Environmental Aspects of Dredging

Environmental Aspects of Dredging

Edited by R.N. Bray

CRC Press
Taylor & Francis Group
Boca Raton London New York

CRC Press is an imprint of the
Taylor & Francis Group, an **informa** business

A TAYLOR & FRANCIS BOOK

Cover: Beach replenishment along the North Sea coast.

Published by: Taylor & Francis/Balkema
                 P.O. Box 447, 2300 AK Leiden, The Netherlands
                 e-mail: Pub.NL@tandf.co.uk
                 www.balkema.nl, www.taylorandfrancis.co.uk, www.crcpress.com

First issued in paperback 2020

ISBN-13: 978-0-367-57753-7 (pbk)
ISBN-13: 978-0-415-45080-5 (hbk)

Visit the Taylor & Francis Web site at
http://www.taylorandfrancis.com

and the CRC Press Web site at
http://www.crcpress.com

Typeset by Charon Tec Ltd (A Macmillan Company), Chennai, India
The book is edited by: R.N. Bray

Library of Congress Cataloging-in-Publication Data
Environmental aspects of dredging / Edited by R.N. Bray
    p. cm.
    Includes bibliographical references and index.
    ISBN: 978-0-415-45080-5 (hardback : alk. paper) 1. Dredging—Environmental aspects.
2. Water—Pollution. I. Bray, R.N. (Richard Nicholas)
    TD195.D72E55   2008
    627'.730286—dc22   2007042703

Contents

Contents

Acknowledgements

Any book of this range and comprehension owes its existence to the efforts of many experts, researchers and professionals. The International Association of Dredging Companies (IADC) and the Central Dredging Association (CEDA) wish to thank both the original authors of the IADC/CEDA *Guides on Environmental Aspects of Dredging* for their contributions and revisions, as well as the recent authors who have added their knowledge to this new updated edition. They are:

- Andy Birchenough
- Jan Bouwman
- Nick Bray
- Neville Burt
- Anna Csiti
- Caroline Fletcher
- Pol Hakstege

- Anders Jensen
- Axel Netzband
- Hans Noppen
- Bo Mogensen
- Eleni Paipai
- Dick Peddicord
- Jos Smits

Without the guidance and vision of the IADC/CEDA Editorial Advisory Board comprising Nick Bray, Marsha Cohen, Anna Csiti, Constantijn Dolmans and Gerard van Raalte, who carefully read and re-read the manuscript, this book could not have been realised.

In addition, the CEDA Environment Commission as well as the EuDA Environmental Committee and numerous IADC, CEDA, EADA and WEDA members were kind enough to review and comment.

Reviewers included: Kees d'Angremond, Jan Brooke, John Dobson, Robert Engler, Tom Foster, Barry Holliday, Polite Laboyrie, Jesse Pfeiffer, and Peter Whiteside. To all those individuals whose efforts made this book possible, and who have not been mentioned by name, we offer our gratitude.

Foreword

Since the industrial revolution, people have been interferring with the environment at an ever-increasing rate. Our past lack of understanding and lack of appreciation of our environment have brought us to a point where this interference is manifesting itself in ways that can no longer be ignored.

Studies conducted on a worldwide basis have shown that, if we are to try to guarantee the future of human existence, we must take a more responsible attitude towards how we live and behave in our environment. Growing public awareness and concern endorse the view that sustainability is paramount.

Dredging is a necessary activity in civilisation's development. Given the right circumstances, dredging may also be a useful tool for remedying past environmental interference. By its very nature, however, the act of dredging is an environmental impact. It is, therefore, of the utmost importance that we be able to determine whether any planned dredging will have a positive or negative impact on our environment. Evaluation of environmental impact should examine both the short- and long-term effects, as well as the sustainability of the altered environment.

In its simplest form dredging consists of the excavation of material from a sea, river or lake bed and the relocation of the excavated material elsewhere. It is commonly used to improve the navigable depths in ports, harbours and shipping channels or to win minerals from underwater deposits. It may also be used to improve drainage, reclaim land, improve sea or river defences or clean up the environment.

Dredging techniques and dredging activities in and of themselves will always make some change in the environment. Although dredging always seeks to improve a given situation awareness of the repercussions of change is essential. For instance,

- alterations to coastal or river morphology may result in enhancement or loss of amenity, addition or reduction of wildlife habitat;
- alterations to water currents and wave climates may affect navigation, coastal defence and other coastal matters, and
- reduction or improvement of water quality will affect benthic fauna, fish spawning and the like.

The question of the duration of such effects, long term or short term, is also important to consider. Environmental effects of dredging may include increases in the level of suspended sediment in the vicinity owing to the excavation process, the overflow while loading hoppers, and the loss of dredged material from hoppers or pipelines during transport. At the placement site there may be disturbance or loss of benthic fauna. Most often, however, these effects will change the environment to a lesser extent in the long term than will be immediately apparent.

Frequently, the level of suspended sediments generated by dredging activities are no greater than those caused by commercial shipping or bottom fishing operations, or even those generated during severe storms. Unfortunately this is often difficult to demonstrate without undertaking comprehensive studies. Yet the investment in these studies may prove to have added value in the long run.

The marine environment is a complex combination of natural features and phenomena, supporting a diverse but largely concealed, underwater population. Because of this complexity, predicting the effects of human-induced changes and short-term operations is extremely difficult. Comprehensive and detailed investigations of environmental characteristics are frequently an essential prerequisite for any planned dredging activity, together with an assessment of all the potential pros and cons. Of course, it goes without saying that re-suspension of contaminated materials poses special problems and demands rigorous scientific analysis.

Neglecting environmental issues in the past has resulted in the present situation in the industrialised nations where many rivers, ports and harbours contain soils that have been contaminated by undesirable levels of metals and chemical compounds. When dredging in these soils, contaminants may be released into the water column and thence into the food chain. Thus, the environmental effects of dredging and relocation of the dredged material may be more severe than when dredging clean material and will require closer scrutiny. In certain cases the very existence of the contaminated soils has led to dredging: by removing the contaminated soils and placing them in a more secure location, the environment is improved. The treatment and storage of contaminated soils is a highly complex subject and requires detailed study. Long-term improvement does, of course, depend ultimately on preventing pollution at its source.

Nowadays an international framework of legislation relating to dredging and the management of dredged material at sea has been developed. This contains regulations, which must be implemented by national authorities. A number of European countries are also developing legislation to control the placement of dredged material on land. Such legislation is constantly changing, as scientific knowledge increases and implementational frameworks evolve. All promoters of dredging works need to be aware of current legal requirements.

An additional positive environmental aspect of dredging, one that is actively encouraged by the controlling authorities, is the reuse of dredged material, even including some that is contaminated. Typical uses include beach nourishment for sea defence, the creation of wetlands for recreation and wildlife sanctuaries, reclamation of land for commercial and industrial development, and the improvement of agricultural land.

This one-volume IADC/CEDA publication *The Environmental Aspects of Dredging* has been created by integrating the seven stand-alone guides, which formed the series of the same name published between 1996 and 2001, and by updating and adding to the original material to reflect advances made in recent years.

The Editorial Advisory Board
Nick Bray (CEDA), Anna Csiti (CEDA), Constantijn Dolmans (IADC)
and Gerard van Raalte (IADC)
January 2008

CHAPTER 1

Consciousness, Concern and Cooperation

1.1 INTRODUCTION

What are the environmental aspects of dredging? And why have they become increasingly important? In the past, dredging focused on nautical, hydraulic and mining applications, that is, the maintenance of navigation channels, the discharge of water and the extraction of minerals such as sand and tin ore. Nowadays dredging clearly has a far wider range of applications. These include building islands and developing other land reclamation areas, such as those in the Middle East and Asia. Dredging is also used for cleaning up contaminated areas or restoring the environment or (re)creating habitats.

1.2 GLOBAL AWARENESS

The environmental aspects of dredging go well beyond mere theoretical musings. The fact that major banking organisations and multinational companies refer to environmental matters in their annual reports and attach environmental requirements to projects that they propose to finance or promote illustrates their importance.

In the fiscal year 1994 the World Bank formally included "environment" as a funding sector. Now, under its Operational Policies, it encourages and supports borrowing governments to prepare and implement an appropriate Environmental Action Plan (see Text Box on page 2). In addition, funded projects will undergo Environmental Impact Assessments and will proceed with an Environmental Management Plan. Other multinational banks show the same awareness.

1.3 THE DEVELOPMENT OF ENVIRONMENTAL CONCERN

A country's attitude towards the environment and its awareness of the fragility or robustness of components of the natural world are constantly evolving. Generally speaking, in the early stages of industrial development, a nation tends to be preoccupied by the pressing needs of its populations in terms of improving income and standard of living. As these goals are attained and a nation becomes wealthier, more time and money become available to consider the wider aspects of life. These can be local factors such as maintaining a comfortable and fulfilling lifestyle as well as global concerns about the earth as a whole. People begin to recognise the value of

World Bank and the EAP

1. The World Bank (www.worldbank.org) encourages and supports the efforts of borrowing governments to prepare and implement an appropriate Environmental Action Plan (EAP) and to revise it periodically as necessary. Although the Bank may provide advice, the EAP is the country's plan and responsibility for preparing and implementing the EAP rests with the government of the country.

2. An EAP describes a country's major environmental concerns, identifies the principal causes of problems, and formulates policies and actions to deal with the problems. In addition, when environmental information is lacking, the EAP identifies priority environmental information needs and indicates how essential data and related information systems will be developed. The EAP provides the preparation work for integrating environmental considerations into a country's overall economic and social development strategy. The EAP is a living document that is expected to contribute to the continuing process by which the government develops a comprehensive national environmental policy and programmes to implement the policy. This process is expected to form an integral part of overall national development policy and decision-making.

3. The Bank draws on the EAP for environmental information and analysis to plan its assistance with appropriate attention to environmental considerations. The Bank encourages each government to integrate its EAP into sector and national development plans. The Bank works with each government to ensure that information from the EAP (a) is integrated into the Country Assistance Strategy, and (b) informs the development of programme- and project-level details in a continuing process of environmental planning.

the natural world and the environment in which they exist (see Figure 1.1). This awareness or consciousness of the environment can then translate into "concern".

The growth of this concern is one of the fundamental driving forces behind the development of legislation aimed at protecting the environment. It explains why the so-called "green" organisations are so vocal in their opinions. They are promoting environmental concern. Nowadays, with the development of global communications and the Internet, countries are becoming aware of their own environmental heritage as well as of global concerns at an accelerated pace.

As a country begins to assimilate the messages being promoted by the environmental lobby, it initiates steps to do two things. Firstly, it endeavours to determine the real threat to the environment posed by whatever human activity is being studied. Secondly, it initiates legislation to counter the perceived threats to the environment.

Figure 1.1 Waterfront development in Japan at Laguna Gamagori, situated between Tokyo and Kyoto on Honshu's southern coast, has created a large recreation complex focusing on marine activities.

Perceived is a key word here because the science needed to study the matters in question is often also a work in progress, still being developed, and the initial answers to the questions posed are frequently tentative.

Impetus to implement legislation providing some measure of environmental protection generally is motivated by two factors; disaster (either caused by humans or by nature) and precautionary science. The mode of introduction of legislation arising from these two causes is often markedly different.

In the first instance, when a disaster occurs, political will often results in strong steps to prevent re-occurrence with legislation that tends to be restrictive. This has happened in the United States with dredged material disposal, and to some extent in other parts of the world with the adoption of inappropriate suspended sediment level controls around working dredgers. As the negative effects of this restrictive legislation become apparent, those affected are inclined to fund more definitive and focused science. This new research in turn provides more reliable answers. Legislation is then able to be appropriately adapted. During this process, because the whole situation is being resolved, environmental concern begins to abate.

Where environmental concern has been raised by precautionary science, political will to introduce strong legislation tends to be lacking (unpopular measures lose votes!) and a greater resistance to its proper implementation may arise. In such circumstances, the legislation is often full of loopholes. The process of correcting this becomes lengthy and thus there is a long lead-time to arriving at satisfactory levels of control and environmental protection.

Obviously it would be more efficient if a country could adopt the most appropriate legislation from the outset. In fact, nowadays, in some circumstances, this may be achievable, because many of the industrialised countries have been through the process described above and are now striving to reach an acceptable balance. So as new issues arise they are more attuned to taking appropriate steps to reach adequate legislation the first time round.

On the other hand, new considerations must be taken into account. One of these is climate and its effect on species type and diversity. Hopefully the publication of this book will add sufficient material to the worldwide data bank to enable both industrialised and emerging countries to arrive at the desired end point without excessive expenditure.

1.4 COOPERATION: A FRAMEWORK FOR INTERESTED PARTIES

In many places throughout the world the social costs of environmental degradation have risen. These include growing health costs, increased mortality rates, loss of productivity, reduced output in natural resource-based sectors and deterioration in the overall quality of the environment.

Not everywhere in the world, however, is there a consensus about these problems. Many other views, opinions and interests play an important role, and the interrelationships amongst the various standpoints are constantly changing, sometimes converging, sometimes colliding. Whatever the case, the environment is clearly no longer a matter of concern for the environmentalists and conservationists alone.

First of all the people living in the area subject to an environmental impact must be considered – they are probably the ones most directly affected by a project's challenges. Additionally, governments, financial institutions, port-related organisations, dredging firms, local authorities, private enterprises, and other members of society at large are all potentially interested parties.

Take, for instance, the presence of contaminated sediment in a navigable river, which is also used for drinking water production and for commercial fishing. The implications are diverse:
- in maintaining navigability, dredging and soil treatment/disposal expenses will be incurred. These may be costly and, as well, there may be difficulties in locating suitable sites for relocating the dredged material;
- the purity of the water system may be impaired, which may lead to the need for more sophisticated and more expensive treatment technologies for supplying drinking water;
- fish can accumulate contaminants through their food chain, which will in turn impact adversely on fish sales, and
- the need could arise for the introduction of special (and perhaps more costly) dredging techniques.

To promote a better understanding of such complex matters, a basic framework is required within which all interested parties (or so-called "players") can have their concerns and needs properly considered on their own merits. Such a framework will allow all players to have a fair chance of expressing their interests and of cooperating in the decision-making process. Only in this way will all players have a sense of "ownership" of the project; only in this way will the environment and society as a whole benefit from the results.

A framework such as this is a practical first step towards developing a systematic approach for successfully managing the environmental issues related to a dredging project, whether of sea- or riverbeds, whether it involves habitat loss or creation.

1.5 THIS BOOK AND ITS BEGINNINGS

This book in its present form has evolved from the seven volume IADC/CEDA series of guides entitled *Environmental Aspects of Dredging*. When the guides were originally conceived, they were perceived to have a number of uses:
● to provide a balanced view of the whole subject of dredging and environment;
● to lead the reader through the state-of-the-art environmental evaluation process;
● to provide a source of information on dredging and environmental matters, and
● to provide references for more detailed study.

In seeking these objectives the authors amassed a wide range of information on the environmental aspects of dredging culled from case histories, research results and practitioners in the field. In its present form the book continues to strive to fulfil the goals of the previous edition. The single volume format provides a way of transmitting this essential information in a more streamlined form.

This publication promotes particular approaches to environmental evaluation and procedures for implementation:
● *It provides the reader with a vast amount of relevant background reading and many sources for further research.*
● *It is unbiased and favours neither the promoter of dredging works nor those who are opposed to dredging activities.*
● *It does not, and never set out to, prescribe specific controls for dredging works on a generalised basis.*
● *It is not intended to be used, nor should it be used, in this manner.*
● *It is intended to be an aid to planners, engineers and environmental scientists when they are making their own individual assessments.*

The book commences with a look at the many people or "players" who become involved in the development of a dredging project and it considers their various perspectives. Chapter 2 promotes a systematic approach to this process, and in so doing attempts to shed more light on the roles of various interested parties. Those involved in the contractual, planning and legal aspects of starting up dredging programmes may find it particularly useful.

Chapter 3 is the operational control centre of the entire book. It is intended to give the reader a "blow by blow" account of how to go about characterising the environmental attributes of a site, defining the project to be executed, and discovering and ranking the potential benefits and shortcomings of any proposed construction works. This chapter also points the reader to specific locations in the book where more detailed information is available.

Chapter 4 offers an in-depth description of how a project may influence a particular environmental regime. This is essentially how the finished works will affect the environment over the short and long term. Here, as in other parts of the book, environment means the whole spectrum of natural and human regimes and activities.

Chapters 5, 6 and 7 are support chapters. They cover matters that may be running in parallel within the overall environmental and project development processes. Chapter 5 presents matters relating to the collection of data and its interpretation. Chapter 6 describes the main dredger types and explores their environmental effects, mitigating measures that may be taken to make them less intrusive, and specialised machines that have been devised to work in sensitive areas. Chapter 7 describes the various uses for dredged material as well as the processes for determining the optimum methods for managing dredged material.

Chapter 8 explains monitoring methods and processes. These may be either to establish baseline conditions on a site or to monitor the impacts of projects or dredging activities on the local environment and apply controls as appropriate.

Finally, Chapter 9 looks to the future and discusses some of the more philosophical aspects of environmental assessment and evaluation.

In the Annexes information will be found about typical legislative conditions and controls imposed by international conventions and regional and national agencies around the world for the placement of dredged material, both in the sea and on land. Further additional Annexes cover case histories, general descriptions of environmental regimes and describe dredged material properties in detail.

CHAPTER 2

Players, Processes and Perspectives

2.1 INTRODUCTION

Recognition of the "players" in any field is an important step towards implementation of a project. It provides a means to identify all those who are likely to be interested in or affected by the project. It is the eventual consensus of these "players" that will determine whether or not a project is allowed to proceed. Most promoters of a dredging project would not see their initiative as a "problem". Rather at best as a contribution, and at worst as a challenge. Increasingly, however, someone somewhere will nearly always view the project as a problem. Consequently, we are all likely at sometime in the future to confront the "problem-solving" or "challenge-solving" process.

The focus of this chapter is on the uses of dredging and ways of finding a solution to "a challenge". The challenge presented here as an example is to deepen a channel for navigation in an area where the existence of contaminated material in a riverbed has caused deterioration in the environment and its ecosystems. The chapter illustrates how every so-called "challenge" goes through certain predictable "life cycle" phases: awareness, solution searching, solving and managing. It also introduces the concept of "players" in this field, people and organisations sometimes referred to as "stakeholders".

2.2 CHALLENGES, PLAYERS AND THEIR PERSPECTIVES

In the field of dredging there are numerous players, each with a specific role and each dependent upon one another in a complexity of interactions and communications. They pursue their own, often conflicting, interests in their roles as either proponent, opponent, facilitator (e.g. lawyers, financiers) or bystander. Some members of a particular group (e.g. politicians) may belong to the proponents, while other members of the same group to the opponents. This may change during the life cycle of the project.

The players are either individuals or organisations (public or private) with clear objectives and opinions concerning the environment: its quality and value, its maintenance, its financing, and acceptable technologies to name a few. Usually they each have their own perception of a project, and see their roles as either cautionary, advisory, directorial, informational or as providers of finance. Their viewpoints may be either global or regional, commercial or altruistic.

2.2.1 The life cycle of an environmental challenge

As pointed out before, most promoters of a dredging project would not see their initiative as a "problem". The project may however be seen by others as this, especially if the material to be dredged is contaminated.

An environmental problem does not often occur suddenly and obviously, but usually develops and is recognised over a long period of time. A small amount of pollution is usually not enough to draw attention and may initially even be denied, especially by those groups that can be held responsible. Over time, however, as the consequences of the problem accumulate, citizens groups who are affected will demand change.

In practice, the way in which an environmental matter is approached is not dissimilar to the way in which society deals with all kinds of different challenges. This has become known as the "life cycle" approach and is based on the fact that the way in which a challenge is met is greatly influenced by the ability to secure and maintain political attention.

With such attention, there is every expectation of reaching decisions and of making available the means (e.g. finance, standards and legislation) with which to implement a solution. During the life cycle of a challenge, political attention will rise and fall, increase and decrease over time.

A life cycle can be divided into four phases (see Figure 2.1):

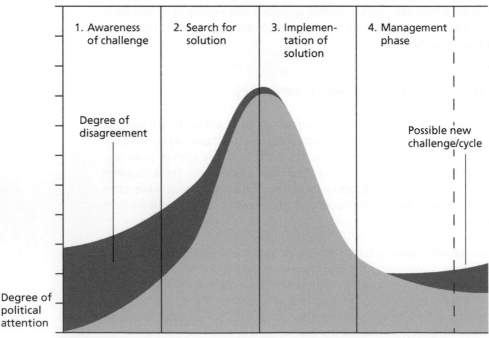

Figure 2.1 Life cycle of an environmental challenge.

Awareness of the challenge . . . . Phase 1
Many groups with different interests and different degrees of knowledge are involved at the outset. Discussions usually start with significant disagreement about the causes and nature of the challenge, its continuation and effects, areas of responsibility, possible solutions, and ultimate results.

This stage is essential even though it can give rise to some heated arguments. Public opinion is mobilised, people arrange campaigns, and debates take place in the media between experts and laypersons alike. The outcome of this first phase is the conclusion that something has to be done.

Search for a solution . . . . Phase 2
Recognition of the challenge is followed by the search for a solution. Relevant information is collected, modelling carried out and data analysed. Options are investigated and alternative solutions developed. There is consensus on the challenge, but not yet on the solution. Consequently, discussions must continue.

Take, for example, the question of what to do with contaminated silt after it has been dredged. Most people agree that the material can no longer simply be disposed of at any site. Soil cleaning remains very expensive, which indicates that well-controlled storage, for the time being at least, is the best solution in most cases. There is, however, disagreement concerning questions such as "how" and, not least, "where" to store the silt. The NIMBY – Not In My BackYard – attitude is much in evidence (see Section 2.5).

Political attention has reached its peak level. The decision-making process has now begun and continues into the next life cycle phase.

Implementation of the solution . . . . Phase 3
Once consensus about the improvement is reached, people tend to focus on carrying out the work efficiently and effectively rather than debating principal issues. Technical and economic matters now play an important role, as do rules and legislation.

To illustrate how this phase works in practice, consider how public attitude has changed towards the recycling of household waste. Initially there was considerable reluctance by people when asked to sort their waste at home before dumping it in the trash can. Nowadays, in many parts of the world, sorting trash for recycling has become commonplace, even required by law, and people are increasingly compliant.

At this stage, most parties involved agree on the challenge or problem and, in very broad terms, on the solution. Some people undoubtedly may continue to feel the need for further debate on the necessity and efficiency of the proposed measures.

Without going into more detail about the decision-making process itself, suffice it to say that decision-making is more interactive than linear in nature. Consequently, during the cycle-time there may repeatedly be switches back and forth between Phases 2 and 3.

Management . . . . Phase 4

Finally there is the management phase. The challenge now appears to be under control and political attention has diminished. The danger exists, however, that attention to the challenge will be lost when people mistakenly think "the challenge has been met" and deem it no longer necessary to be concerned. In that event, or when new challenges emerge, a new awareness phase begins. This can also come about as a result of unforeseen side effects or from changing demands by society.

The management of traffic is a case in point. Growth in traffic levels has been adequately dealt with in the past by extending or upgrading the infrastructure. In fact the source of the challenge, traffic growth, has not been dealt with. Traffic growth is being allowed to continue to grow and since expanding infrastructure is not always possible, this way of dealing with the challenge is now being questioned.

2.2.2 The players

In dealing with the life cycle phases of an environmental challenge, the existence of distinct interest groups, the so-called players, has been identified. That these players have different or even conflicting interests and may play different roles in different phases of a project is clear.

The following players are identified:
- Politicians: at local, national or international level;
- Financiers: bankers at, once again, local, national or international level;
- Activists: those who wish to increase awareness, such as pressure groups;
- Contractors: the organisations which undertake the dredging;
- Consultants: the advisers who contribute to the pool of knowledge;
- Administrators or Standards Institutes: governmental bodies, London Convention of 1972, Oslo and Paris Commissions, ANSI, ISO, JIS, DIN, BSI and others;
- Owners: the individual or organisation, private or governmental entity responsible for the considered area, and
- Polluters: the source of contamination, and also source for payment for remediation.

The Politician may be a proponent or opponent of the project and is concerned with:
- growing awareness (owing to e.g. activists, calamities or literature search);
- validation and balancing of interests;
- current laws, rules and standards, and
- national and international conventions.

These concerns will lead to updated legislation, standards and guidelines.

The Financier is relatively neutral with respect to the project's progress and is concerned with:
- balancing interests;
- bank continuity;
- (inter)national laws and conventions;
- viewing future developments, and
- investments, loans.

Addressing these concerns results in the financing of some projects.

The Activist is often, but not always, an opponent of the project and is concerned with:
● calamities, accidents (actual, potential or perceived);
● gaining an understanding of potential challenges, and
● agitating and campaigning for public awareness.

These concerns result in a greater consciousness/awareness of the challenge, with other groups taking an interest.

The Contractor is traditionally a service provider during the construction phase of the project. More frequently nowadays the contractor works alongside the owner and, sometimes with the assistance of consultants, assists in the development of the project in an environmentally friendly manner. This cooperation may start at quite an early stage of the development process. In this role the contractor is a proponent of the project or initiative.

The Consultant is concerned with:
● providing technical/managerial support to other players;
● applying standards, codes of practice and legislation;
● capitalising on his or her experience for solving challenges, and
● gathering information from previous projects.

These concerns increase knowledge of the matter amongst the other groups and assist the parties involved. Although consultants are engaged by one of the players, as an "independent" they are not really a proponent or an opponent of the project.

The Administrators and Standards Institutes are also neutral with respect to the project and concerned with:
● developing new standards, codes of practice and regulations;
● issuing permits and verifying conformance;
● registering state-of-the-art technology;
● developing standards of good practice, and
● taking account of health and safety considerations.

Two other types of players have not yet been mentioned: One is the "Owner" of the project. This player can be any individual or organisation or a combination of these, but is most often a private or governmental entity which "owns" or is responsible for the considered area.

The Owner of the project, the main proponent, is concerned with:
● making others aware that the challenge exists;
● being prevented from carrying out business (i.e. not being allowed to dredge the channel);
● researching ways of solving the challenges;
● conforming with rules and regulations (quest for licences, etc.);
● contracting out the dredging operation;
● supporting the operation, and
● registering the measures taken and the work carried out (compliance monitoring).

These activities eventually result in a dredged area and relocated or processed contaminated material, enabling the "owner" to continue its primary (business) activity.

Having mentioned contaminated material, the other player, perhaps the least identifiable player, comes to the fore – *the Polluter*. In cases dealing with contaminated sediments, the "polluter" is an important player in terms of taking responsibility for paying for remedial work, as well as preventing further contamination. Often however in cases of historic contamination, this player is not easily identified.

2.2.3 The field of play

Clearly all players have their own particular interests, which may be elements of a much larger entity. Therefore knowing from which point of view each player approaches the issues is very useful. The position of the reader or the player must be clearly understood in order to gain insight into this complex matter.

The interrelationship of all these groups of players and elements is depicted in the diagram (Figure 2.2). The diagram illustrates the complex "field of play" when considering a dredging project with contaminated materials in the bed of a river. Identify a particular player and trace the lines to the result of the processes in which that individual or organisation is involved. This will also reveal the interrelationship that takes place with the other interest groups.

The reader may possibly identify with one or more of the parties in the diagram. Since the diagram thus provides an overview of the entire field of play, it also indicates where the reader fits into the whole and where the reader's principal interests lie.

The illustration also shows the area of focus. Subjects are, in particular, seen from the viewpoint of the project's owner and the contractor. However, other roles can be played where appropriate.

2.3 DEVELOPING A SYSTEMATIC APPROACH

2.3.1 Combination of players and life-cycle phases

The previous section gives an impression of the broad area in which the subject "environmental aspects of dredging" is positioned. The major players are identified and defined. In the next section, each of the life-cycle phases – awareness, the search for a solution, implementation of a solution, and management of the project – are related to the particular player(s).

While all players adopt one role or another throughout the phases of the life cycle, some players are more dominant than others. The combination of players and phases is presented in the form of a matrix in Figure 2.3, where the vertical axis shows the players and the horizontal axis the life-cycle phases. The cross-sections show in which phase a player has a dominant role.

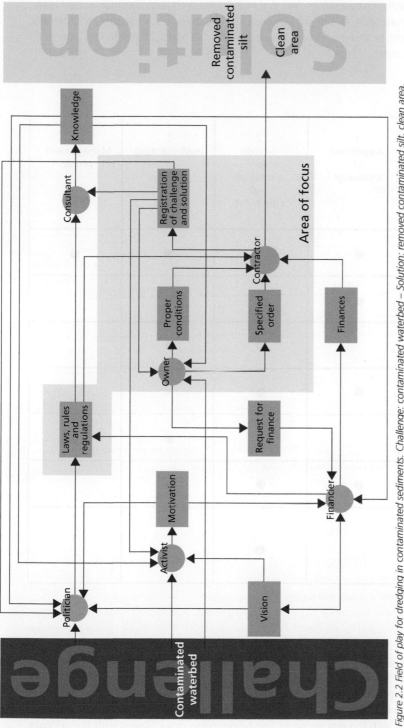

Figure 2.2 Field of play for dredging in contaminated sediments. Challenge: contaminated waterbed – Solution: removed contaminated silt, clean area.

For the matrix illustrated here (see Figure 2.3), the dredging of contaminated material has been taken as the subject area. On the basis of this matrix, a systematic approach can be developed with which to tackle the particular challenges associated with this project.

Phase / Player	1 Awareness of challenge	2 Search for a solution and consensus	3 Implementing of solution	4 Management phase
Owner	●		●	●
Politician	●			
Financier		●	●	
Activist	●	●		
Contractor		●	●	●
Consultant	●	●	●	
Administrations/ Standards Institute		●		

Figure 2.3 Matrix of phases and players.

The first essential step to be taken is for the players to ask themselves, "Who are we?" and "What phase are we in?". The answers to these questions will reveal whether or not those players currently have a role to play. If they do, then they must take action by making use of the information other people provide, by acting in accordance with the requirements and by providing other players with the information or products they require.

Each phase calls for different actions and requirements, and may be identified by one dominant or key player:

Phase 1. . . . Awareness of the challenge
This phase may be dominated by the "politician". Alerted by the public or the media, the politician is pursuing the adoption of (international) legislation, which can be the first step in formulating boundary conditions upon which national legislation can be based.

Such legislation creates working space within which politicians, government agencies, environmentalists, and others can discuss and (re-)formulate national, regional and local policies. The creation of legislation and the ensuing environmental policies are based upon a fundamental understanding of the contaminants and their effects on the environment.

Phase 2 . . . . Search for a solution
Here the "consultants" play an active role. They will have to address three important questions:

1. Do we have to use absolute criteria? (e.g. are the rules imperative and do we have to work according to them?)

2. Do we have the option to go for the best available technology and are the rules for guidance purposes only?

3. How important are economic considerations? Should we seek the lowest cost-effective solution? Or are economic investments in new technologies possible?

In this phase – the definition and/or specification of the project's challenges – the interests of all parties concerned and the scope of work should be clarified. Risks must then be identified by, for example, a systematic analysis of parts and processes (fault trees) to determine causes, consequences, probabilities and so on. The risks must also be quantified on the basis of historical data, information from expert sources, calculations, and other parameters.

Furthermore, sensitivity analyses should be performed, and mitigation measures and alternatives regarding their impact on the solution or course of action should be proposed and evaluated. On the basis of all the information collected on the subject, decisions can be made.

Phase 3 . . . . Implementation of the solution
Phase 3 may be dominated by the "contractor". The preparation and execution of dredging works involving contaminated material is often different from normal dredging operations. The pre-dredging investigations required to locate, classify and quantify the contaminated material may be more elaborate than conventional pre-dredging investigations. They may demand a higher degree of accuracy, more detailed laboratory testing of samples and additional safety measures.

Whereas most nautical dredging projects entail working to a predetermined dredge profile, the execution phase of dredging contaminated material involves a more selective removal process. Because of the high costs of treating and storing contaminated material, every cubic centimetre of dredged material increases costs substantially. Dredging accuracy is therefore extremely important. In order to meet all requirements, special equipment and techniques have to be applied or may have to be developed.

The transport of contaminated material from the dredge site to the treatment depot/disposal area demands special attention as well. Owing to the contaminated nature of the material, transport whether by barge, pipeline, hopper, or other means should be organised in such a way that it does not contaminate the environment.

Finally, it should be borne in mind that often even contaminated dredged material can have significant value if properly used for beneficial purposes. To realise its full value detailed planning is necessary and activities must be efficiently coordinated between all parties concerned, especially the potential material users.

Phase 4 . . . . Management
No player dominates in the "management" phase; guidelines take the lead. Guidelines and procedures have already been established for the execution and monitoring of the dredging operations which must be undertaken in a controlled and environmentally acceptable manner.

Each critical activity must be covered by a procedure that details how, when and by whom questions that may arise should be answered. This, for example, goes for design and execution activities, as well as for inspection and maintenance which form part of this fourth phase.

There will also be procedures covering safety (loss of human life/injury/long-term consequences), the environment, natural and social values, and multi-criteria analyses. This calls for expertise in the fields of data acquisition, risk analysis, geotechnics and hydraulics.

2.3.2 A step-by-step approach

Using the framework described above, a step-by-step approach can be followed. This, however, is not essential for all players. Listed below, as an example, are the steps to be taken by an "owner". Those for the "politician" and other players will of course be somewhat different.

Step 1 – As an owner, locate your position in the complex field of play. Look at the other players from your point of view. Define what others expect from you and what you as an owner expect from the other players.

Step 2 – Find out what other people/owners in similar situations have done and how. This may be achieved by comparing relevant facts and figures of important (pilot) projects that have been or will be carried out worldwide.

Step 3 – Find out, especially in your locality, but also worldwide, what standards and legislation have been created to control, monitor and maintain the skills of dredging in locations containing contaminated material.

Step 4 – Carry out pre-dredging investigations according to pertinent rules and conditions. Before dredging can commence, all sources of contamination must be identified and classified. Once this has been done, the concentrations, quantities and spread of the contaminants can be determined. Check out procedures that show how the contamination can be identified and quantified, and what techniques are available for boring, sampling, laboratory testing, hydrographic and seismic surveying, and other procedures.

Step 5 – Select from the various techniques and equipment, offered by contractors, which may be specially designed for this type of work. Compare these techniques with similar techniques used for "normal" dredging operations to clearly identify the differences between the various methods and to highlight the specific (technical) measures to be taken when dredging and transporting contaminated soil in an environmentally acceptable way.

Step 6 – Choose from the available options the techniques and methods for treating and disposing of the dredged material in an environmentally acceptable manner and identify the requirements. Investigate possibilities for re-using the material following processing.

Step 7 – Obtain political consensus and financial agreement on the way to execute the work.

The steps given in the approach above relate to dredging works, where the material to be dredged is contaminated, but a similar process may be applied to any environmental aspect of dredging. It is recommended that when any dredging works are proposed, the interested parties should:
- identify the players in the field,
- determine what the environmental issues are,
- locate the issues in the context of the life cycle diagram (Figure 2.1), and
- follow a step-wise approach to implement the works.

2.5 NOT IN MY BACK YARD (NIMBY)

For any proposed project there will often be a number of local people who feel that they will be directly affected in some way by the development. Some of them will indeed be affected, but many will not. This does not stop them from feeling that a potential development will have a negative effect on their locality. If the development had been somewhere else, many of these people would have positive or, at worst, neutral views on the desirability of the project. Unfortunately, as a project moves closer to their "backyard", their feelings and views grow more acute. This phenomenon is known by the acronym, NIMBY, meaning "Not in my backyard".

NIMBYism is a phenomenon that causes considerable difficulties for potential developers. It can only really be overcome by ensuring that every opportunity is taken to engage with local people and organisations, quashing wild rumours and disinformation, providing everyone with quality information, genuinely listening to and addressing objections. This process will foster understanding and hopefully get these players on board with the development at stake as early as possible.

In all dredging projects someone's backyard is likely to be effected. It is for the politicians, supported by all available expertise, to decide whose backyard has to be "sacrificed" in the general interest and with the least "damage" to society. This process requires very careful deliberation. Provision of a form of "compensation" to the owner of the backyard being affected may sometimes be a key part of good practice.

CHAPTER 3

Control, Coherence and Coordination

3.1 INTRODUCTION

From an environmental perspective, every dredging and reclamation project consists of two inter-related elements; the Dredging Process and the Constructed Project. The dredging process is generally short term and has a number of short-term effects, some of which may lead to long-term effects. The constructed project is the finished article, the culmination of a study, design, construction and completion process. It is the part of the environment that has been intentionally altered by humans and which will most likely have long-term environmental effects, both positive and perhaps to some degree negative. All these effects in both the construction process and the constructed design are considered here.

3.1.1 Definitions: Acceptable versus unacceptable

Environmental effects are in essence impacts, commonly considered either positive or negative, good or bad. But in fact the issue of environmental effects is not quite so polarised. The promoter/initiator of a dredging and reclamation project is attempting to improve a given economic, social or environmental situation and the considerations involve nuanced, carefully weighed distinctions. Environmental effects can also be seen as neither good nor bad, but rather acceptable or unacceptable. A promoter of a dredging project is striving for a situation where the proposed project has no "unacceptable" environmental effects. The following definition may clarify this statement (see Section 5.7):

"Unacceptable effects are changes of such a nature and of sufficient magnitude, duration, and/or a real extent that society deems the benefits accrued from the project not worth the environmental impacts."

The important point to be made is that all or most projects will make changes to the environment and that some changes may be adverse, but that these adverse changes *per se* do not necessarily constitute an unacceptable impact.

This chapter describes how the appropriate information can be assembled, analysed and presented for society to make the judgement, acceptable or unacceptable. It describes the various tools used in the environmental evaluation process. It also illustrates how some of the material obtained and activities carried out during this process are linked to the implementation of a permitted project. In short, this chapter is intended to be the implementation manual for a complete holistic environmental evaluation process.

3.2 THE OVERALL FRAMEWORK

The overall project implementation process is illustrated in Figure 3.1. This is a simplified diagram. Though the overall process is more complicated than this, for purposes of clarity, some of the other activities and linkages are illustrated in subsidiary diagrams, which are explained later in this chapter. Figure 3.1 is essentially a road map for the reader to see how the chapters in this guide are integrated into the whole environmental and project development process. The diagram also illustrates:

- Sequencing: The logical steps taken as the reader/manager works through the implementation process;
- Linkages: How matters considered at particular stages of the process are linked to other subjects, and
- How the environmental assessment and the design are related.

The relationship with stakeholders has intentionally been omitted. Stakeholders are so important that they are given a section of their own, and in addition they are included as part of the partnering process (see Sections 3.7 and 3.10.1).

3.3 PROJECT OBJECTIVES (M1/D1)

Project objectives encompass both the socio-economic benefits to be gained from the project, combined with the physical changes that have to be designed to reap those benefits. These are covered by boxes M1 and D1 in Figure 3.1. Regarding dredging and reclamation, the socio-economic benefits may include: deeper channels and berths to receive larger vessels; expanded industrial working areas; additional land for commercial and domestic activities; improved beaches, coast defences and other such maritime facilities. At this stage of the process, a simple description of the project is required, one that can be presented to interested stakeholders and will answer three fundamental questions:

- What do we want to do (generally in a physical sense)?
- Why do we want to do it (what are the social and economic benefits)?
- When do we want to do it?

As a minimum requirement, the physical changes that are to be made to achieve the objectives of the project should be defined. This will mean producing conceptual designs of the initiative to include details of the new water depths in berths, manoeuvring areas and channels, dimensions of these areas and broad outlines of reclamation areas (where applicable) or places where dredged material is to be relocated. To achieve this some of the activities described in Chapters 4, 5 and 7 (boxes M4, M5 and M6), particularly with respect to use/placement of dredged material may need to be carried out in a preliminary way.

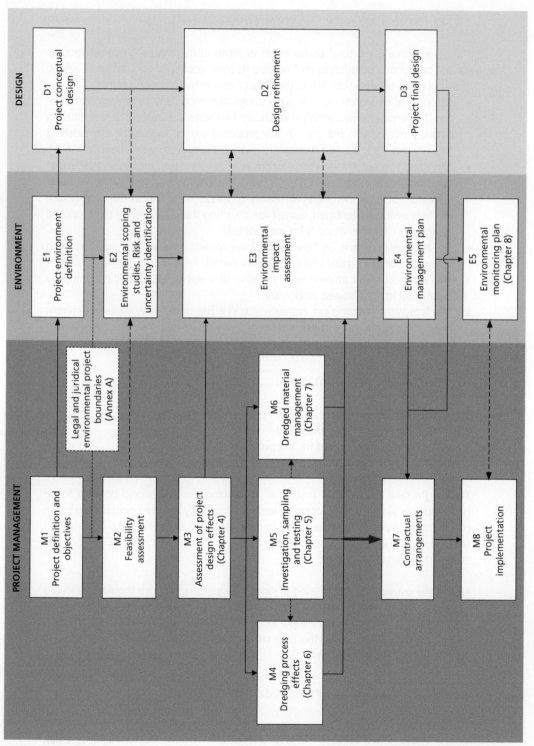

Figure 3.1 The overall project implementation process: M = Management controls; E = Environmental controls; D = Design controls.

3.4 THE PROJECT AREA (I.E. THE TOTAL ENVIRONMENT) (E1)

The project area is the total environment in terms of its physical characteristics, its natural or human operation or regime, and its socio-economic and aesthetic values. It is, therefore, necessary to define the project area in the following terms:

- The natural environment, with such things as morphology, hydrology and ecology. These may be reasonably well known in a general sense, but less defined in detail, particularly in the area of any proposed works. Tools such as modelling (see Section 3.11) may be required to provide the necessary definition;
- Any identified natural long-term trends (e.g. sea level rise, morphological evolution, land mass movement etc.). Of these, morphological evolution is particular important because natural evolutionary processes occurring in the project area are frequently of significant magnitude dwarfing those caused by the proposed works (see Chapters 4 and 5 for more detail);
- The built environment, i.e. how people have already impacted the area and what temporary or permanent changes have occurred;
- The socio-economic environment, in terms of stakeholders and all those having an interest in the planned works, and
- The legislative environment within which one has to operate, including current regulations, permits required and so on. (Annex A contains a vast amount of information on current [2007] regulations around the world).

3.5 CONCEPTUAL DESIGN (D1)

The conceptual design should be accompanied by a description of:
- Why is the project being proposed?
- What benefits does it hold for the promoter of the project?
- What are the wider benefits to the local community and the country as a whole?

This might take the form of, first of all, explaining the operational benefits to be gained by carrying out the proposed works, followed by a more general description of the social, economic and aesthetic benefits accruing from the works. The overall time frame of the project should be given, together with any key deadlines or milestones that are to be achieved.

Finally, the criteria that will be used to judge whether the initiative should be allowed to proceed should be presented. This is the basis for evaluating the environmental acceptability of the plan (i.e. "triple bottom line" approach or other criteria, see Section 3.11.1.). Once again the plan must include the completed construction as well as the construction process.

3.6 INITIAL PROJECT FEASIBILITY ASSESSMENT (M2)

The initial project feasibility assessment (see box M2 in Figure 3.1) is an important part of the development process, because it assists in determining where and when resources have to be made available for further investigations, design and study (see Figure 3.2). This is a period in the development process where stakeholder participation is crucial (Section 3.7). However, to assist in determining who the relevant stakeholders are, the project area must be defined and characterised (see box E1 in Figure 3.1).

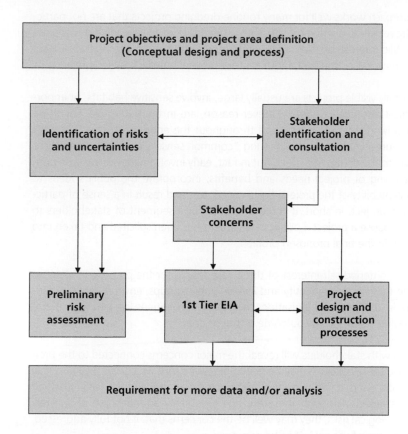

Figure 3.2 Stakeholder role in Risk Assessment.

3.7 STAKEHOLDER INVOLVEMENT

Stakeholder participation in the early stages is illustrated in Figure 3.2. Stakeholder identification will be facilitated by carrying out the project area definition described above. The players and processes described in Chapter 2 may also suggest additional stakeholders. Finally, local and regional publicity given to the project may attract responses from stakeholders, individuals or groups, not already caught in the initial net.

Public interest in any particular dredging project will vary from non-existent to highly focused. The degree of interest will depend on the socio-political structure of the country where the project is located and local perception of potential environmental impacts associated with the dredging project.

There are two basic approaches to managing public involvement in dredging projects:
- A relatively small team plans the project then informs the public and solicits their comments.

- The public is brought in during the project planning process and participates throughout the project.

The first approach works well for many routine dredging projects that are not particularly controversial. It is the most efficient expenditure of usually limited resources and the public interest is often adequately represented using the first approach. However, it may not be the best approach for environmentally "visible" projects.

Environmentally visible projects are usually large, involve sensitive habitats, or important biological resources, or, for whatever reason, are in the public eye. For those projects, involving the public early and throughout the entire process is best. This may seem counter-intuitive to engineering "common sense", especially when there is opposition to the dredging project. But in fact, early involvement will increase public understanding of project needs and benefits, incorporate the public's ideas to increase acceptability of the project, build consensus, and result in a sense of participation in the project. In short, it is more difficult for a segment of stakeholders to successfully oppose a project if public involvement has been solicited and taken into consideration in the final proposed project.

It is in the enlightened self-interest of those responsible for the proposed project to take it upon themselves to identify and involve public groups, environmental organisations, trade and labour organisations, and any other entities likely to view themselves as having a stake in the outcome of the project.

Consultation with stakeholders will reveal the major concerns connected to the project proposals. These concerns may be real or perceived, but whichever they are, they need to be addressed. It is important to identify the magnitude of the feeling behind the concerns raised. Although these are not necessarily an indication of the real physical or ecological risks, they may well be the concerns that, if not fully addressed, can lead to failure of the project to be approved.

During these early stages, the main stakeholder groups will become apparent. It should be possible to identify which groups should be represented on a permanent or intermittent basis in consultation exercises, both during the project development phase and during implementation.

At this stage, identifying where resources should best be placed to carry out further investigative work, additional studies and design should also be possible.

3.8 OUTLINE OF THE DESIGN (D2)

Figure 3.2 shows that the project design and the construction processes are influenced by the first tier Environmental Impact Assessment (EIA) (see also Chapter 4) and the stakeholder concerns. This is because what one is trying to achieve at this stage of the development process is a project design that suits the project definition, but that also causes the least amount of environmental impact. At this stage of the project it may be possible to "engineer out" some of the environmental effects of the completed project. It may also be possible to engineer out some of the effects of the construction process.

A key feature of the outline design stage for any dredging project (see Box D2, Figure 3.1) is the question of where the dredged material is to be placed. This is covered in full in Chapter 7, but it is important to note here that the ability of the project developer to find good uses for the material dredged may be the key to the whole project's success, particularly if the material is being used rather than being merely relocated. This is often where the "win/win" outcome is conceived.

Once again, detailed knowledge of the project area (see Section 3.4) is fundamental in assessing both the effects of the finished design (Chapter 4) and the effects of the construction processes on the natural environment (Chapter 6). At the end of the outline design stage, one should be able to characterise the project in terms of:
- environmental effects of the completed project and any alternatives;
- construction processes involved;
- environmental effects of the construction processes, and
- potential impacts of the noted effects, both positive and negative.

One should be able to use risk analysis to identify the severity, consequence and relevance of each impact and to focus on where additional detailed studies/ investigations are required to:
- reduce the number of alternative schemes;
- reduce the chances of the identified risks being realised;
- reduce the severity of the consequences of the risks;
- eliminate those risks that can be shown to be of very low chance or negligible severity, and
- reduce as many areas of uncertainty as possible.

This first evaluation then shows where investigations and studies are required. These are carried out so that the whole of the above process can be repeated later in much finer detail when more information is available.

3.9 DETAILED DESIGNS AND ALTERNATIVES (D3)

Bear in mind that the detailed design of the project is an iterative process and cannot be completed in its entirety until all the investigations and studies have been undertaken (note the interconnections between Boxes E3 and D2 in Figure 3.1). As each piece of information is received and analysed the project may become more highly refined and alternatives may be eliminated, but the final design cannot be "cast in stone" until all data have been processed. Thus, everyone involved should acknowledge that the team developing the project must be prepared to make changes right up to the end of the design process.

The establishment of a data collection programme in the early stages of the project is clearly of great importance. This includes beginning the collection of data at the earliest moment and being prepared to continue its collection until the construction process has been completed and post project monitoring has been undertaken (see Chapters 5 and 8). A considerable amount of environmental monitoring data exhibits seasonal characteristics, therefore, the data should be collected over at least a full year, and preferably a few years, to establish the seasonality and the variability within a single season. Note also how the pre-project, project and post-project monitoring are linked.

All in all, the final design and construction process should strike a balance amongst all the various pressures being applied to the development process. This balance is illustrated in Figure 3.3. At the conclusion of this stage the project promoter should be able to identify:
● the preferred scheme;
● the preferred construction process (methodology);
● the methods of monitoring the construction process and the effects of the project on the locality, and
● the methods by which environmental control and mitigation may be applied during and after the construction of the project.

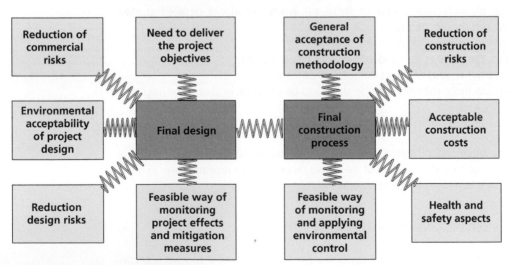

Figure 3.3 The pressures influencing the Final Design and Final Construction Process.

3.10 LINKAGES

This section demonstrates how some of the actions taken in the early stages of the development process are linked to matters that are generally to be considered in detail at a later stage. For this reason it is imperative to have a broad appreciation of all the aspects covered in Chapters 4, 5, 6, 7 and Annex A, before embarking on the development process in earnest. Examples of the linkages between the various stages of the project are described below.

3.10.1 Stakeholders

There will be many stakeholders associated with any large development and the identities of some of these have been described in Chapter 2 (the players). There is a consistent link between the actions of the stakeholders and the development and subsequent use of the project to be developed. This will be readily apparent if the stakeholders' involvement, as would be appropriate in developments with open consultation, is described:

1. In the early stages of project development the stakeholders will be consulted and their relationship with the project will be assessed. Their views will be collated as an aid to defining a project that is as acceptable as possible, within the constraints of the project objectives, and their concerns about the development, and its construction phase, will be noted. Action groups that may appear to be actively hostile to the project should also be involved in this consultation process.

2. During the preliminary and detailed risk analyses processes, the views of the stakeholders will be used to help in identifying those concerns which are, or are perceived to be, of major importance. None of these concerns should be dismissed out of hand. Concerns that are perceived to be of major importance have to be treated as just that – as matters of importance. During this phase a full and frank exchange of views about the development is crucial. This will help some stakeholders to "buy in" to the project.

3. As the project moves forward towards the construction stage, major concerns relating to the effects of the construction operations on the local ecology usually arise. Major stakeholders should be encouraged to take an interest in the collection and analysis of monitoring data, so that they may see how the environment is being protected during the construction stage and during any subsequent post-project monitoring period. During this phase the stakeholders will begin to "buy in" to the development and encouraging them to do so is essential. Key stakeholders can be asked to join a Technical Advisory Group (TAG) that meets regularly to discuss the progress and effects of the project.

4. Finally, after completion of the works, the TAG can be continued as a "partnership" of interest organisations, to discuss matters of mutual concern regarding the aftermath of the project and its relationship with adjacent activities. In the long run this benefits all stakeholders, because stakeholders who were originally reluctant, once having had a positive experience, may now be encouraged to develop their own activities in the future.

3.10.2 Win-win potential

One of the main difficulties in dredging-related developments is often deciding where to relocate the dredged material. Dredged material can be used in many ways (see Chapter 7 where box M6 may be linked to box D1), depending on the type (and level of contamination, if any) of material, the volume being moved, the rate of removal and the location of the site relative to potential relocation sites.

Keeping an open mind with no preconceived ideas is essential as the optimum place for relocation or use of the material may not be obvious at the outset. Ideas for possible uses may be suggested during stakeholder consultation (see Section 3.7) or may be more obvious when the material has been fully characterised.

In some respects, many dredging projects should be viewed not as dredging design tasks (which are often relatively easily accomplished), but as material management tasks. The art, therefore, is to treat the dredged material as a resource, as recommended by the London Convention (see Chapter 7) and to find the most appropriate use for it. In many cases, this may involve splitting the project down into separate parts to reflect the differing materials in the site. It is in this phase of the project that innovation comes to the fore.

The essence here is to bridge the needs of the project promoter with the aspirations of some of the stakeholders, following environmental guidelines, and create as many win-win solutions as possible.

3.10.3 Environmental control

The collection of data to characterise the dredging site in terms of the natural environment might be termed "environmental measurement" (see Chapter 5, where box M5 and box E1 may be linked). The proposed works on the site are then assessed to determine the nature and severity of their effects on the natural environment. This is usually called an "Environmental Assessment" (EA) or an "Impact Assessment" (EIA). From this assessment a determination can be made whether or not it is necessary to introduce any "environmental controls" during the works. The formalised system of controlling the works on the site and monitoring to demonstrate that the control is having the desired effect is called the "Environmental Management Plan" or EMP (see Chapter 8 and Annex A, Appendix 5).

The evolution of all the above items takes place at different times during the project development process, but they are all inextricably linked. It is, therefore, useful to take account of the following:

1. Think hard about what environmental measurements are going to be made and how they are to be used. Remember that they may be used in the establishment and verification of models of the site, but they may also be used as baseline data (i.e. before works data) and if so, the same type of measurement and analysis should be performed at each stage of the project.

2. The impact assessment may demonstrate (through modelling or otherwise) that suspended sediment is the key water quality parameter that is going to be affected by the works. How can it be measured during the works? How will this relate to baseline data? Can this data be measured in real time and used to control the operations?

3. How variable are the site conditions? How do the effects of the works compare to natural variability?

4. Can baseline conditions actually be defined or are they too variable? How are they to be defined from the measurements? How are extraneous (define?) effects identified and removed?

5. Can the effects of the works really be measured in the same way as the natural effects? Can extraneous effects be separated from these?

6. Can the works actually be controlled on the basis of the measurements made or is it possible to mitigate the effects in some other way?

These questions clearly demonstrate that there are strong linkages running through the whole environmental assessment and control processes, leading to the EMP. The moral is, therefore, that one should not wait until the final moment before developing the EMP. The EMP should be in mind from the moment that environmental measurements are commenced in the early stages of the project. To do this some small iterations of the whole project development process, at least in outline, should be run in the first development stage.

3.10.4 Risk assessment, investigation and study focus

Risk assessment is a management tool (see Sections 3.11.2 and 4.7) that is used to highlight, analyse and manage risk over the whole project development process. It also has certain features that make it useful in focusing work during the development process.

One of the great quandaries in the development of a project is deciding where, and to some extent when, to put money into investigation and analysis. During the early stages of the project existing data is collected and analysed. From these analyses certain areas of the environment that are largely unknown and others that may only be partially understood become apparent. The question then arises: How do we prioritise any additional investigative and analysis work?

At this point Risk Assessment (RA) may emerge as a useful tool to prioritise the work. The advantage of using RA to do this is its ability to identify not only the actual risks, whose probability of occurrence and severity of effect are of consequence, but also the perceived risks, which may have a low probability and severity, but whose existence may be of disproportionate concern to stakeholders. Allaying the fears of the stakeholders is as much a part of the process as reducing the more urgent risks of the project. Thus, RA may be able to point the project development team towards those areas of the unknown that give the greatest benefit in moving the project towards completion.

3.11 THE BASIC TOOLS

The tools needed to move through the implementation process include the various technical devices that are available for making predictions, assessing results and making evaluations. They include environmental assessment, risk assessment and modelling. These tools are applied to control and coordinate the process in a coherent manner.

3.11.1 Environmental assessment

Evaluating the benefits and effects of a project is a highly complex matter with many competing pressures (see Figure 3.3). The ability to define how to evaluate the merits and disadvantages of a project and to display these for others to review is difficult and crucial.

In effect, small "mini evaluations" are often being made throughout the development process. The way these evaluations are made must be as transparent as possible, and the basic data on which decisions are based should not be obscured. Whether the evaluations are small or large, the basic balancing act depicted in Figure 3.3 is at work in the evaluation process.

A common evaluation process is called the "Triple Bottom Line" approach, also referred to as People-Planet-Profit. In this, the attributes of the project are grouped into three categories; social, environmental, economic (see Text Box). For each of these categories the positive and negative effects of the project are listed and scored or ranked (weighted), allowing the environmental, economic and social aspects of the project to be evaluated.

Clearly, the characteristics of a project and its construction processes may be viewed from a range of perspectives. Each of these will vary depending on the viewpoint of the person carrying out the evaluation and weightings will also be made in this light. In any evaluation the following should be provided:
● A clear description of how the scientific background to the project has informed the evaluation process;
● The point at which judgement, rather than science, has been used in the evaluation process;
● A description of all scientific bases for judgements so that others may see if they agree with the judgements made, and if necessary make their own, and
● The evaluation criteria that lead to any conclusions.

In particular, judgements made at an early stage of the project, and cast into the process, should not become invalidated by later changes in the project or its construction processes. Additional information relating to Environmental Assessment is given in Chapter 9.

3.11.2 Risk assessment

Risk Assessment (RA) is a relatively modern management tool for handling all aspects of risk and uncertainty connected to project development. It is addressed in detail in Section 4.7. It is sufficient to say here that it may be used to:
● identify all project risks and uncertainties, both in connection with the design itself and the construction process;

Triple Bottom Line: A methodology for assessment of dredging technology and placement options for a major dredging project taking account of the environment

Dredging technology

The short-listed dredging technology options are compared for each dredging area using multi-criteria analysis. The multi-criteria analysis involves comparison of options against relevant environmental, social and economic criteria (the triple bottom line). Options are scored for each criterion on a scale of 1 to 10 with the highest value, indicating highest suitability. The analysis is performed separately for dredging of uncontaminated and contaminated materials. The criteria address the key technical, operational and environmental factors that might influence the selection of the preferred dredging technology.

Placement options

The three options are compared using a multi-criteria analysis against the following criteria covering environmental, social and economic aspects, in a similar approach to that of dredge technology selection. Criteria are selected giving consideration to:
- relevant environmental, social and economic factors (including shipping safety)
- findings from the project risk management process
- assets, values and uses in the vicinity of the locations of the short-listed options
- indicative dredging schedule.

To compare the alternatives, the three options are scored in relation to the identified criteria. To investigate the sensitivity of the outcome to different criteria weightings, three weighting approaches are used in the analysis as follows:
- Non-differentiated: each criterion is given equal weight.
- Balanced: each category (economic, social and environmental) has equal weight with weightings within categories assigned to reflect importance.
- Green: the environmental, social and economic categories are given weightings of 60%, 25% and 15% respectively.

- focus the design team on those areas where additional investigation and analysis needs to be carried out (see Section 3.10.4);
- provide a basis for evaluating the environmental effects of the entire project, and
- provide a framework for managing risk throughout the project.

3.11.3 Modelling

Models are used in all branches of scientific study. They may be physical, chemical, biological or numerical, but all have the same purpose, to replicate the natural environment and/or human interference with it in some manner. Models are used essentially as predictive tools. For that reason, they are extremely useful in the environmental assessment process, as they can be used to predict the changes brought about by the planned works and the construction processes.

For all models a number of golden rules apply for their use:
- A model must be based on some demonstrably scientific theory or empirically derived relationship(s);
- A model must be capable of calibration using prototype field data;
- A model must be validated using a totally different data set from that used for the calibration process; and
- Any restrictions or limitations on the use of the model must be stated when the results of the modelling are given.

Additional information on modelling is given in the final section of Chapter 4.

CHAPTER 4

Effects, Ecology and Economy

4.1 INTRODUCTION

The potential environmental and socio-economic effects of a "constructed project" resulting from dredging and reclamation works are numerous. The variables in any project are site specific and cause and effect relationships vary considerably. No chapter can present a detailed overview but the information given here attempts to present as complete a picture as possible.

To create a structure for discussion, a combined overview of possible physical, eco-logical and socio-economic effects is presented, with concise lists in the tables of the possible positive and/or negative effects related to different types of projects. The subject is presented in a matrix-like structure, highlighting selected examples from actual projects. The purpose is to allow readers to easily select the types of dredging project most similar to their own and to get insights about possible effects.

As a first step, the Environmental Assessment (EA), also referred to as Environmental Impact Assessment (EIA), is described because as a tool for planning new projects, EA has grown in importance in recent times. In some countries the terminology EA or EIS or EES are also used to indicate an environmental evaluation and sometimes the terms may have a formal or legal specificity.

In the second section, risk assessments and risk management, including materials characterisation, in various stages of a project – from feasibility to final design and construction – are discussed extensively.

The third section introduces the traditional classification of dredging and reclamation projects which, in principle, is divided according to the purpose of the project: Capital, maintenance and remedial dredging. To facilitate a more systematic description of effects especially when making references to actual case studies, a further subdivision of these main groups has been made.

In the fourth section the potential effects of dredging and reclamation projects are introduced and defined. The types of effects are categorised into physical, ecological, socio-economic and political, with the emphasis on effects that endure beyond the actual construction period.

Sections five, six and seven examine the typical effects for each of the three main project categories (capital, maintenance and remedial dredging). Selected case studies, representing examples of each of these project categories, are described in Annex B. When focusing on capital dredging, a further division by type of aquatic environment of the project site (tidal inlets and estuaries, open wave-exposed coasts, and rivers) is distinguished. Short descriptions of these aquatic environments and characteristic ecosystems are given in Annex C.

In section eight various techniques for the prediction of environmental impacts are described and future trends in evaluation tools are briefly discussed.

4.2 CLASSIFICATION OF DREDGING AND RECLAMATION PROJECTS

To understand the significance of the effects of dredging and reclamation works, a full understanding of the different types of dredging and reclamation works is necessary. Dredging is a general term covering a wide variety of different activities, which are traditionally divided into three main groups: Capital, maintenance, and remedial works.

4.2.1 Capital dredging works

Capital dredging involves the creation of new or improved facilities such as a harbour basin, a deeper navigation channel, a lake, or an area of reclaimed land for industrial, residential or recreational purposes. Capital dredging is also involved in large infrastructure projects, such as bridges and/or tunnels, beach nourishment, mining and aggregate dredging, i.e. exploitation of energy, mineral resources or construction material.

Such projects are generally, but not always, characterised by the following features:
● relocation of large quantities of material;
● compact soil;
● undisturbed soil layers;
● low contaminant content (if any);
● significant layer thickness; and
● non-repetitive dredging activity.

Clearly, a negative environmental effect of such dredging or disposal action is often the destruction of natural habitats (e.g. reclamation of wetlands, disposal of excavated material in biologically sensitive zones, disappearance of inter-tidal flats). On the positive side, however, additional wetlands or inter-tidal flats can be created and important sites can be protected from erosion. How exactly the environment is affected is largely determined during the design phase of a capital dredging project. The spread of chemical contaminants adsorbed on the dredged material is generally of little concern as the material to be excavated is virgin.

The environment is best served when a full analysis of the environmental implications is integrated in the design process of capital dredging works. It is here that the main remedial actions can be taken, namely informed site selection, improved design (e.g. of the location, depth, erosion protection), and suitable selection of dredging plant. Furthermore, use of the dredged material can also have a large positive effect on the overall environment at the project site.

4.2.2 Maintenance dredging works

Maintenance dredging concerns the removal of siltation from channel beds, which generally occurs naturally, in order to maintain the design depth of navigation channels and ports.

The main characteristics of maintenance dredging projects are:
- variable quantities of material;
- weak- to well-consolidated soil;
- contaminant content possible;
- thin to variable layers of material;
- occurring in navigation channels and harbours;
- repetitive activity;
- dredging in a dynamic environment; and
- sedimentation/erosion is on-going while dredging.

Since maintenance dredging occurs mainly in artificially deepened navigation areas, the dredging activity is, in itself, not necessarily damaging to the natural environment. The main potential for environmental impact is from the disposal of the dredged material and by the increasing quantities of suspended sediments during the dredging process (possibly inducing dispersion of contaminants). Suspended sediment problems can, however, be readily controlled by careful choice of dredging equipment and procedures as discussed later. These problems are compounded by the need to repeat maintenance dredging regularly, since siltation is a never-ending story.

The contaminant content of the material to be dredged can also have an important bearing on the environmental impact. Many cities have for a long time allowed their sewage and industrial waste to spill out into navigation channels, while the silt on the bottom of our rivers has over the years soaked up numerous contaminants that have entered the water stream. Dredging can spread the particles to which the contaminants are attached and increase the speed with which they spread. The extent of this phenomenon depends upon the type of dredging equipment used and the attentiveness of those engaged in the dredging work.

4.2.3 Remedial dredging works

It should be appreciated that both capital and maintenance dredging can have a beneficial or remedial side effect of removing contaminated material from the dredging location. But this is not their primary purpose. Remedial dredging, purely for the purpose of cleaning the dredging location, should be recognised as a separate type of dredging, with distinctive characteristics. It is carried out in an effort by society to correct past actions, which have, in some cases, resulted in heavily contaminated sediments. Remedial dredging requires the careful removal of the contaminated material and is often linked to the further treatment, reuse or relocation of such materials. Its characteristics are:

- small dredged quantities;
- high contaminant content;
- weak- to well-consolidated soil; and
- a non-repetitive activity (if the problem is effectively controlled at source).

Given that the aim of remedial dredging is to remedy an existing adverse situation, the main environmental effect is bound to be positive, provided the dredging is done with great care and does not significantly damage the environment in any other way. A prerequisite for a successful remediation project is the removal of the source of the contamination prior to the start of any remedial dredging.

4.3 GENERAL EFFECTS OF DREDGING AND RECLAMATION PROJECTS

Clearly, each of the three categories of dredging described above has different goals, but none has dredging as a goal in itself. Dredging is a means to an end, such as the deepening of a harbour, the removal of contaminated material, the creation of wetlands or the construction of a safe place for industrial or residential development.

In order to attain these objectives, specific requirements must be met regarding, in general, the dredging equipment, and, in particular, the acceptable level of environmental impact. In attempting to meet these requirements, the dredging industry has developed a wide range of equipment, each with its own specific characteristics and abilities to deal adequately with the challenges of a particular project. These are described in Chapter 6.

This chapter reviews a number of typical dredging projects, focusing on their environmental implications and socio-economic impact. The projects have been selected, covering the most common and frequent projects on a worldwide basis separated into the main groups of dredging activities mentioned above; capital, maintenance and remedial dredging. Note that the selection of examples has been based on relevance, but also was limited by the availability of information open for publication.

4.3.1 Definitions and scales

This section contains definitions and general descriptions of possible effects related to dredging and reclamation projects. That is, some of the potential positive or negative effects on the surroundings brought about by projects in which dredging and reclamation form a part are highlighted here.

The effects of these projects on the surroundings include:

- the consequences of handling dredged material in the aquatic environment;
- re-shaping the bathymetry; and
- the use of new areas or facilities for which they have been designed, e.g. as part of an infrastructure improvement.

The types of effects thus encompass those related to the possible alterations of:
- the physical-chemical conditions;
- the biological-ecological conditions;
- the socio-cultural conditions; and
- the economic and operational conditions.

As illustrated with the basic impact chain (see Figure 4.1) obviously once a decision is made to initiate a project, environmental impacts will automatically follow at both the dredging and placement sites. These impacts may be positive or negative. However, the decision to proceed or not is ultimately made in the best overall interest of society, after balancing the potential environmental impacts with the benefits of the project and the cost to construct and maintain.

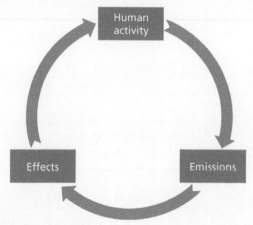

Figure 4.1 The basic impact chain.

Longer-lasting effects
The focus here is effects that last longer or that give rise to related effects which have a duration longer than the actual construction period. Shorter-term effects caused by the presence of dredging and reclamation equipment, such as noise and disturbance of navigation, are discussed in Chapter 6.

An exception to this is the impact of phenomena such as increased turbidity and sedimentation. These are often considered temporary effects strictly related to the construction activities, but are included here because the consequences of these effects may be of a longer duration than the construction period. When they initiate other effects, for example on ecology, their effects may last for a considerably longer period. However, turbidity (see Text Box and Figure 4.2 on page 38) is also a natural phenomenon.

Four main groups of possible effects
The possible effects caused by dredging and reclamation projects are divided into four main groups: the first two are directly related to the natural environment and the remaining two groups are related to society (see Table 4.1).

Turbidity

Turbidity is a description of how clear water is, or in other words, the degree to which water contains particles that cause cloudiness or backscattering and the extinction of light. Turbidity occurs naturally and high turbidity may be caused by a high content of fine sediments and/or of organic particles, or by low concentrations of material with high light absorption. Suspended sediment concentration (TSS or Total Suspended Solids) is the measurement of the dry-weight mass of sedimentary material that is suspended in the water per unit volume of water . A more detailed description of turbidity is given in Chapter 5.

Figure 4.2 Turbidity can be caused by dredging, but it also exists naturally. Shown here, the Pearl River (Hong Kong) turbidity front where the river meets the oceanic waters.

Table 4.1 Classification of possible dredging- and reclamation-related effects

Possible effects	Examples
1. Physical environment	Waves, current, water level, turbidity, sedimentation, coastal morphology, geology and so on
2. Ecology	Vegetation, fish, mussels, corals, birds, marine mammals and so on
3. Economy	Infrastructure, industry, fishing, tourism and recreation, farming and so on
4. Political and social	Environmental awareness, heritage, welfare, changes in labour opportunity and so on

The possible effects tend to appear as a chain of related effects where effects from group 1 influence group 2, which in turn could influence groups 3 and 4. Moreover, there is often a close interaction between effects within the groups 1 and 2 and again within groups 3 and 4. The interaction amongst the nature-related effects and society-related effects would normally be restricted to a one-way chain of reactions. If the nature-related effects give rise to unacceptable effects on society's activities, feedback actions may be initiated to readjust the project in order to reduce the undesirable effects. An example of a general chain reaction of effects is outlined in Figure 4.3. However, the extent to which this chain reaction occurs is highly variable and, in some cases, may hardly occur at all.

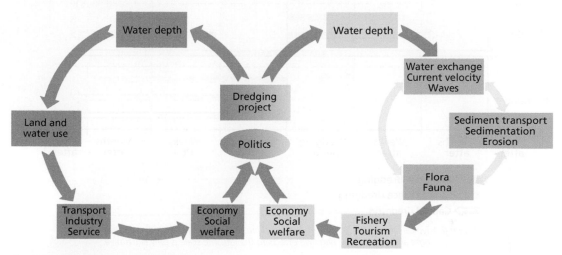

Figure 4.3 General chain of possible effects.

Figure 4.3 illustrates the general chain reaction of effects, which, as a result of a considered development project, might occur when either dredging or reclamation introduces new water depth conditions. The loop on the right side reflects the effects on the existing hydrographic and ecological conditions, and the socio-economic effects that might be related to a change in these conditions. For example, changes in hydrographic and ecological conditions, may affect fishery and tourism sectors that based their income on the particular area or they may influence the recreational value. The left-hand loop considers the effects on the "new" land-and-water-use potential, which the development plan is thought to introduce. It shows how this contributes added value by improving conditions for industry and trade. The two loops meet where the political decision is taken with a balanced view to whether the effects on the existing conditions are acceptable compared to the benefits likely to be introduced with the future conditions.

Time scales
The duration of the effects is defined here as being the time span from the moment when the effect could possibly have been detected (or measured) to the moment when the specific effect can no longer be detected or measured. Some effects may

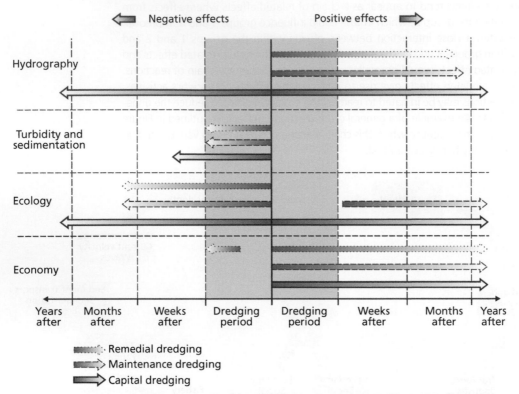

Figure 4.4 Typical time scales. The figure is divided into 2 parts, the possible positive effects to the right and the possible negative effects to the left. The scale is only illustrative.

be temporary and diminish soon after finalisation of the project. Other effects, especially those related to the effects on the ecosystem, may last for several years upon completion of the project. Still other effects may be of a more permanent nature, lasting at least as long as the lifetime of the actual project.

The duration of specific possible positive or negative effects depends very much on the type of environment and the type of project initiating the effect. Typical time spans of dredging and reclamation-related effects are presented in Figure 4.4. But note again that some of these effects may not occur at all.

Horizontal ranges
Like the time scales, horizontal ranges of the possible effects vary considerably depending on the project, the local physical and biological environment, and the economic and political situation. Typical horizontal ranges are presented in Figure 4.5.

The scaling terminology applied in Figure 4.5 is, as mentioned, arbitrary – near-field effects may accommodate effects on the specific construction site including an inner and outer impact zone adjacent to the site. Within these zones different criteria for allowable environmental impact may be defined. The far field may cover a larger

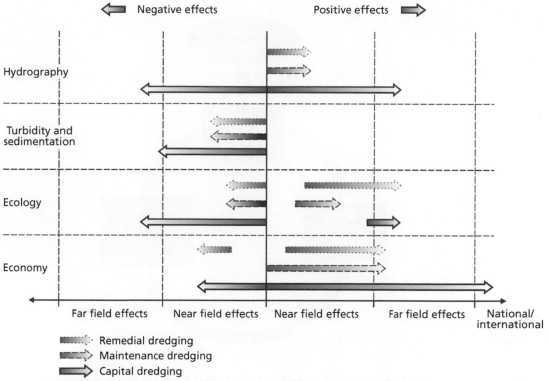

Figure 4.5 Typical horizontal ranges. The exact scale depends on local conditions. The figure is divided into a positive and a negative side. The scale is only illustrative.

regional area or even extend across national borders and will normally involve very low tolerable impact levels or may even have to obey so called "zero-impact" criteria.

4.3.2 Effects on the physical environment

Dredging projects very often lead to changes in the hydrographic conditions because dredging or reclamation projects may change the bathymetry in the development area. Changes in hydrography may also have an effect on the sediment transport, which may again influence the bathymetry.

The possible physical effects of dredging and reclamation projects are related to each other as a chain of related effects where the entrance point is the change in bathymetry (see Figure 4.6).

The effects on the physical environment caused by altering the bathymetry depend on the following parameters:

● existing bathymetry;
● shape and location of dredged or filled area relative to wave and current direction;
● hydrographic conditions (tide, waves, currents) low or high energy, and
● sedimentary regime (silt, sand, rock) sediment transport, sedimentation rates.

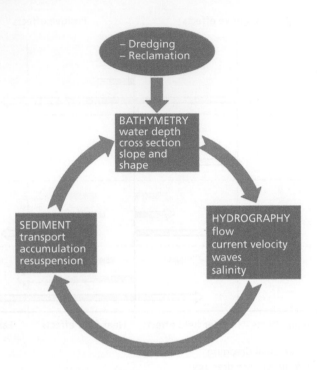

Figure 4.6 Chain of possible effects caused by changing the bathymetry.

The chain of effects caused by dredging may initiate feedback on the bathymetry either by counteracting the dredging (sedimentation) or by enhancing the dredging (erosion). The latter is less usual. Similar feedback on the bathymetry may be induced in the vicinity of a reclaimed area. In the case of beach nourishment, where the purpose is to maintain or improve the recreational value or the protective function of a coastline by feeding material to the beach, the intended effect is to counteract erosion. However, like the maintenance of water depths in navigation channels, beach nourishment is normally required to be performed at regular intervals.

As can be seen from the above list, the potential physical effects of a dredging and/or reclamation project can be very complex. In all cases they are greatly dependent on the local conditions and the shape and size of the changes in bathymetry. For this reason the more specific chains of physical effect of projects are described under the main project categories in Sections 4.4 to 4.6 in the examples given in Annex B. In this section the fundamental relationships are highlighted.

The energy of the water links the chain of physical effects together. The energy comes from the movement of the water and is dispersed and absorbed because of friction on the seabed and coastline. By altering the shape of the water pathway, the transmission of the energy in the water is altered. Widening and/or deepening the natural pathways of the water allows the energy in the water to be transported with less loss from friction. This situation is illustrated in Figure 4.7 where the increased cross-section reduces the current velocity in the dredged area but normally increases the flow capacity of the dredged area. If instead the natural pathway is limited by reclamation, the friction will be increased and the energy will be dispersed at the reclamation area, sheltering the

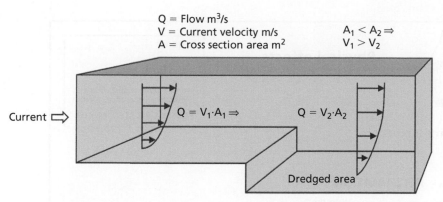

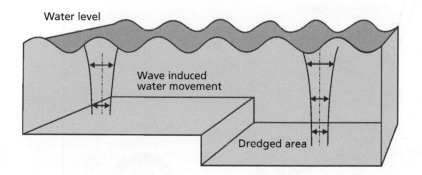

Figure 4.7 Relationship between cross-section area A, current velocity V and flow Q in a channel.

areas landward of the filled area. Locally increased current velocities can often be seen around reclamation areas as the available water pathway is narrowed. The effect on the wave conditions for the same situation as in Figure 4.7 is shown in Figure 4.8. The increased distance from the surface reduces the wave impact in the dredged area compared to the situation before.

Figure 4.8 Water movement under waves at different depths. The reduced wave action at the seabed inside the dredged area may increase sedimentation.

The increased water depth also allows faster propagation of larger waves, e.g. tidal waves, as the velocity in shallower water is a function of the depth. This allows the tidal wave to move further inland, with more amplitude than before, owing to the reduced friction.

On a larger scale the resulting physical effects depend very much on the type of environment. Figure 4.9 presents an overview of possible effects in various types of systems. The degree of effects will normally be proportional to the amount of available energy in the system. The potential effects of changing the bathymetry are likely to be largest in areas with large tidal ranges and/or high waves.

The sedimentary regime also plays an important role. Dredging and reclamation in areas where there is a naturally large sediment transport can potentially cause far greater effects than in areas where the natural sediment transport is limited because of, for instance, lack of loose sediment. A basic introduction to hydrography and sediment transport is given in Bearman (Ed.) 1989.

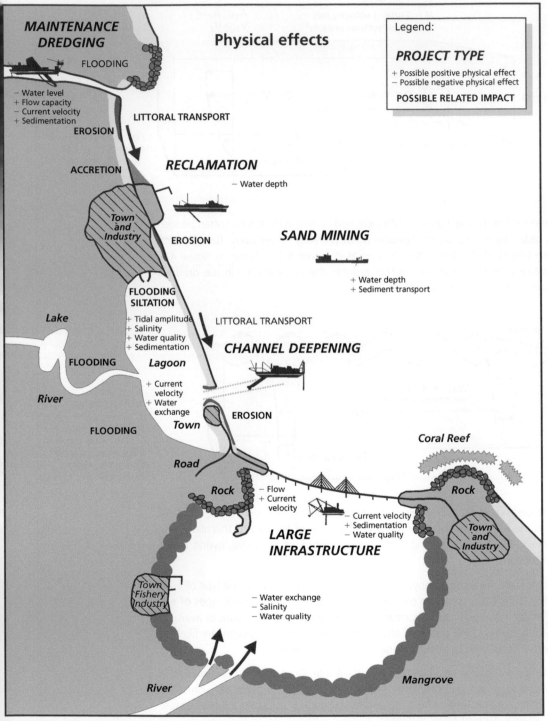

Physical effects

MAINTENANCE DREDGING

FLOODING

− Water level
+ Flow capacity
− Current velocity
+ Sedimentation

EROSION

LITTORAL TRANSPORT

ACCRETION

RECLAMATION

− Water depth

Town and Industry

EROSION

SAND MINING

+ Water depth
+ Sediment transport

FLOODING SILTATION

Lake

+ Tidal amplitude
+ Salinity
+ Water quality
+ Sedimentation

LITTORAL TRANSPORT

CHANNEL DEEPENING

FLOODING

Lagoon

River

+ Current velocity
+ Water exchange

FLOODING

Town

EROSION

Road

Coral Reef

Rock

− Flow
+ Current velocity

Rock

− Current velocity
+ Sedimentation
− Water quality

Town and Industry

LARGE INFRASTRUCTURE

Town Fishery Industry

− Water exchange
− Salinity
− Water quality

River

Mangrove

Legend:

PROJECT TYPE

+ Possible positive physical effect
− Possible negative physical effect

POSSIBLE RELATED IMPACT

Figure 4.9 Possible physical effects, which may be caused by dredging or reclamation projects in various environments.

Physical systems

The hypothetical map in Figure 4.9 depicts a variety of marine, estuarine and fluvial environments. Within these aquatic environments different types of geomorphologic forms appear, forms that are determined by the processes taking place and material present. These processes have different time scales and may vary in strength and frequency of occurrence. Some are directly visible like the action of waves and with a bit of patience the water level variation in tidal areas may be observed as well. Others can hardly be noticed even if observed over several years – these are the water level variations associated with climate changes and the geological forces associated with the dynamics of the crust of the Earth.

When considering the physical effects of dredging and reclamation, an understanding of the processes that take place at the particular site is required in order to be able to determine or forecast how the system will develop as a consequence of the changes occurring with or without the project. A particular project area often consists of different types of physical "systems", or the project may affect a larger area with different types of systems present. However, a preliminary assessment of the type of system present in the project area may tell what kind of typical effects one can expect as a result of a specific type of dredging or reclamation project.

Figure 4.10 shows a possible classification of coastal types controlled by tidal range. If the tidal range is less than 2 m it will often be the wind waves that provide the dominant coastal process, and beaches, spits and barrier islands will be the dominant features of the coast. Areas that experience tidal ranges above 4 m are dominated by tidal landforms such as tidal flats and salt marshes. Wind waves may still occur in these areas but the effect on the coastal landforms (Pethick, 1984) will only be minor, if any.

A large proportion of the seabed in these different coastal landforms is in motion (Bray *et al.*, 1997). Equilibrium is often characterised by a balance of material transported into and out of a site. If a hole is excavated it normally fills quickly with material;

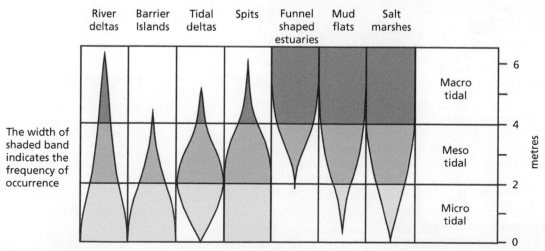

Figure 4.10 The type and frequency of occurrence of a wide variety of coastal land forms are related to the tidal range (cf. Pethick, 1984 modified after Hayes, 1973).

likewise a heap of deposited material tends to flatten out and disperse. The state of any site can be defined by the net transport of material into and out of it, taking into account any seasonal variations. To understand the regime of any site, the natural mechanisms of sediment movement in the area should be examined.

Historical maps, aerial photos, coastline profiles and mathematical sediment models are important tools in the prediction of effects from dredging and reclamation projects.

4.3.3 Effects on the ecosystem

Marine ecosystems and habitats are described in detail in Annex C.

The effect of a dredging or reclamation operation on the ecosystem can be divided into:
- direct effects caused by the construction activities, and
- indirect effects caused by:
  - the release of chemical substances from the dredged or disposed sediment
  - changes in the hydrographic regime, and
  - changes related to changes in land (sea) use.

A simplified model of possible ecological effects of a dredging operation is presented in Figure 4.11.

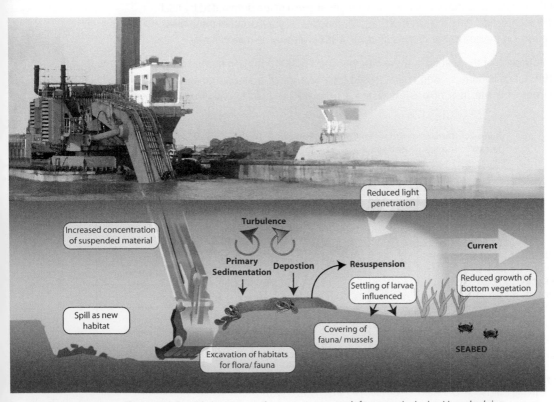

Figure 4.11 Possible impacts on the ecosystem caused, for example, by backhoe dredging.

Direct effects
Removal or burial of habitat: The removal and/or burial of habitat are the most likely direct effects of dredging and reclamation projects. Any direct loss is often neutralised by the establishment of new habitats, either directly in the dredged area as a result of improved bed conditions or by the introduction of new habitats on the slopes of a reclaimed area.

The change of water depth in a dredged area may, however, be of such magnitude that the original ecosystem in the area is replaced by a new system suitable for the new conditions. Slow-growing and/or very fragile systems such as corals may only reoccupy the dredged area if the new conditions are totally suitable for both the corals and the algae with which they have a symbiotic relationship. Nevertheless, even though reoccupation of the areas involved often takes several years or even decades, these direct effects tend to be temporary rather than permanent.

Turbidity: Another direct effect may be caused by turbidity generated by a project. Turbidity can be harmful to the bottom vegetation and fauna owing to shading and burial by the released sediment. This effect only occurs when the turbidity generated is significantly greater than the natural variation of turbidity levels and sedimentation rates in the area (see Figure 4.2). Although the natural ecosystem adapts to the natural light and sediment conditions over long periods, changing these parameters, even for short periods, can lead to serious effects in areas where the system is not used to large quantities of particles in the water.

Corals are especially vulnerable to increased levels of turbidity, mainly because they are only able to survive if the rate of settling of suspended particles is relatively low. There are a number of management techniques such as tidal dredging, physical barriers and such, which may be used to mitigate the effects of dredging on sensitive organisms (Annex A). Seasonal- and area-determined spill rate allowances, as implemented in the Øresund Link project, are also a possible management technique (Annex B).

Upon the completion of a dredging and reclamation project, visibility in the water may be improved, for example, by removing easily resuspended sediments or improving sediment conditions. In this case even a minor operation may have a significant positive effect on the surrounding ecosystem. For more details about direct ecological effects caused by dredging and reclamation works see e.g. Sten and Stickle (1978), Donze (1990), Hodgson (1994) and Øresund Proceedings (1999).

Indirect effects
Indirect ecological effects are:
• those generated by the release or removal of chemical (toxic) compounds from the dredged or disposed sediment, and
• those caused by permanent changes in the overall hydrography.

A major environmental focus with regard to the dredging industry has been related to the handling and disposal of contaminated sediments. Even though the dredging industry itself has never been the primary polluter, it has been confronted with authorities and interest groups in projects involving contaminated sediments.

Acute effects from contaminated sediments may arise in the immediate vicinity of the actual dredging operation. Acute toxic effects are seldom seen in areas exposed to currents and waves, because the amount of acute toxic substances is small compared to the volume of water into which the substances are diluted. The release or exposure of oxygen-consuming, but otherwise harmless, substances in the sediment can also initiate an acute effect in the water, such as oxygen depletion. This effect is also diminished by water exchange and significant oxygen depletion caused by dredging is a rare event.

The risk of acute effects occurring is related to the degree of water exchange, with the highest possibility of occurrence in stagnant water. Knowledge of how different chemical compounds with acute toxic effects behave when either directly released into the aquatic environment, or contained in the seabed material being disturbed by a removal operation, is essential for the evaluation of the potential impacts. Some constituents may decompose quickly, causing the effect to be limited both in time and space. Others may be more persistent.

The release of chemical substances, which might be absorbed into the food chain, is the most severe ecological effect of dredging or reclamation projects in industrially contaminated areas. Levels of these substances must be carefully examined to see if they are high enough to negatively affect human health or are toxic to aquatic organisms. The compounds, e.g. heavy metals, are often not degradable in nature or are very slow to degrade and the concentration tends to increase with the level in the food chain. The effect on the ecosystem will often become evident in the form of reduced fertility or reduced resistance to infectious diseases at the top of the food chain several years after the release of these substances. More information about chemical contamination in sediments and the dredging industry can be found in Donze (1990).

Permanent effects
Dredging and/or reclamation projects, which have permanent effects on the hydrography in a large area, may also result in permanent ecological effects in coastal areas. Very often projects, such as deepening of channels, improve the water exchange in the affected area. In coastal areas this may cause the water body located landward of the dredged site to become more marine because of increased salinity. This will benefit marine species and negatively impact brackish species, which are pushed further landward than before.

A shift of the saline border can have an impact at water intakes with consequences for the operational lifetime of different types of processing plants and on groundwater flows (intrusion in lowlands). Channel deepening projects may not change the hydrographic conditions on the seaward side of the project because of the large volumes of water involved. The restoration of river systems and wetlands naturally improves the diversity of the ecosystem involved.

The possible direct and indirect effects on the ecosystem are even more dependent on the existing conditions than the physical effects. Figure 4.12 highlights possible

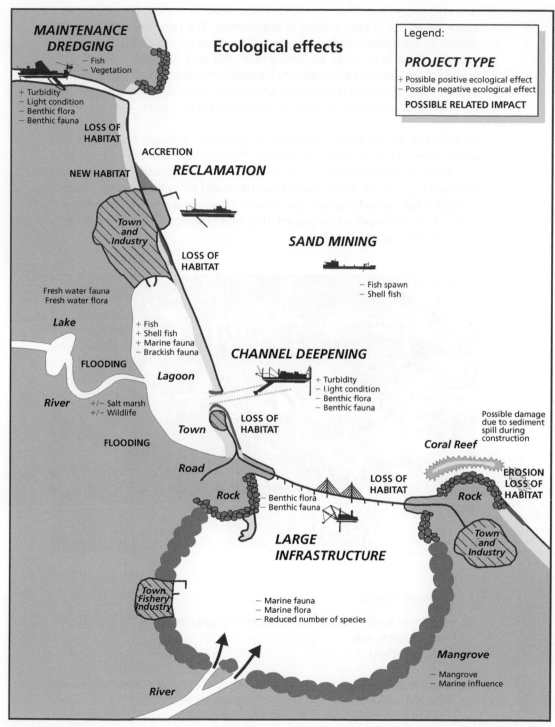

Figure 4.12 Possible ecological effects which may be caused by dredging or reclamation projects in various environments.

ecological effects in various types of environment. The potential effects will also depend on the climatic conditions in which the projects are carried out. For instance, in arctic or sub-arctic zones the slow growth and long biological turnover times would enhance any bioaccumulation of contaminants released by dredging projects. In these regions, fish and sea mammals belonging to the high end of the food chain are often an important part of the daily human food consumption.

In tropical zones, the presence of coral reefs makes the ecosystems very vulnerable to high levels of turbidity and sedimentation. The coral reef itself forms an important ecosystem, which hosts numerous other species that are affected if the corals are damaged (Figure 4.13). In some places coral reefs also act as natural wave breakers, protecting the adjacent coastline from erosion. Corals are slow-growing species and recovery from damage is a slow process. Thus where damage occurs, it causes significantly reduced species diversity.

Figure 4.13 Tropical marine environments are complicated ecosystems which are vulnerable to outside disturbances.

4.3.4 Economic effects

The economic effects of dredging and reclamation projects can be divided into:
- tangible effects, and
- intangible effects.

Tangible effects: are those that can easily be valued in monetary terms. This includes the increased value of new and existing investments in e.g. harbours, channels, property, ships, machinery, which benefit from the dredging and reclamation project.

The tangible effects can again be divided into direct and indirect effects. The direct economic effects normally include the benefits to those stakeholders with an economic interest in the project. When a port decides to increase its capability to service larger ships, the project economy is evaluated, based on the direct outcome of the investment for the port. Indirect effects include spin-off of the investment to other parties, e.g. companies servicing the port, increased employment opportunities, lower transport costs and cheaper goods in the market.

Intangible effects: are those that cannot as easily be valued in monetary terms. This includes nature conservation, tourism, or stimulation of business climate and increased trade area for a harbour or city. Other benefits might be increased security, recreational opportunities and aesthetic appeal. These effects are difficult to measure, but the effects may easily be larger and last longer than the direct economic outcome of the project. Improving inland shipping capacity by dredging channels can reduce road traffic problems by reducing the amount of heavy traffic and thus costly wear on roads. These kinds of interactions may influence the entire economy of a region, a state or a country.

The economic effects of dredging or reclamation projects can be positive or negative depending on the point of view. What is positive for one town or port may turn out to be negative for a neighbouring town or port, which loses traffic and commerce when a neighbouring harbour is improved. This competition between ports usually has the effect of maintaining a country's international competitiveness and is a net benefit to the country as a whole. On the other hand, the economic growth in general may leave plenty of room for neighbouring ports to extend, specialise and further compete. Dredging projects that increase the tidal range locally, may have a positive spin-off on fishery in the area, but may also have a negative effect on local farms, which become more exposed to flooding and saline water intrusion.

Part of the economic evaluation of a new development plan such as a tourist reclamation area, or an extension to a port, comprises consideration of the pressure on existing hinterland facilities. Examples of this are the impact on existing fresh water resources, power supply, communication systems, waste-water treatment, solid waste collection, roads, navigation channels (capacity and safety) and so on, both during the construction phase and after completion of the project. The costs of the operation and maintenance of these utilities form part of this evaluation. The total costs of the project must be balanced by the potential benefits to the economy in the region and increased welfare to its inhabitants. The effects on the regional economy may comprise attraction of investments, influence upon payment balance, taxation base, competition for customers and so on.

The effects on socio-economic conditions may involve:
- considerations of employment opportunities likely to be created both during the construction phase and in the operation phase with identification of age group, sex and profile of the required categories of the work force,
- the possible benefits/disadvantages to the local community resulting from an eventual shift in employment structure and the influence upon local salary levels,
- the value of the potential sectors, e.g. fishery, impacted by the project and impact on the employment situation in affected sectors,

- the demands for housing and urban development, and the impact on the cost of living for the local population, e.g. property values and food prices, and
- the influence on public health, e.g. noise, dust, air and water quality.

The example of the Delta Plan (see Text Box) shows how economic analysis can assist in considering a problem and in posing the right questions.

How ecological considerations changed a plan

An interesting historic example of valuation of effects in a project that comprised dredging and reclamation activities was made in the benefit-cost analysis applied in the planning of flood prevention constructions for part of the Rhine Delta (Oosterschelde, the Netherlands) in 1974, the so-called the Delta Plan (Hufschmidt, 1983).

In 1953 in the southern part of the Netherlands, as a result of severe storms and high tides, the sea broke through the dike system and caused widespread damage, including the loss of 1,853 lives. Intensive research into ways of preventing similar disasters began. Initially a dike closing off the estuary was thought as the best solution. Public objections, however, led to reconsideration. The estuary, in its natural state, provided valuable environmental benefits: as an important water resource for commercial fisheries and shell fish production; as the major sport fishing area and a popular site for other water-based recreational activities, and as a nursery for small fish of significance for North Sea fishery and a habitat for migratory and breeding birds.

The first closure design would have destroyed all these attributes. A commission was established to report on all safety and environmental aspects of the Eastern Scheldt closure. It recommended that the dike closing the estuary should not be built. Instead a special dam with large gates should be constructed. Its gates would be left open to permit normal tidal flows and only be closed to block off the sea when dangerous storm conditions prevailed.

Five basic plans with an array of variants were considered and detailed benefit-cost analyses were carried out on six selected plans. Benefits from reducing flood damage were estimated as the value of property and agricultural production that no longer would be lost through expected storm damage. The direct construction and operating costs for the original proposal were the lowest, but the indirect monetary costs owing to loss of income from fisheries would have been substantial. Analysis of the various plans was made and costs and benefits established at various discount rates. The final selection was based on a combination of priced elements and unpriced adverse impacts on environmental quality, not explicitly stated in the monetary analysis, as well as non-monetary costs in loss of life and human suffering.

4.3.5 Social and political effects

The social and political effects caused by a dredging or reclamation operation are influenced by the economy and the political situation. The effect of the project on the economy will also affect the political situation in the vicinity. This is comparable to the interaction pattern in the physical and ecological systems where the chain of effects is complex and depends very much on the local and regional situation. An operation, which will boost the economy in one town at the expense of another town, will, as a rule, evoke political resistance from the neglected area.

Political resistance may also be expressed directly or more indirectly as environmental concern. The eventual decision on whether to proceed with the project or not may well have to be taken at a national, or even international level, depending on the overall benefits of the whole development.

The effect of a project on local socio-cultural conditions may depend on the potential impact on resources valued by the local community, e.g. ecological assets such as beaches, marine and terrestrial ecosystems, and on the visual impression of the project and its impact on adjacent areas. Socio-cultural attitudes may also be affected if historical/archaeological sites or cultural traditions may be influenced by the project.

Many countries, young or old, industrialised or emerging, have a rich historical heritage manifest in archaeological sites. Sites that previously have been "terrestrial" may today be flooded owing to land subsidence. If dredging takes place in such areas and excavation reveals evidence of old settlements, the works are stopped and archaeologists called in to make an assessment of the significance of the discovery. If the discovery seems to possess the potential to elucidate historical aspects that are little known or controversial, the find is surveyed and "rescued".

In the 2006 Dutch guidelines for EIAs for the development of offshore wind farms (http://www.noordzeeloket.nl), there are questions raised to the developers regarding the effects of the activity on the cultural history and archaeology. Based on existing knowledge (Indicative Map of Archaeological Values of the North Sea, geogenetic and hydrographical knowledge), information should be provided on sites with low, moderate or high expectation of shipwrecks, but also special attention should be given to old (sub-Atlantic) channel depositions. Also the expected impact of the activity on prehistoric landscapes should be described.

Geology/geomorphology are also seen as having a certain (historical) value, which might be affected by activities such as dredging. How to quantify and describe "historical value of geo-morphological features present in the seabed" is not clear, but it has been introduced in The Netherlands as an issue in recent EIA procedures.

The possible socio-economic effects of various dredging and reclamation projects are highlighted in Figure 4.14. Such effects will depend very much on the actual political and economic system.

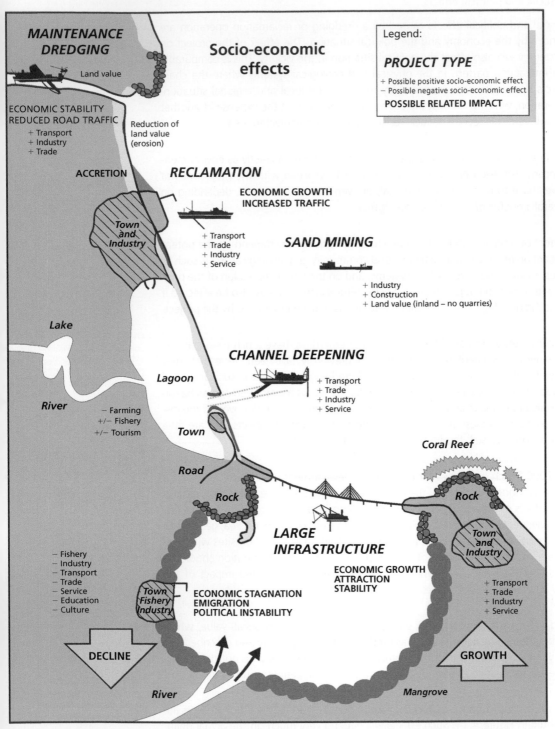

Figure 4.14 Possible socio-economic effects that could occur due to dredging and reclamation projects in various environments.

Economic analysis and environmental assessment
Economic analysis of proposed projects involves considerations related to macro- as well as microeconomics. The purpose of the analysis is to determine the overall economic benefits and compare these with the costs, and to guide or possibly redesign such that the project is given the best conditions to produce a satisfactory economic rate of return. Adverse environmental impacts are part of the costs of a project. Positive environmental impacts are part of the benefits.

Economic valuation of the consequences of changes in environmental state is part of the economic analysis and environmental assessment. The valuation relies on:
● first of all, critically understanding and measuring the physical, chemical and biological effects of the project, and
● secondly, the methods for placing monetary value on non-market goods and services.

Although this is an evolving science, advances have been made in the understanding of these "effect" relationships. Progress in the valuation technique shows that the uncertainties are reduced. This ultimately results in fewer decisions being made on a subjective basis making it easier to choose appropriate plans and technologies.

One of the difficulties in determining "effect" relationships and economic valuation is understanding how aggregate effects relate to one another and are integrated. The Pressure-State-Response Framework (see Figure 4.15) is at first glance causal: Human activities place pressure on the environment, leading to changes in the

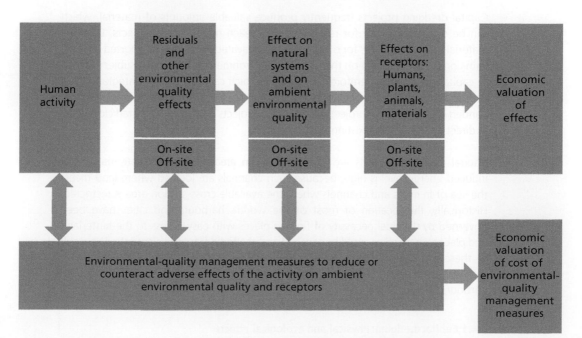

Figure 4.15 Pressure-State-Response Framework: Valuation of natural systems, environmental quality effects and environmental quality management measures (after Hufschmidt et al.,1983).

environmental state that ultimately produce human responses. But this causality breaks down as soon as there is aggregation of indicators or sufficient complexity in the causes of environmental change (Dixon, 1999).

4.4 EFFECTS OF CAPITAL DREDGING

The results of capital dredging are an increase of the existing water depth and sometimes the creation of new land. The purpose can be either to improve navigational conditions, to extract raw materials and/or for construction purposes.

Capital dredging may be located:
- in shallow near-shore environments, for navigational and some construction purposes,
- in deeper waters offshore when dredging for raw material (mining) and pipe or cable-laying activities occurs, and
- in inland waterways.

Globalisation and growing trade volumes are drivers for port expansion and upgrading of the navigation infrastructure requiring capital dredging works port regions. The increasing pressure from transporting goods on the road and rail networks, especially in Europe, has led to increasing interest in upgrading and extending the inland waterway systems.

Capital dredging projects frequently produce valuable amounts of material, which can be used beneficially for reclamation or beach nourishment projects. Dredged material often substitutes for other material excavated from and transported by land, with considerable impact on the terrestrial environment as a result. A subject related to capital dredging is therefore reclamation, which can either be an integral part of the project or completely separate. In the second case the dredged material is sold or handed over to another project. Reclamation projects are typically located close to or in direct connection to existing land.

Capital dredging projects are often located in areas where the potential water-induced energy level is high, because such channels are located where land meets the sea or in rivers and channels where the available cross-section area is restricted. Historically, the location of most of the world's harbours and cities have been governed by the dual necessity of finding places with easy access to the hinterland and places where ships could be serviced and sheltered from the elements. The result of these functional needs is that almost all of the largest harbours and cities in the world are located in or near river entrances, straits or other suitable "natural" harbours.

4.4.1 Capital dredging: Physical and ecological effects

The effect of a dredging or reclamation operation depends very much on the type of physical environment in which it is located. For this reason the possible effects of

dredging and reclamation have been described in the context of the following typical environments:
- tidal inlets, estuaries, bays, straits and other tide- or current-dominated areas;
- open wave-exposed coasts or shelf areas; and
- rivers.

Each type of environment responds to any human-caused alterations of the bathymetry with typical chains of effects that, from a general point of view, are independent of geographical location. The resulting positive or negative impacts on individual flora and fauna species depends of course very much on local conditions, but the principal chains are still universal and independent of geographical location.

The magnitude of the effects depends, of course, very much on the relative size of the operation and the natural level of energy as discussed in the section on Rivers below. A special variant of capital dredging is the construction of embankments for bridges, dams and trenches for submerged tunnels. These large infrastructure projects often include capital dredging and major reclamations in the same project. One of these projects, the Øresund Fixed Link, has been chosen as an example of capital dredging (see Annex B), because this project has, in many ways, generated new standards for assessment and control of environmental effects.

In support of the following discussions on the potential effects of dredging activities, some of the physical processes that take place and govern the characteristics of the different aquatic ecosystems considered in this section are briefly described in Annex C. The social-economic importance of these areas is highlighted as well.

Tidal inlets, estuaries, bays and straits
Dredging
As a result of the morphological nature of inlets, straits, bays and estuaries, the limiting water depth is often found close to the actual dredging site where bars or sills form underwater obstacles. Deepening channels or lowering such obstacles allows more saline water to enter the landward sea area. How the sediment regime will be affected by this intervention depends primarily on how much the tidal prism (the total volume of seawater that moves in and out of the area over a tidal cycle) is changed. If significant changes are introduced either in the tidal prism or in the water flow or with regard to the relationship between fresh and seawater volumes, the effects of the project may be quite considerable.

For example, the deepening of a channel allows the tidal wave, which is very long, to penetrate faster and further landward than before because tidal wave propagation is a function of the water depth. This mechanism increases the water level fluctuations landward of the dredged area, which in some cases may lead to a higher risk of flooding. The tidal pumping, which acts as a concentrating factor for fine-grained sediment in estuaries, may extend and in some situations may lead to increased siltation rates landward of the dredged area.

Any increase in tidal activity increases the landward flow and water exchange of the dredged area. This may further develop the marine influence in the area, which often improves water quality with respect to oxygen levels, nutrient load and concentrations of contaminants. The marine fish species and their spawn, as well as the marine fauna, may thus experience more favourable conditions. The flora, however, may be reduced as a result of higher levels of energy and/or increased turbidity. The area of salt marshes, if any, may extend because of the increased tidal range, which again may favour bird populations and increase the overall biological diversity. These effects will, of course, only take place if the tidal influence is allowed to act further inland than before and if the hydraulic circulation pattern is not disturbed significantly.

Capital dredging normally includes the removal of consolidated materials varying from consolidated marine sediments to bedrock. Generally speaking the materials are, as a rule, uncontaminated, as they have never been in contact with the water during the time of industrial development. Dredged material is often usable as fill or construction material.

Table 4.2 Capital dredging in an estuary, strait, bay or tidal lagoon

Parameter and figure	Possible effects after dredging	Assessments to be made
Water levels Figure 4.9	Increased cross section ⇒ increased tidal wave velocity and less loss of energy ⇒ increased tidal range	Flooding risk landward of the dredged area
Water exchange Figure 4.9	Increased tidal range ⇒ increased water exchange ⇒ increased salinity ⇒ increased marine influence (see below)	Assess future progression of salt front. Assess water quality and ground water quality
Current conditions Figure 4.9 and 4.12	Increased tidal range ⇒ increased current velocity ⇒ change in sediment transport ⇒ possible damage to bottom life	Assess future current pattern and possible effects on sediment transport (erosion and siltation) including sensitive bottom life
Sediment spill Figures 4.9, 4.11 and 4.12	Increased turbidity during and after dredging ⇒ shading and burial of flora and fauna ⇒ loss or formation of habitat	Assess sediment spill and dispersion including background turbidity. Check for vulnerability of local marine environment
Removal of existing seabed Figure 4.12	Temporary or permanent loss of habitat ⇒ increased erosion ⇒ increased turbidity. Change in sediment type after dredging	Check significance of habitat to be removed. Assess possible recolonisation of species, rate and extent
Increased marine influence Figure 4.12	Extension of the marine ecosystem. Reduction of the fresh water ecosystem	Check importance of existing fresh water species to be affected
Disturbance and risk for accidental contamination Figure 4.12	Increased human activity ⇒ disturbance of sea mammals, birds and fish ⇒ reduction of population (migration, death) Contamination: Oil, chemicals nutrients and so on	Check existing conditions. Assess future conditions including availability of alternative habitat. Assess risk, possible remedial actions (way out corridors and so on)

The spill released from capital dredging projects can vary considerably, depending on the type of material to be removed, the type of equipment used and current velocities. The background turbidity in these areas is often of the same magnitude or higher than the turbidity created by dredging. If this is the case, increased turbidity of uncontaminated material should not necessarily be regarded as an environmental problem.

A checklist of parameters to be considered, possible effects of capital dredging and recommendations for impact assessments is given in Table 4.2.

Reclamation
Constructing new land or structures, which decreases the available flow cross-section, will generally have the opposite effect on the natural environment to that described under dredging.

Reduction of tidal action, water flow and water exchange decreases the marine influence in the area located landward of the construction site. The effects can be limited or, in some cases, even completely eliminated by carefully planned compensation dredging. The purpose of compensation dredging is to restore or rather maintain the natural hydrographical conditions in the affected area by adjusting the remaining cross-section to such an extent that the natural flow is maintained.

The effects caused by sediment spill and increased human activity will be similar to those described under dredging. A checklist of parameters to be considered, their possible effects and impact assessments to be considered is given in Table 4.3.

Open wave-exposed coast
Dredging
Increasing the water depth in the open sea has an effect on the wave regime if the water depth is less than half of the wavelength. This is the approximate water depth where waves start to be influenced by the seabed:
- if the water depth is greater, the wave regime will not be affected, and
- above the wave depth limit, the dredging-induced change in water depth has an increasing importance as water depth decreases, until the breaking zone is reached.

For example, increased water depth in a navigation channel allows the waves to move faster and with less loss of energy than before. However, the waves will tend to turn away from the channel as they are refracted by the edges of the channel. Increasing and concentrating the wave energy may cause increased coastal erosion in some locations and less in others. The resulting effect is controlled by:
- the available wave energy,
- the refraction pattern, and
- the geology.

Coastlines formed by loose sediment or fragile rock, facing large and deep free stretches where long waves can be generated, are more vulnerable to adverse effects caused by dredging than low energy sheltered coasts, or coasts formed by hard bedrock.

ble 4.3 *Reclamation in an estuary, strait, bay or tidal lagoon*

Parameter and figure	Possible effects after dredging	Assessments to be considered
Water levels Figure 4.9	Decreased cross-section ⇒ decreased tide wave velocity and loss of energy ⇒ decreased tidal range	Flooding of reclamation area Check extreme water levels
Water exchange Figure 4.9	Decreased tidal range ⇒ decreased exchange of water ⇒ decreased salinity ⇒ decreased marine influence (see below)	Assess water quality
Current conditions Figure 4.9	Decreased tidal range ⇒ decreased current velocity ⇒ change in sediment transport ⇒ possible damage to bottom life	Assess future current pattern and possible effects on sediment transport (erosion and siltation) including sensitive bottom life
	Increase in current velocity around reclamation area ⇒ change in sediment transport ⇒ possible damage to bottom life	Check erosion and revetment stability
Sediment spill Figures 4.9, 4.11 and 4.12	Increased turbidity during and after dredging ⇒ shading and burial of flora and fauna ⇒ loss of habitat	Assess sediment spill and dispersion including background turbidity. Check for vulnerability of local marine environment
Burial of existing seabed Figure 4.12	Permanent loss of habitat	Check significance of habitat to be removed. Assess possible recolonization of species, rate and extent
Decreased marine influence Figure 4.12	Shift towards a more brackish eco-system ⇒ less diversity in aquatic life	Assess impact on marine species. Check for possible mitigating actions (e.g. compensation dredging)
Disturbance and risk of accidental contamination Figure 4.12	Increased human activity ⇒ disturbance of sea mammals, birds and fish ⇒ reduction of population Contamination: Oil, Chemicals nutrients and so on	Check existing conditions. Assess future conditions including availability of alternative habitat Assess risk, possible remedial actions
Drainage Figure 4.12	Improved drainage capacity ⇒ improved water quality at fresh water intake/increased volume of fresh water resources	Establishment of design flows in drainage channels, backwater effect in the drains, optimisation of drainage reserve

On coasts formed by loose sediments (sand) the coastline is normally in some kind of dynamic equilibrium with the prevailing wave climate. The sediment transport on the coast is initiated by the wave action, which moves the grains in an onshore or offshore direction. The onshore/offshore movement generated by the waves is modified in a long shore direction by coast-parallel currents, which are usually generated by tide and wave action. The coast-parallel transport is normally referred to as littoral drift.

Removal of sand either by dredging or by trapping it in trenches and channels may disrupt the natural balance with a risk for increased erosion at some point on the coast.

Offshore sand removal

Removal of sand located offshore of the coastal sediment circulation does not have the same effect on the coastline. Sand deposited on the shelf is normally lost to the coastal circulation either because it was left behind during the sea rise after the latest ice age or because it has been transported out of reach of normal wave action, e.g. by sliding. The effect caused by dredging the seabed outside the high dynamic coastal zone is, from a physical point of view, very small, especially when compared to nearshore dredging projects.

Offshore dredging normally results in very small relative changes of the bathymetry with little or, more often, no effect on the wave climate, but the direction of the offshore drift should be checked to see if it takes part in the littoral sediment balance.

Dredging of the seabed removes the flora and fauna located inside the dredged area. The removal is only temporary in most cases. The new seabed may recolonise within a few years after dredging has stopped unless the change in water depth or the change in sediment characteristics is of such a magnitude that the original flora and fauna are unable to exist under the new conditions:

- if the seabed geology is changed permanently by the dredging operation, e.g. by removal of sand or alteration in bed mobility, the composition of species living on the seabed may change accordingly;
- if the dredging activity creates isolated deep depressions in the existing seabed, areas with stagnant and oxygen-depleted water may periodically form, resulting in less favourable conditions for the marine life, and
- If dredging is carried out at the wrong time of year, fish populations may be affected by removal or destruction of eggs, especially if spawn grounds are affected either directly by dredging or by increased sedimentation caused by spill.

Effects of spill generated by dredging projects in the open sea or on a wave-exposed coast are normally of a temporary nature. Light-sensitive flora is not often found in such areas and the dispersion of the spill is often so high that the resulting turbidity will be very low. Increased sedimentation caused by spill may locally affect mussels, fish eggs and other benthic fauna.

Coral reefs – the largest biological constructions on earth are, unlike many other marine species, very sensitive to increased turbidity and sedimentation. Coral reefs form important ecosystems in tropical seas and often form natural barriers against the sea. Destruction of coral reefs either directly by dredging or by increased sedimentation may lead to increased coastal erosion and to a significant reduction in fish population.

Cold-water coral reefs are also susceptible to turbidity and sedimentation. These reefs (see Freiwald *et al.*, 2004) are found in dark, cold, nutrient-rich waters off the coasts of at least 41 countries, in fjords, along the edge of the continental shelf and around submarine banks and seamounts in almost all the world's ocean and seas. They are found at depths of hundreds of metres up to just 40 metres.

Table 4.4 *Assessment checklist: Capital dredging including sand and gravel mining on an open wave exposed coast*

Parameter and figure	Possible effects after dredging	Assessments to be made
Wave conditions Figure 4.9	Increased depth ⇒ increased wave height and velocity ⇒ change in refraction and reflection pattern ⇒ erosion	Future wave conditions inside the dredged channel and land ward hereof to be estimated
Littoral transport (coast parallel transport of sand) Figure 4.9	Increased depth ⇒ reduced transport capacity ⇒ increased sedimentation and possible down drift coastal erosion ⇒ possible damage to bottom life	Existing and future littoral transport to be assessed. Estimate of sedimentation rate and coast stability – see maintenance dredging. Assess sensitive bottom life
Sediment spill Figures 4.9, 4.11 and 4.12	Increased turbidity during dredging ⇒ shading and burial of flora and fauna ⇒ loss of habitat. Damage to corals may be permanent	Assess sediment spill and dispersion including background turbidity. Check for vulnerability of local marine environment
Removal of existing seabed Figure 4.12	Temporary or permanent loss of habitat. Removal of coral may be permanent	Check significance of habitat to be removed. Assess possible recolonisation species, rate and extent
Disturbance and risk of accidental contamination Figure 4.12	Increased human activity ⇒ disturbance of sea mammals, birds and fish ⇒ reduction of population Contamination: Oil, chemicals nutrients and so on	Check existing conditions. Assess future conditions including availability of alternative habitat Assess risk, possible remedial actions
Change in bottom relief Figure 4.12	Possible creation of isolated deep depressions ⇒ lack of oxygen ⇒ increased local mortality	Assess final depth of dredged areas in relation to wave/current action. Assess of stagnation occurrence

A checklist of parameters to be considered, their possible effects and recommendations for impact assessments are given in Table 4.4.

Reclamation
Reclaiming land in wave-dominated areas such as open coasts reduces or removes the wave action completely from the area landward of the reclamation. This results in calmer conditions in the sheltered areas, leading to the same effects regarding the sedimentation regime and water quality as described for current-dominated environments.

The wave conditions seaward of reclaimed areas may be affected by waves reflected from the reclaimed area:
● if the revetments are steep and not constructed to absorb the wave energy, the wave condition immediately seaward can be very rough and dangerous for small boats and ships, and
● increased erosion may also be expected, as the wave energy is concentrated in a smaller area.

Table 4.5 Assessment checklist: Reclamation projects on an open wave exposed coast

Parameter and figure	Possible effects after reclamation	Recommended assessments
Wave conditions Figure 4.9	Creation of sheltered areas ⇒ change in sedimentation pattern	Assess sedimentation in artificial lagoons (if any)
	Creation of a new surf zone	Check navigational conditions Check possible erosion in front of reclamation
Littoral transport (coast parallel transport of sand) Figure 4.9	Blocking of transport ⇒ reduced transport capacity ⇒ increased sedimentation up drift and down drift coastal erosion ⇒ possible damage to bottom life	Existing and future littoral transport to be assessed. Estimate of sedimentation rate and coast stability. Assess sensitive bottom life
Sediment spill Figures 4.9, 4.11 and 4.12	Increased turbidity during dredging ⇒ shading and burial of flora and fauna ⇒ loss of habitat	Assess sediment spill and dispersion including background turbidity. Check for vulnerability of local marine environment
Burial of existing seabed Figure 4.12	Permanent loss of habitat	Check significance of habitat to be removed. Assess possible recolonization species, rate and extent
Disturbance and risk for accidental contamination	Increased human activity ⇒ disturbance of sea mammals, birds and fish ⇒ reduction of population (migration, death)	Check existing conditions. Assess future conditions including availability of alternative habitat
Figure 4.12	Contamination: Oil, chemicals nutrients and such	Assess risk, possible remedial actions (way out corridors and so forth)

The impact on the littoral drift is caused by the blocking effect on the sediment transport. A naturally sandy wave-exposed coast is in a delicate state of equilibrium between the supply and removal of material. Temporary or permanent blockage of the supply leads unavoidably to erosion downstream and sedimentation upstream of the filled area. The upstream area will undergo sedimentation, which in the long run may create a bypass for the littoral sediments, which are then transported on the outside of the filled area.

If a navigational channel is required in connection with the reclamation operation, the channel may undergo rapid sedimentation starting years before the completion of the project or after completion depending on the sediment bypass.

A checklist of affected parameters, the possible effect and recommendation for impact assessments is given in Table 4.5.

Rivers
Enlarging the water depth or the cross-section of a river may increase the flow capacity of the actual reach, if the dredged increase represents a significant proportion of the cross-section and the dredged area extends down the whole reach:
● upstream of the dredged area, a reduction of the water level may thus result unless dams regulate the water flow, and

- downstream of the dredged area, the river may experience higher risk of flooding, because high water surges travel faster in the dredged area than before.

This effect is enhanced considerably if the water slope after dredging is steeper than before, for example, by removing meanders or point bars, thus allowing the water to travel faster than before. Again, dams or artificial pools or basins that allow storage of the excess water can regulate this effect. These, however, will also capture sediment, which may then have to be regularly dredged.

Deepening a river can also reduce the risk of flooding by improving the flow capacity and by removal of obstacles which otherwise may block all or parts of the river by catching and piling up drifting objects.

In dynamic rivers, if the water surface slope of the river is maintained, the dredged area is subject to increased sedimentation owing to a reduction of the mean current velocity. If dredging increases the water slope, the mean current velocity and the sediment transport capacity increases. This effect might cause erosion problems along the riverbanks inside the dredged area and sedimentation problems downstream of the dredged area, where the current velocity decreases and the excess sediment yield starts to settle.

In low energy rivers, dredging an existing riverbed reduces or completely removes morphological elements, which are important for the biological system in a river. Point bars, riffles, pools and dunes create areas where fish and other organisms can find shelter from the current or where increased turbulence creates easy access to food. Areas with coarse sediment can be important spawning areas for fish.

Isolated depressions and trenches created by dredging in the riverbed can lead to stagnant water with increased sedimentation of fine and organic-rich material where oxygen depletion might develop.

Improving navigational conditions on a river might also include building of dikes and dams, which prevent flooding of neighbouring wetlands. This increases the risk for flooding downstream owing to the removal of temporary water storage areas. The reduction of flooding of the wetlands reduces wildlife habitat in the affected areas. These effects can be avoided by allowing controlled flooding of the wetlands either for river regulation purposes or for maintaining the wetlands.

Currents may rapidly disperse increased levels of turbidity created during dredging projects in rivers, and the levels normally do not exceed the natural annual fluctuations seen in rivers. Increased turbidity can have temporary effects on sensitive fish species such as salmon and trout. River sediments may contain contaminants, which could be released to the water by dredging. A checklist of affected parameters, the possible effects and recommendations for impact assessment is given in Table 4.6.

4.4.2 Capital dredging: Economic, social and political effects

Infrastructure improvements and utilisation of natural resources are the objectives for capital dredging projects. Of these two objectives, seen from an economic, social

Table 4.6 Assessment checklist: Capital dredging in a river

Parameter and figure	Possible effects after dredging	Assessments to be made
Water levels Figure 4.9	Increased cross-section ⇒ increased flow capacity	Flooding risk down stream Reduced water level up-stream
Water surface slope and current conditions Figure 4.9	Increased cross-section ⇒ decrease of current velocity in dredged area.	Assess future current pattern and possible effects on navigation, river banks and sediment transport (erosion and sedimentation)
Sediment transport Figure 4.9	Changes in current velocities ⇒ changes in transport capacity and river bank stability	Assess future sedimentation and erosion pattern Assess river bank stability
Sediment spill Figures 4.9, 4.11 and 4.12	Increased downstream turbidity during and after dredging ⇒ shading and burial of flora and fauna ⇒ loss of habitat	Assess sediment spill and dispersion including background turbidity. Check for vulnerability of local environment. Check for content of pollutants
Removal of existing seabed Figure 4.12	Temporary or permanent loss of habitat ⇒ increased erosion ⇒ increased turbidity	Check significance of habitat to be removed. Assess possible recolonisation of species, rate and extent
Changes in current and river morphology contamination Figure 4.12	Temporary or permanent loss of habitat ⇒ reduction of flora and fauna	Check possible changes in riverbed morphology. Assess mitigation actions

and political point of view, the improvement of infrastructure is by far the most important. World trade has long been, and will continue to be in the foreseeable future, the backbone of the world economy. Capital dredging projects not only benefit navigation and shipping, but can also be advantageous for both land and air transport sectors as seen in Hong Kong (Chek Lap Kok) (see Annex B), Singapore (Changi), Japan (Osaka/Kansai) and Australia (Sydney).

Improving and upgrading waterways and harbours is not only beneficial to the shipping industry, but also reduces the pressure on the land transport system, which in some places, especially in Europe and North America, is increasingly becoming a bottleneck for further economic growth.

As available land for industrial development and growth, especially in the industrialised world, is becoming limited, the creation of new land by reclamation projects becomes increasingly important for the future. Creation of new land close to suitable transport infrastructure facilities, such as navigable rivers and the sea, has not only measurable economic effects. It may also have more indirect, mostly positive effects on the environment by reducing the need for land transport and reducing the pressure on existing land.

The benefits of large infrastructure projects, like bridges and other types of fixed links crossing rivers and straits, are likewise not only restricted to road and train traffic but may also have effects on a number of other sectors including tourism and the shipping industry. Bridges and other fixed links normally reduce the need for ferry traffic, but they can also create new or larger trade districts for existing harbours. Seen from an environmental point of view, such projects may even reduce emissions to the air by reducing the distance between two centres and/or by reducing emissions from ferries.

Utilisation of raw material from the sea saves valuable landscapes from being scarred by numerous sand, gravel and stone quarries, from which the material would otherwise have been extracted and transported over land.

The political and social effects of capital dredging projects are normally difficult to distinguish from other impacts. Steady economic growth is one of the most important parameters for social and political stability. Seen in a broader perspective a capital dredging project is one of many parameters, but nonetheless an important one, supporting modern industrialised society.

The local social and political effect of a capital dredging project depends very much on the economic effects and to some degree on the environmental effects. Large projects are always met with scepticism in the beginning, but as the positive effects of the project start to materialise, or become visible to the population, the attitude towards the project often grows more positive. More permanent resistance can be expected from those sectors and communities that directly experience negative (economic) effects from the project or are remotely situated from the project area.

A checklist of possible economic, social and political effects is presented in Table 4.7. The effects here are grouped according to types of projects, rather than to parameters as in the previous tables describing the physical and environmental effects, because the chain of these effects is more directly linked to the type of project.

Table 4.7 Checklist: Possible economic, social and political effects

Purpose of project	Areas that might be affected	Assessments to be made
Improvement of harbour capacity	Harbour and shipping industry Existing transport industry and service City growth	Existing infrastructure capacity Available land for harbour and city expansion Labour force capacity
Improvement of inland water ways	Shipping industry Harbour industry Land transport industry	Existing infrastructure capacity Shipping capacity Harbour capacity
Reclamation	Available alternative land. Land prices. Existing infrastructure. Value of neighbouring and alternative land	Land use Regional planning Infrastructure capacity
Large infrastructure projects	Existing transport industry Industrial and service sector economy	Land use Regional planning Infrastructure capacity
Energy and mining	Cost of energy and raw materials Transshipment of energy. Land prices. Existing land-based industry	Sustainable energy resources Harbour and transport capacity

4.5 MAINTENANCE DREDGING

Maintenance dredging is the activity of keeping existing waterways, harbour basins and other such waterways at the required nautical and/or hydraulic depth by removing recently deposited (up to several years) sediment. The environmental effects of such an operation are usually minimal and limited to the effects of the dredging and placement operations. Environmental issues increase in importance when the material is contaminated.

Fine-grained materials deposited naturally in harbour basins, navigation channels, river mouths or other areas are often contaminated by the discharge from industrial waste outlets. In such circumstances, the major environmental concerns associated with maintenance dredging are related to placement of the dredged material.

Maintenance dredging always includes the removal of sediments, which have been deposited and often sorted by the water. This means that the material removed is likely to be sorted and relatively unconsolidated, unless the maintenance is irregular. Maintenance dredging is normally carried out at regular intervals, but even in such circumstances fine materials may well have consolidated.

4.5.1 Maintenance dredging: Physical and ecological effects

Maintenance dredging includes the dredging of materials ranging from sand to silt and clay. Materials coarser than sand do not normally appear in this context, because these materials are only rarely transported by nature in navigable waterways. However, larger particles may sometimes be found in channels in the littoral zone where the wave forces are able to transport them.

Maintenance dredging of fine material in industrialised areas includes, as a rule, handling of some quantities of contaminated material. The level of contaminants in the sediment is not only a function of local sources but also of the time between maintenance dredging campaigns. In areas where sediments accumulate quickly and thus require frequent maintenance, there tends to be smaller concentrations of contaminants than in sediments from areas with a lower sedimentation rate where there is less frequent maintenance dredging.

The placement of contaminated sediment is strictly regulated in many countries both by national and international regulations (see Chapter 7 and Annex A). The effect on almost every parameter caused by the need for the safe placement of contaminated sediments will often be considerably larger than the effects caused by the dredging operation itself.

From a hydraulic point of view, maintenance dredging can be regarded as having effects similar to capital dredging, but the effects, if any, will normally be much smaller. This is because maintenance dredging does not increase the natural cross-section but only maintains or restores existing conditions seen over a long time scale.

Maintenance dredging prevents natural processes from adjusting natural channels and openings to the dominating hydrodynamic forces. This effect is especially important in tidal inlets, where the tide and the littoral drift balance each other by forming a system of bars and channels allowing the littoral drift to pass the inlet. Maintenance dredging in such areas disturbs the natural balance in favour of the tidal processes. Unless the dredged material is relocated downstream in the littoral zone (by the use of bypassing techniques), severe erosion may occur on the downstream coast. Maintenance dredging in estuaries, tidal inlets, lagoons and rivers tends to maintain the status quo in the affected areas by counteracting the natural processes, which would otherwise tend to close or silt up these types of environments.

Maintenance dredging also functions as a clean-up agent by removing contaminated sediment from the aquatic environment when the dredged material is disposed in confined areas or cleaned. As environmental awareness increases, especially in the industrialised world, the discharge of hazardous compounds decreases. This has already led to a general decrease in the levels of contaminants in many areas. In some places, ongoing maintenance dredging combined with good placement practices has reduced the amount of contamination in the aquatic environment.

From an overall ecological point of view, keeping existing waterways and harbours open means that the shipping industry is able to compete with land-based traffic which is putting more pressure on the global environment than water-based traffic.

Finally, the act of maintenance dredging can be used to obtain equilibrium of the sediment transport regime in an area by proper maintenance of sediment cells. More details about this are given in Chapter 7. The effects of and the assessments needed before dredging contaminated sediment can be found in Chapters 5 and 7 as well as in various PIANC Guidelines.

4.5.2 Maintenance dredging: Economic, social and political effects

The positive economic effects of keeping existing waterways and harbours open are worth mentioning. Large investments in infrastructure are maintained and kept attractive for further investments. From a national or a regional point of view, economic investments in maintenance dredging are positive investments, although the economic balance for the individual owner may be more troublesome because of the increasing costs of environmental considerations and regulations.

The overburdened land traffic infrastructure systems in Europe and USA are increasingly becoming bottlenecks for further increases in goods traffic. Maintenance of existing infrastructure for ship-borne goods traffic is therefore becoming increasingly important for the economy.

There is a need to identify suitable disposal locations close to dredging areas. These are often difficult to find because of the normally high utilisation of port or industrial areas for other purposes, which may, at first, seem more profitable and acceptable than depots for sediment.

4.6 REMEDIAL DREDGING

Remedial dredging is a clean-up operation where, for the sake of the environment, contaminated sediment is dredged and deposited in isolation from the surrounding environment. Remedial dredging is carried out preferably in areas and water systems, where the contaminating sources have been stopped or reduced to such a level that further contamination of the sediment will be negligible. Remedial dredging always takes place in fine-grained sediment and in sediment with a high or very high degree of contamination.

Sometimes situations may occur where removal of non-contaminated sediments is required in order to preserve the environment. This type of removal of unwanted sediments is here called "conservation" dredging. In Annex B an example of conservation dredging is given in which a mangrove area in Indonesia that had experienced a large accretion of sediment has now been restored.

4.6.1 Remedial dredging: Physical and ecological effects

The effects of remedial dredging on the hydrodynamics of a site are usually small or non-existent. The areas affected are usually areas of natural accumulation for fine-grained sediment, where the water depth is of minor importance for the hydrodynamic processes. In the long term, the dredged material will be replaced by new and hopefully cleaner material, especially if the contamination is stopped at source.

The removal of heavily contaminated material from the aquatic environment is obviously beneficial for the ecosystem over a longer time span. The short-term effects, however, may not be positive, as some of the contaminants will unavoidably be dispersed into the surrounding waters. Still, when proper precautions are taken during the dredging process, the spill of contaminants into the surroundings can be contained or eliminated.

The positive effects of such an operation, i.e. a reduced content of contaminants in the biological system, must be seen in the longer view since the long turnover time in these systems can vary between a year for lower organisms to decades for the higher organisms such as sea mammals. This means that the improvements may not immediately be evident and long-term management and monitoring are required.

4.6.2 Remedial dredging: Economic, social and political effects

The profit obtained from a remedial dredging project cannot be measured over a short period of time in purely economic terms. Governmental or regional authorities, therefore, often finance these projects. In principle the polluter (if identified) can be charged for the cleaning operation, but in most cases the sources of the contaminants are so diffuse that it is difficult to charge a single polluter.

The benefits of remedial dredging projects should be considered as a long-term investment in the environment, which will have economic and social spin-offs for many sectors. In many large harbours, especially in industrialised countries, old docks and other harbour-related property are transformed into residential and office areas. The value of the investment in new buildings and infrastructure can be increased considerably if the water quality in the old harbour basins is improved by removing contaminated sediments.

The relatively large investment in a remedial dredging operation increases environmental awareness and responsibility in the community. The overall attitude of the population towards such projects can be quite positive initially, but this supportive attitude can be reversed at the local level by other concerns, for instance, the necessity of constructing a disposal site or a cleaning plant for the contaminated sediment. Neighbouring landowners may regard trying to determine a location for a disposal site or a cleaning plant facility as a threat to the environment as well as an economic threat. This phenomenon is known by the acronym NIMBY – not in my backyard (see Chapter 2). And indeed, if the quantities of material involved are large and/or if local disposal sites are unavailable, the costs of transporting material and the transport-related impacts on the environment can jeopardise a remedial dredging project.

4.7 IMPACT ASSESSMENT AND RISK MANAGEMENT

Fundamentally, all human activities cause impacts which may be positive or negative or neutral. Performing an activity implies a change to the supply or consumption of natural resources. This may cause an improvement to or degradation of the resources, which again may affect the capability of performing the same activity in the future. Although the scope of this book is to describe effects that arise from "human activities" covering entire projects in which dredging and/or reclamation works take place, one soon realises that the list of "entire projects" seems to be endless.

Take, for example, a power plant built on a reclaimed area made from dredged materials. The list of potential impacts to consider may involve:
- those induced at the borrow area;
- those at the project site;
- those which may affect adjacent areas as a consequence of changes in the physical shape and utilisation pattern at the project site;
- those connected with the operation of the plant, for example, ecological effects as a result of excess water temperature generated by the discharge of heated cooling water; and
- those from emissions of gaseous and particulate substances.

How will human beings, animals and plants respond to eventual changes in the environmental state? Who will be positively or negatively affected and what are the economic benefits of the whole scheme?

Complete answers to these general questions, whether with regard to a new power plant on reclaimed land or any other such developmental projects, require an insight into a vast number of sectors and disciplines, including the methods used by the different professionals to analyse positive and negative effects and mitigate any problems. Although the main focus here is on the environmental effects directly coupled to dredging and reclamation, social and economic effects must also be considered as part of the entire project planning and implementation.

4.7.1 Environmental impact assessment

The EIA, Environmental Impact Assessment, is internationally recognised as the framework within which evaluation of positive, negative or neutral effects associated with proposed development or maintenance projects takes place. In some countries the terminology EA or EIS or EES are also used to indicate an environmental evaluation and sometimes the terms may have a formal or legal specificity.

Starting as an instrument focused solely on contamination assessment it has developed to cover a broader ecological assessment, and nowadays is used as an all-encompassing assessment of almost any aspects affected by almost any proposed development, including socio-economic impacts. Growing public awareness of environmental issues as well as the general need for insight into matters which may affect the local community have motivated many local governments to adopt a more transparent and open information policy. This means that the available literature on possible physical, ecological and socio-economic effects of projects is expanding, both with regard to the more general descriptions of what should be considered, e.g. in regulatory procedures, as well as in more specific descriptions of project evaluations at various geographical locations. Completed EIA documents may thus prove more and more useful as a source of background information on effects from specific types of projects, in specific geographical areas.

During the last two decades a number of countries, regional groupings and multilateral agencies have adopted its use into planning regulations as can be seen, for example, in the EC directive EC 1985/97 (see Text Box).

EIAs are required for the majority of projects that involve dredging and reclamation activities since they may have a significant impact on the environment by the nature

---

" 'EC 1985/97, Article 3:

The environmental impact assessment shall identify, describe and assess in an appropriate manner, in the light of each individual case and in accordance with Articles 4 to 11, the direct and indirect effects of a project on the following factors:

– human beings, fauna and flora;
– soil, water, air, climate and the landscape;
– material assets and the cultural heritage, and
– the interaction between the factors mentioned in the first, second and third indents.';"

of these activities. More than 60 percent of the world's population live within 60 km distance from a coast. Add the population living along rivers and lakes, and this figure may well exceed 75 percent of the population living near water. As the world population continues to grow, the number of individuals living in these areas is expected to increase dramatically in the future. Consequently, unless careful environmental management and planning are implemented, severe conflicts over coastal space and resource utilisation are likely to occur. Ultimately, the degradation of natural resources will limit social and economic development options.

As a process (Sorensen et al., 1990), EIA is usually imposed by governments in order to force public agencies and private developers to disclose environmental impacts, to co-ordinate aspects of planning, and to submit development proposals for review. Chapter 2 gives an example of the typical "players" involved in dredging projects where material is contaminated. Multilateral development and lending institutions, such as the World Bank, also require that the borrowing country implement institutional measures and assess the environmental impacts of projects. This requirement recognises that good investment policy requires an accounting of projects in terms of both economic and environmental feasibility if sustainable development is to be achieved.

As an analytical method, EIA is used for predicting a proposed project's potential effects on both the sustained use of resources as well as the quality of the human environment. Thus EIA may be regarded as part of a project optimisation method based on an integrated view of a variety of different interests.

In order to ensure a satisfactory outcome of EIAs (and to avoid them being used as a cosmetic exercise, resulting in only minor and maybe insignificant fine tuning of project design), they must be fully integrated into the progression of the various project phases. This ensures not only that the findings of the various EIAs be used to adjust the final design and work plans during the implementation of the project (see Chapter 3), but also that EIAs do not delay the processing of the project.

Finally, EIAs may provide answers to the many questions raised during the public debate and thereby contribute to ensuring a safe landing and broader acceptance of the project (see also Chapter 8, Monitoring, where implementation of EIAs, reflected in Environmental Monitoring Plans (EMPs), is described in more detail).

Economic valuation of the consequences caused by changes in environmental state is part of an environmental assessment. The valuation relies first of all on understanding and measuring the physical, chemical and biological effects of the project. Secondly, it depends on the methods for placing monetary value on non-market goods and services (see Chapter 9).

The complex combination of natural features and phenomena, predicting the effects of human-caused changes and short-term operations on the marine environment is difficult as has been pointed out. Indeed, the prediction of environmental responses to any possible impact is an evolving science, and gaps in our knowledge exist. Still, much progress has been made in methods and tools, which makes the evaluation and optimisation of new development projects more and more reliable.

The accuracy and time required to identify and assess potential impacts depends, however, not only on the tools, but also on the available data. In many countries the amount and quality of data are steadily improving, but are still insufficient in many areas. Field surveys are often required as part of project preparation, both for obtaining information on the present state of the environment (baseline conditions, which are a prerequisite for the assessment of the potential impacts for development proposals) as well as for design and construction purposes.

Descriptions of effects on a global basis, which at the same time can be of assistance for local use, require that a systematic approach be applied. Categorisation into "systems", in which processes and responses to impacts are the same, is one alternative. For example, by dividing the *physical environment* into different types of coastal forms, the *ecological environment* into habitats and characteristic ecosystems, and the *socio-economic environment* into industrialised versus emerging countries or main income sources, it may be easier to compare and evaluate environmental impacts. Still, although some relationships may be placed in a general framework, every local site has its own characteristics requiring special attention.

Another consideration is that a proposed project may spawn some additional smaller projects. For instance, in the "win-win" scenarios, the use of dredged material may possibly generate another project. In such cases, two or even more EIAs may be required; one for the main project and one for each of the subsidiary projects.

4.7.2 Risk assessment and management

Risk
Risk can be defined as the probability of the occurrence of an adverse event; risk assessment is a method of determining that probability. The role of risk assessment in the overall project development process is described in Chapter 3, section and is also discussed here because it provides a valuable framework for determining:
- what data need to be gathered in the pre-dredging and materials characterisation phase of the project,
- for evaluating that data, and
- for management and communication of risk.

While minimal data may be required for preliminary hazard assessment, substances that do not pose a hazard need not be carried any further in sampling and analysis. A preliminary risk characterisation may be sufficient to show low potential risk associated with many substances and allow more detailed sampling and analysis to focus on those potential stressors shown to have higher risk potential for the conditions of the specific project being investigated.

Overview of the risk assessment process
The risk assessment process consists of four steps: Hazard Identification, Effects Assessment, Exposure Assessment and Risk Characterisation.

1. *Hazard identification* is the planning stage where the problem is formulated, i.e. each risk scenario (each potential stressor – pathway – receptor) is evaluated to determine if it can cause an adverse effect. If a risk potential is determined, the stressor is referred to as a hazard, and is carried further in the risk assessment process. Types of evidence considered here include laboratory toxicity studies, epidemiological studies, clinical analysis studies, quantitative structure-activity relationships and so on.

2. *Effects assessment* determines the conditions necessary for the effects found in the hazard identification to occur for the species (receptors) of concern. This may typically be expressed in close-response curves obtained from experimental tests in the laboratory.

3. *Exposure assessment* determines the conditions expected to result from the proposed project in terms of the magnitude, duration and frequency with which the receptor(s) of concern would be exposed to the stressor.

4. *Risk characterisation* brings together the results of the effects assessment and exposure assessment to yield an estimate of risk. By definition, all risk assessments are uncertain. Discussion of these uncertainties is an important element of risk characterisation. It provides the risk manager with a level of confidence that can be placed in the results and also points to important data gaps that could greatly reduce the risk uncertainties.

Risk management and risk communication
If risk assessment identifies a risk of unacceptable adverse effect in the proposed project, then that risk must be managed. Risk management alternatives for a dredging project theoretically range from acceptance of the risk to extensive modification of the project design to abandonment. For instance, risks caused by chemical or physical stressors are almost always managed by controlling the potential for exposure because the intrinsic toxicity or dose-responsiveness of a chemical can rarely, if ever, be altered. In reality, less drastic risk management approaches are often effective when an environmental risk from a proposed dredging project is identified.

In developing a plan to reduce risk, the risk manager considers not only the results of the risk assessment, but factors such as applicable laws and regulations, engineering feasibility, potential benefits, costs, economic impacts, the socio-political decision environment, and so on. Clearly, this process is very similar to the one presently undertaken in evaluating the potential environmental impacts of dredging operations.

Risk communication is a dialogue, not a monologue, that occurs at two levels. The first is between the risk assessor and the risk manager. In practice, this usually occurs during risk characterisation when the assessor communicates technical findings to the manager.

The manager must be provided with a clear and accurate picture of the environmental risks of the proposed project, including an appreciation for the limits and uncertainties.

If this does not occur then the next level of risk communication (risk manager to the public) will be unsuccessful. Here, the public includes not only the general public but also all other interested parties or "players" such as regulatory agencies, special interest groups, and any human population that may be at risk.

There is clearly an important interplay between risk management and risk communication. If the risk manager does not clearly understand the magnitude of the estimated risks and the uncertainties associated with those estimates, the public cannot be expected to do so. In such cases, the public is likely to push for draconian project alterations that may well result in different and greater overall risks, unnecessary project delays, and greater costs borne directly or indirectly by the public.

Benefits of the risk assessment process in evaluating dredging operations
A fundamental and legitimate question is why risk assessment should be utilised in evaluating dredging projects. The incentives for a risk assessment approach are found in the following realities:
- regulatory decisions are always based on incomplete and uncertain data;
- achievement of zero risk is impossible;
- achievement of near-zero risk may be cost-prohibitive, and
- everyone accepts a certain amount of risk.

Using a risk-based approach in this decision-making environment has political, managerial, and technical value. Some of these advantages are:

1. Risk assessment is the only approach currently available for quantifying environmental risks that has broad acceptance in the scientific and regulatory communities. It is not perfect, but it is the most logical, technically sound approach to objective evaluation of potential environmental impacts presently available. The approach is gaining an ever-increasing acceptance within the scientific and regulatory communities worldwide.

2. The risk assessment process treats uncertainties explicitly. This eliminates the need for worst-case testing scenarios. When properly designed and conducted, risk assessments yield a continuous solution as opposed to a discrete yes-or-no answer. This solution is expressed in the form of probability distributions. While some may be uncomfortable with this type of technical output, it reflects the real world and provides a basis for professional judgement about incremental risks when an outcome is close to some regulatory threshold.

3. Regulators are charged with making decisions, not finding scientific truth. The risk assessment process is compatible with this charge because it deals with probabilities, not absolute truths.

4. Risk assessments are purposely designed to be highly conservative, yielding upper bound risk estimates. Therefore, if projected project risks are found to be acceptable, the risk manager and his constituency can be reasonably assured that the actual risk is quite low.

5. If projected risks are not acceptable, the risk assessment process offers a means of identifying where the problems are and how they can be corrected. Sensitivity analysis is the manager's tool for pinpointing these critical elements. Once the important determining factors have been identified, supporting data and assumptions may be more closely scrutinised. If the data associated with these elements is poor or non-existent, the risk manager has the option of collecting additional information. If the knowledge domain is sufficient, sensitivity analysis can help focus risk management activities on the most critical elements.

6. Large uncertainties can be partially ameliorated by conducting comparative or incremental risk assessments. In this approach, the quantitative difference in risk associated with various scenarios is examined rather than the absolute risk of each. Any conservative assumptions or large uncertainties common to each scenario do not contribute to the difference between the calculated risks. For example, one could calculate the incremental risk associated with placing dredged material at an aquatic site versus placing it at an upland site. Conservative assumptions and large uncertainties common to both actions become irrelevant. What is important is the difference in risk between the two sites.

7. Finally, utilising a risk-based approach has a distinct managerial advantage. Risk assessments identify what is important, unimportant, and unknown. This permits managers to allocate critical, and usually limited, resources to areas of greatest need. It provides an objective way for managers to identify knowledge gaps and direct resources in such a manner that will facilitate the future conduct of the job.

The Environmental Impact Assessment process in Victoria, Australia for the proposed Channel Deepening Project for the Port of Melbourne Corporation (PoMC) incorporates the Australian Standard AS4360 for Risk Management (see Annex B).

Risk management and risk assessment for dredging projects comes almost always into play if it is known that the sediment to be moved is contaminated with a chemical substance. The exposure of a substance or compound when sediments are disturbed as a result of dredging, transportation and relocation may have a pathway that gives rise to a risk of affecting human and ecological health. The formal procedures applied for determination and quantification of consequences of these risks are described below as they would be applied to materials characterisation.

Materials characterisation
The collection and evaluation of pre-dredging information for materials characterisation are not conducted in isolation. These activities are part of a larger continuum of environmental information generation and evaluation relevant to the overall project (see Annex D).

The pre-dredging investigations for materials characterisation are important not only for evaluating the proposed dredged material, but also for selection of the most appropriate placement site and determination of any special management techniques that might be imposed for environmental reasons.

The pre-dredging investigations for materials characterisation are essentially predictive in nature. Once the project is completed, environmental monitoring should be designed to determine the appropriateness of the topics addressed in the pre-dredging investigations to test or verify the hypotheses about environmental effects. A monitoring programme can also identify unexpected impacts and indicate measures that may be effective in controlling them. If a properly evaluated continuum consistently reveals:
- environmental impacts that were not predicted, or
- no environmental impacts when impacts were predicted (i.e. hypotheses are not verified),

then the scope, procedures, and/or interpretation of the pre-dredging investigations, the site selection process, dredging and placement operations, the monitoring process, and/or the formulation of hypotheses should be revised to improve them for future dredging operations.

Environmental risk assessment and dredging evaluations
Pre-dredging evaluations for materials characterisation may include chemical analysis of the sediment proposed for dredging, or tests to determine the toxicity of the proposed dredged material to appropriate sensitive animals and the bioaccumulation potential of sediment-associated contaminants. In most instances, this approach yields a simple "pass" or "fail" answer. That is, dredged material is found to be either acceptable or unacceptable for placement at the proposed site. Current procedures do not allow the manager to quantify how "acceptable" or "unacceptable" the dredged material is.

If the material is found "acceptable", this is not usually problematical, but if the material is found "unacceptable", the consequences can be substantial. A risk assessment approach can be valuable in such circumstances by providing a basis for evaluating the incremental risks associated with a material that was just over the threshold of "acceptability". Acceptance of a slightly greater risk may be considered more acceptable than the consequences of altering the project to avoid that slight increase in risk.

Historically, most sediments have been found to be "acceptable". Those considered marginal or "unacceptable", while representing a small volume of total material dredged, consume a disproportionately large share of limited resources. These costs, expressed as time, money and productivity, are initially borne by the project proponents. Ultimately, they are often passed on directly or indirectly to the consumer and taxpaying public.

The lack of technically sound procedures for assessing the probability of adverse impacts associated with dredging operations is a major reason extensive testing has been requested. Risk-based procedures provide a basis for balancing potential environmental impacts with other factors (e.g. costs) in a more technically defensible manner. These procedures also provide a quantitative means of comparing the risks associated with different placement options (e.g. open water, upland) including the "no action" alternative.

This weighing and balancing goes on under the present evaluative approaches, but the process is subject to criticisms of subjectiveness, distortion and inconsistency. Use of risk assessment to determine the degree of unacceptable impacts of dredged material placement can help mute these criticisms, and constitutes a significant technical step forward, and as well as encourages an increased credibility among regulatory agencies and the public.

Risk assessment provides an objective basis not only for evaluating placement of dredged material at a pre-selected site (site evaluation), but also for determining the environmentally preferable site (site selection). In the latter application, it offers a promising approach for objective across-media comparisons (e.g. open water versus upland) of the potential consequences of dredged material placement.

For additional guidance on assessing human health and ecological risk in relation to dredging and dredged material placement see PIANC EnviCom 10 (2004).

4.8 MODELLING AND INFORMATION MANAGEMENT

This chapter has focused on highlighting and presenting possible chains of effects on nature and human activities caused by dredging. This long list of possible effects may look overwhelming, but please keep in mind that this chapter is a catalogue of *possible effects*. Not all possible effects are actually probable. In real projects the effects and chains of effects caused by dredging as a rule will be simpler and more manageable.

4.8.1 Modelling

Prediction of the behaviour of spatially distributed dynamical processes is a complex activity and the following tools may be used (often as a mix of) to quantify the nature and extent of environmental interactions (Modified from: Challenges in managing the EA process – EA Sourcebook UPDATE, The World Bank, December 1996, Number 16):

- Mathematical models (such as formal air and water dispersion models, income multipliers);
- Physical models (such as hydraulic models of, for example, stability of mounds and dikes);
- Field experiments;
- Structured or semi-structured approaches to produce a mix of qualitative and quantitative predictions (for example, landscape change and social impacts); and
- Scientific experience and judgement.

Mathematical modelling plays an increasingly important role in the modern quantitative assessment of potential impacts due to the fast paced development both in modelling methodologies and computing. It is an efficient tool in the planning and decision-making process. Mathematical models have been developed to describe and predict physico-chemical, biological and socio-economic processes.

Mathematical models can be classified as

- deterministic (the uncertainties of input and output variables and model parameters are not considered) or stochastic (at least one of the input or output variables is probabilistic);
- static (temporal dynamics is not taken into account) or dynamic (time-varying interactions amongst variables are taken into account), and
- one-dimensional (1-D) or multidimensional (2-D, 3-D) to simulate complex spatial conditions.

Today mathematical models with varying complexity are available or being developed to predict a broad array of processes related to dredging and dredged material disposal. These include models to predict

- changes in the hydrodynamic regime (water levels, currents and waves),
- sediment release from dredging plant and the dynamic behaviour of the sediment plumes,
- sediment transport and associated morphological impacts,
- contaminant release from confined disposal or placement facilities (CDFs) (such as effluents, surface runoff) and during aquatic disposal and effects on ground and surface water quality; and
- responses of flora and fauna to changes in water column parameters and deposition of sediments.

The more sophisticated the models become, the more data is needed for set-up and validation. Common to all models is the fact that the quality of the results is dependent on the quality of the input data. The lack of good quality data is often a major obstacle for developing using numerical models. Furthermore the complexity of the interactions between abiotic (physical and chemical) and biotic parameters challenges modellers worldwide in producing realistic and acceptable predictions functional in the assessment process.

An additional problem for ecological models is that most organisms have the option of making choices and, as such, are unlike air or water, which are governed by the laws of physics. Having organisms that make choices, like postponing reproduction or putting less effort into that aspect of life, or switching to an alternative food source, makes ecological models more prone to uncertainties, especially when dealing with multi-celled, complex life-forms. Such small changes will not usually result in sudden or dramatic changes in populations. They may, however, deplete populations, since reproduction or survival, in some cases, only needs to drop a little for numbers to decrease. For a sensitive species, the time needed to find out whether decline, once noticed, is from natural or human causes may be too long. Certain high quality models are available as freeware. The commercially available models are generally more user-friendly (menu driven user interface, user support and good graphical presentation facilities).

4.8.2 Information Systems

Decision-support software is also available in a quality that enables the decision-makers and contractors to use it for screening analyses and assessments, especially in the planning phase of a project. As the demands on a contractor to have still more comprehensive and complex control of a project and its possible impact on the surroundings increases, so does the need for information and decision-making systems.

The future will bring more sophisticated project control and management software on the market. The Geographical Information System (GIS) is now widely used in dredging projects to filter and present the vast amount of data produced during a dredging operation. Artificial intelligence in the form of neural network computer systems is currently being developed for use as an interpretation tool for marine geology and seabed mapping studies. In the future similar systems will probably be used as a helping tool to manage and optimise the work of large sophisticated dredgers.

4.8.3 Future trends

The development of more reliable and user-friendly models makes it easier and more economically feasible to assess the real effects of dredging operations. It helps to separate natural occurring fluctuations from actual dredging-related impacts. This helps the dredging industry and project owners to refute the often widely exaggerated expectations of environmental impact, which are commonly raised by NGOs and other opponents to dredging projects.

Field survey methods and monitoring are also changing because of developments in electronics and telecommunications. Less expensive and more reliable monitoring equipment with online data transmission will be available in the future. The development of survey equipment for bathymetric and seismic surveying is on the other hand moving towards more complicated and more expensive equipment, for example, multi-beam echo-sounders.

The more sophisticated equipment is considerably more effective, but requires highly specialised teams of surveyors to operate them and highly qualified engineers to interpret the enormous amounts of data generated. These very advanced survey methods make it theoretically possible to dredge or fill with very little tolerance. Unfortunately they are sometimes also adopted as a requirement in areas where the need for such accuracy is unnecessary and thus they only serve to increase costs.

Indeed, the dredging manager and the project owner should be aware of the pressure from consultants and manufacturers of survey and monitoring equipment who may try to make dredging operations and the associated investigations more complicated and thereby more expensive than necessary. To distinguish between "nice to have" and "need to have" is a difficult, but crucial question to be resolved at the very beginning of a project.

Keep in mind that the mitigation of any possible negative effects, especially on the environment, is of essential concern and in the best interest of the dredging industry. The degree of environmental awareness in almost all sectors of society is growing rapidly and environmental issues in many countries are now weighty political factors.

Political pressure against the dredging industry, seen in many places, can be balanced with increased levels of information and increased control over potential effects.

4.9 ECOLOGY AND ECONOMY

The title of this chapter announced that ecological effects *and* economic effects of dredging would be addressed. What the title does not indicate is the very strong synergy between "ecology and economy". Economic growth and environmental quality have traditionally been viewed at best as alternatives, at worst as antagonistic. Conventional wisdom holds that to obtain rapid economic growth, environmental quality must be sacrificed. The impression one gets about many past development activities is that there was insufficient concern for the environment.

Today, this view is misleading and as a business attitude has long ago been abandoned.

In today's world the aim is to develop a sustainable society in which the long-term productivity of natural systems is preserved. In addition, potential environmental impacts must be minimised so that, for example, recreational value is maintained or even improved in order to avoid undermining human welfare. Ultimately, the objective of any developmental dredging project is the betterment of human welfare. The role of dredging in achieving or regaining a positive synergy between ecology and economy is demonstrated in many projects, several of which are briefly described in this book. Clearly, dredging helps create a stronger economy in a cleaner environment.

This chapter has provided the reader with an overview of the most common positive and/or negative effects which may arise as a result of dredging in an aquatic system. The knowledge of potential effects, and how they may interact, is of paramount importance to a dredging industry operating in a more and more controlled and regulated world.

CHAPTER 5

Investigation, Interpretation and Impact

5.1 INTRODUCTION

This chapter presents a brief but comprehensive overview of environmental aspects of pre-dredging investigations for materials characterisation. This information is applicable to any dredging project, although not all the information will be equally useful for every project.

Why is material characterisation necessary? Adequate assessment of sediment is necessary to predict behaviour during and after placement and to evaluate the environmental acceptability of management options.

A distinction is made between evaluating potential environmental impacts for new work construction (capital dredging) versus that for routine maintenance dredging. A new project may need to proceed carefully through a rather extensive series of assessments, recognising that precedents may be set for future evaluations, whereas routine maintenance dredging might require minimal new information, especially if existing data demonstrate the absence of unacceptable effects from past dredging activities. More detailed information about dredged material and its characterisation can be found in Annex D.

When dredging in countries where environmental evaluations are not yet the standard practice, attempting to begin with detailed implementation of all aspects presented here is unnecessary and ill advised. On the other hand, delaying initiation of a programme of environmental evaluation is also ill advised. Instead, a programme could begin with general surveys and advance to routine implementation of more complex evaluations as environmental experience and sophistication develop. Doing basic evaluations now is far better than doing nothing because of a current inability to conduct sophisticated evaluations.

For the sake of clarity, in this chapter the choice has been made to emphasise maintenance dredging conducted for navigational purposes and placement of the dredged material at aquatic sites. This emphasis is justified for the following reasons:

- Traditionally maintenance dredging operations are the most common by project number.
- Aquatic placement of dredged maintenance material, sometimes contaminated, tends to elicit the highest and most vocal environmental concerns.

- Most countries' statutory requirements were developed to regulate maintenance dredging operations.

The structure of the chapter is illustrated in Figure 5.1 and emulates the process generally followed when conducting pre-dredging investigations for materials characterisation. The process usually takes place by following four steps illustrated in Figure 5.1:

Step 1   Project Planning
Step 2   Initial Evaluation
Step 3   Field Surveys, Sampling and Laboratory Testing
Step 4   Interpretation of Results

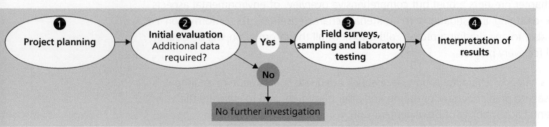

Figure 5.1 Process of pre-dredging investigations for material characterisation.

In Step 1, Project Planning, the following elements are addressed: purpose of the dredging project; regulatory context; and type and location of placement site(s). These factors determine the type of information that is needed.

This information is sought in Step 2, Initial Evaluation.

If it does not exist, it must be generated (Step 3). One can see, therefore, that the nature and scope of pre-dredging investigations and materials characterisation are directly dependent on project planning and existing information.

The results of pre-dredging investigations and materials characterisation are interpreted in Step 4.

5.2 STEP 1, PLANNING OF INVESTIGATIONS

Experience indicates that successful, cost-effective dredging investigations begin with an investment in proper planning. Not surprisingly, thorough project planning is the first step in the pre-dredging investigations for materials characterisation (see Figure 5.1). This planning requires an understanding of dredging and placement options, knowledge of applicable laws and regulations (see Annex A), and understanding of local environmental issues and the public/stakeholders interests. These issues are discussed below.

5.2.1 Placement options

Most dredging projects face two questions concerning dredging and two questions regarding placement options:

Question 1: "I **want to** dig a hole in the sediment. What are the consequences for the existing environment?"

AND

Question 2: "I **have to** dig a hole with minimal impact to the environment resulting from the operations. What systems and measures do I have to consider to achieve this?"

PLUS

Question 3: "I **have to** relocate this dredged material. From an environmental perspective, where is the best place to put it?"

OR

Question 4: "I **want to** place my dredged material at a known location. What's the environmental impact of that action?"

Regarding questions 3 and 4, question 3 leads to investigations supporting placement site selection, while question 4 leads to investigations supporting placement site evaluation. The two scenarios sketched in questions 3 and 4 prompt different suites of pre-dredging investigations and materials characterisations. Thus, it is very important to define clearly what placement options may or may not be available for the project (see also Chapter 7). These two scenarios are discussed in detail below.

5.2.2 Selection of placement site

In this case, the primary purpose of the project is to remove the material from its present location with no predetermined placement location or use other than ensuring that it does not re-enter the dredging site. In practice, placement site selection is often an iterative process. It begins with a broad-brush initial evaluation of placement site options from which one or perhaps a few options are selected for more detailed evaluation. This process eventually leads to the identification of the environmentally preferred placement option. Once the environmentally preferable placement site is identified, it is compared with acceptability criteria as described in the following section on placement site evaluation.

For an example of placement site selection, consider a project whose purpose is simply to remove sediment from a channel, with no particular use envisioned for the dredged material. The options might be:

- An aquatic site might be suitable for the material.
- Constructing a nesting island for birds might also be considered if there were an ecological need for such a site in the area.
- If the dredging site is close enough to land on which a diked (bunded) containment area might be constructed, this option should be considered.

In this situation, the potential physical (e.g. habitat alteration or loss), chemical (e.g. water quality), and biological (e.g. toxicity) consequences of each placement should be evaluated. This evaluation need not always be exhaustive; usually getting a reasonable amount of detail is sufficient to identify the preferable option. The environmental acceptability of this option can then be confirmed through a more detailed placement site evaluation.

5.2.3 Evaluation of a placement site

Placement site evaluation occurs in projects where the dredged material placement location has been identified. This identification could be based on economic reasons (e.g., closest to the dredge site), project reasons (e.g., wharf creation), or environmental reasons (e.g., beach nourishment, remedial dredging of sediment to an approved landfill). Other placement options have already been eliminated as not being compatible with the project purpose, or have already been evaluated as discussed under placement site selection and so they do not require further evaluation. Once a proposed placement option is selected, an environmental evaluation is made relative to acceptability criteria. If the proposed option were found unacceptable, the evaluation process would begin again, with a new search for acceptable options, as in the section on site selection above.

For an example of placement site evaluation, consider a project whose purpose might include upland placement of the dredged material where it would be developed as an industrial site. In this case, neither aquatic nor upland organisms would be directly exposed to the material, and the emphasis of the environmental investigations might be on the dredging and placement activities, potential leaching of contaminants from the material, and loss of whatever habitat might exist at the placement site. In such a case, evaluating potential water quality impacts or toxicity and bioaccumulation by either aquatic or terrestrial organisms might be irrelevant; if so, these need to be excluded from the pre-dredging investigations for materials characterisation.

5.2.4 Regulatory considerations

During the project planning process, laws and regulations that may be applicable to the project must be identified (see Annex A). Appropriate regulatory and funding authorities should be consulted to assist in this determination. Once applicable laws and regulations are identified, a clearer understanding of what is required, vis-à-vis pre-dredging investigations and materials characterisation, should emerge.

Failure to have a clear understanding of regulatory requirements during the planning process can have disastrous consequences on the project budget and schedule. Many pre-dredging investigations (e.g. toxicity/bioaccumulation bioassays) have sequenced procedures that are fixed and time periods that cannot be shortened. Failure to allow time for these procedures can derail other elements in the work schedule (e.g. contracted dredging).

A variety of international, national, and local treaties, laws, regulations and policies may address dredging to varying degrees. In a document of international scope, identifying all the regulatory requirements and considerations necessary for any particular project is not possible. Annex A, on regulations, provides a concise overview of international laws and treaties pertinent to the environmental aspects of dredging and placement operations. Project planners, as well as local experts and authorities, are urged to consult this Annex on regulations early in the project planning process.

The most widely applicable international regulatory instrument for dredged material is the London Convention on the Prevention of Marine Pollution by Dumping of Wastes and Other Matter and the 1996 Protocol thereunto (hereafter referred to as the London Convention or the LC). The Dredged Material Assessment Framework (DMAF) of the London Convention is a carefully crafted and widely reviewed and accepted approach to evaluation of dredged material. Situations may be encountered in which there is little regulatory guidance dealing specifically with dredged material. In such cases the DMAF, even if it is not legally required, may provide useful guidance since it is a technically sound and practical overview of the important environmental considerations.

Under the DMAF (Figure 5.2), dredged material may be excluded from chemical characterisation, and toxicity and bioaccumulation testing, if it meets any one of the following three criteria:
- excavated from a site far away from existing and historical sources of appreciable contamination, so as to provide reasonable assurance that the dredged material has not been contaminated; or
- composed predominantly of sand, gravel and/or rock; or
- composed of previously undisturbed geological materials.

Although the London Convention formally applies only to activities of contracting nations in international waters, these exclusions from chemical and biological testing are reasonable, technically sound, and could be applied to all dredging projects unless doing so would violate national, regional, or local regulatory requirements.

Countries that are not signatories to the London Convention may or may not have their own national, regional, and/or local regulations applicable to dredging and dredged material placement.

Note as well that, even though not regulatory, many international funding organisations (such as the World Bank) have their own environmental requirements. These may differ from the applicable governmental regulations and may be in force even in the absence of governmental regulations.

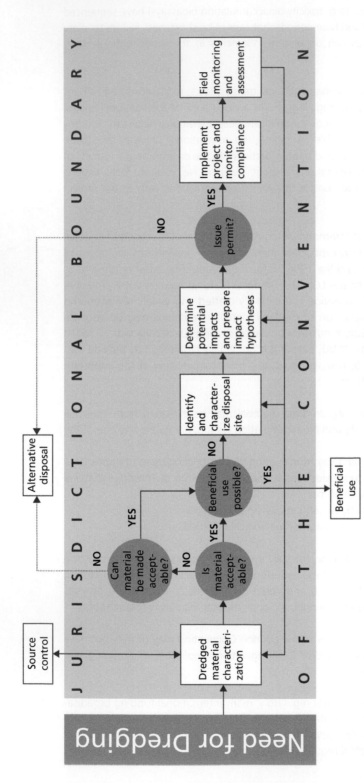

Figure 5.2 Dredged Material Assessment Framework (DMAF) of the London Convention (adapted from LC 1992).

5.3 STEP 2, INITIAL EVALUATION

Once project planning is well underway, the search for information that may assist in the environmental evaluation begins. This second step is called the Initial Evaluation (see Figure 5.1). It is based primarily on a thorough review of existing information. For many dredging projects, the initial evaluation of existing information will provide all that is needed for technically sound environmental decision-making. For example, if one is dredging sand and placing it in a nearby area where biological resources are not a major concern, then the minimal amount of information gathered during the Initial Evaluation should be sufficient. Likewise, for a dredging project that is conducted repeatedly in a similar manner and for which there is no historical evidence of unacceptable effects, information gathered during the Initial Evaluation should suffice. Although the Initial Evaluation is usually based on existing or readily obtainable information, new data may be collected. This may include chemical characterisation of sediment from the dredged site, reference site and placement location.

5.3.1 Determination of sufficient versus insufficient information

The primary purpose of Step 2 is to determine what information is available or readily obtainable for assessing the potential environmental impacts of the dredging project. In addition, the following specific purposes should be achieved during the evaluation of existing information:

- describe the general environs and surrounding land use at the dredging and placement sites;
- describe the physical, chemical, and biological features of the proposed dredging and placement sites;
- identify natural resources at the dredging and placement sites which may be at risk;
- identify and classify sources of contaminants to the sediment proposed for dredging;
- screen constituents of potential concern to determine which (if any) warrant more detailed consideration;
- identify other environmental issues such as noise, air quality, aesthetics, cultural resources, infringement on public use, etc. that might be important, legitimate issues; and
- identify what information is and is not available thereby directing more focused, subsequent information gathering.

The evaluation of existing information also serves as an initial tier in the environmental evaluation process (see Figure 5.3). A tiered approach facilitates decision-making without inefficient generation of more data than necessary to support the decision (see Text Box on page 91).

In most cases, the procedures discussed here provide sufficient information for defensible regulatory decisions about environmental impacts potentially associated with dredging operations. In most cases where decision-making is difficult, the difficulty has more to do with the political, economic, or public acceptance of the decision than

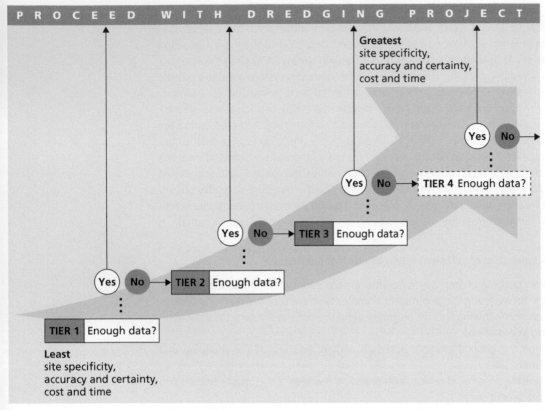

Figure 5.3 Tiered Evaluation.

with the lack of sufficient environmental information. Even though environmental issues often are blamed for the delay, rarely does inadequate environmental information lead to a delay in making a regulatory decision.

In those cases where testing truly does not provide sufficient environmental information to make a decision, more detailed or advanced-tier environmental testing can be conducted. Seldom are advanced-tier tests routine procedures; more often they involve a considerable element of research. Therefore before testing is started, all interested parties must agree:

- which tests will be conducted;
- which criteria will be used to evaluate the results; and
- that the test results will indeed provide the basis for a decision.

If these agreements cannot be reached in advance, there is little likelihood that the increased environmental testing will actually result in a decision.

Phased versus tiered assessments

In a phased assessment, it is generally understood that each phase will be carried out. They may overlap or be conducted sequentially, but all will be completed. In the context of this book, a tiered assessment can be defined as the sequential steps (tiers) of pre-dredging investigations which proceed from the least complex and costly to more complex and costly, producing the correct information in the proper quantities to support technically sound environmental decision making (see Figure 5.3). At the end of each tier, information is examined. If the information available at that point allows a decision regarding potential environmental impacts to be made with acceptable certainty, the tiered assessment stops. If the level of uncertainty is unacceptable, the necessary additional information is gathered or the project plan is re-evaluated (e.g. a different dredging method or placement location is considered). The tiered approach allows managers to more judiciously allocate usually scarce resources (time, money, equipment and personnel). Phased and tiered assessments are not the same. As discussed above, one should exit the tiered process as soon as sufficient information to support a decision has been gathered.

5.3.2 Review of existing information

The type and amount of existing information needed will vary according to the dredging project and local regulatory requirements. Three general categories of environmental information, listed below and discussed in the following sections, are often sought:
- characteristics of the dredging and placement site environs;
- identification of biological receptors of concern; and
- identification of constituents of potential concern in the sediment at the dredging site and perhaps the placement site.

Persons collecting this information should assess the quality associated with this information. This quality assessment will be important when the many lines of evidence are weighed together in the initial evaluation of the proposed project (see Section 5.3.6).

Existing information might be obtained from the following sources:
- prior physical, chemical or biological test results conducted on the proposed dredged material;
- prior physical, chemical or biological tests conducted at the proposed placement site(s);
- monitoring studies conducted at the dredging and placement sites;
- locations of likely sources of contamination adjacent to the dredging and placement sites, including industrial and agricultural areas, storm and sanitary sewer

discharges, surface water drainage patterns and known/suspected/abandoned sites of chemical contamination;
- hydrodynamic conditions including currents, waves and tides; and
- information in the files of environmental authorities or otherwise available from public or private sources.

Examples of sources from which relevant information might be obtained include the national, regional, local, academic or industry equivalents of:
- environmental databases;
- water and sediment data on major tributaries;
- discharge permit records;
- previous environmental evaluations or monitoring of similar projects;
- pertinent, applicable research reports;
- port authority and other industry records;
- college and university research projects;
- records of environmental agencies;
- published scientific literature;
- listings of chemical spill incidents;
- incident records of contamination;
- identification of in-place contaminants;
- hazardous waste sites and management facilities reports; and
- studies of sediment contamination and sediments.

5.3.3 Dredging and placement site environs

As part of the larger project planning effort, a thorough description of the dredging and, perhaps, placement site environs are likely to be available. To ensure the environmental assessment of the project is conducted in the appropriate manner, the following information should be gathered for both sites:
- location, physical size and shape;
- hydrology and seasonal variations; and
- general ecology and habitat type(s).

5.3.4 Receptors of concern

In general when concerns are raised about the impacts a dredging project might have on the environment, the more specific concern is usually directed at risks posed to human beings (i.e. human health receptors of concern) and selected flora and fauna (i.e. ecological receptors of concern). Consequently, existing information must be gathered for both these receptor groups. For ecological receptors, there are several categories:
- commercially important – wildlife, fish and shellfish populations that constitute economically important resource stocks;
- recreationally important – wildlife, fish and shellfish populations that are sought by man for recreational purposes;

- ecologically important – flora and fauna populations whose abundance and/or biomass are important to energy flow and nutrient cycling; and
- special status – individual species whose survival is threatened or endangered.

5.3.5 Constituents of potential concern

The evaluation of the constituents of potential concern in the proposed dredged material is extremely important. To carry out this evaluation, available information may need to be supplemented with additional chemical analyses of the dredged material. Both chemical and microbial constituents may be of concern depending on the source and proposed placement location vis-à-vis biological resources at risk.

Chemical constituents
Identifying specific contaminants that warrant attention for a particular dredged material is dependent on the information collected in the Initial Evaluation. In some instances, sufficient existing chemical information may be available. In other cases, no information will exist but none may be needed. Most likely, however, some amount of chemical analysis will have to be performed.

The initial list of chemical analytes to be considered will be driven by several factors. One is the chemicals associated with known/suspected sources that may input to the dredging and placement sites. Some regulatory bodies may have a "standard list" of chemicals they require to be analysed. The toxicity and bioaccumulation potential of chemicals should also be considered. Some contaminants are toxic but not bioaccumulative, while others are not toxic to initial receptors, but are persistent and bioaccumulative and may exert toxic effects on higher trophic level organisms (e.g. humans). McFarland *et al.* (1989) summarise many of the major factors affecting bioavailability of sediment-associated chemicals and may be helpful in identifying chemicals to consider and in evaluating results.

The issues of bioavailability and potential toxic effects are dealt with quantitatively in higher tier pre-dredging investigations and materials characterisations.

The initial analyte list represents a compromise between cost constraints and what may reasonably be expected to be present in the dredged material. As the list grows, the negative impact on project budget and schedule becomes substantial. Conducting comprehensive analyses on a relatively small number of representative samples may prove advantageous. These results may help reduce the analyte list and enable one to narrow the focus of future studies (if needed) on a much smaller suite of chemicals. Analyte selection is discussed further under 5.6.2 "Chemical analysis".

Microbial constituents
If dredging or placement sites are close to shellfish beds, swimming beaches or drinking water intakes and sediments are suspected of having high levels of microbial contamination, then evaluation of the microbial constituents may be warranted. National, regional, or local health and water quality agencies should be consulted for appropriate methods and interpretive guidance. The ultimate concern with pathogens and viruses, as with any other contaminants, is not their mere presence but their potential to affect adversely humans and susceptible macrobiota.

Physical parameters

The physical parameters most often considered from the perspective of potential environmental impacts of dredging and dredged material placement are turbidity and suspended solids. Although these two water quality parameters are related, they are not the same. While suspended sediments affect turbidity they are distinct properties and one is not quantitatively predictable from the other.

Turbidity is an optical property of water that causes light to be scattered and absorbed rather than transmitted in straight lines through the water. It is caused by the molecules of the water itself, dissolved substances and organic and inorganic suspended matter. Turbidity is measured in various optical units such as Secchi depth, Jackson Turbidity Units (JTU), and Nephelometric Turbidity Units (NTU).

Suspended solids comprise fine particles of inorganic solids (e.g. clay, silt, sand) and organic solids (e.g. algae, detritus) suspended in the water column. Suspended solids are usually measured as Total Suspended Solids (TSS), the dry weight of suspended solids per unit volume of water. TSS is usually reported in milligrams of solids per litre of water (mg/L).

The ability of suspended particles to scatter light depends on the size, shape, and relative refractive index of the particles, and on the wavelength of the light. Thus, two samples of water with equal TSS, but different particles or even different size distributions of the same particles, will produce very different turbidity readings on the same turbidity meter. In addition, two different types of turbidity meters may produce different readings on the same sample. The relationship between turbidity and TSS is not constant, even for different sampling times at the same site (Thackston and Palermo, 1998).

Turbidity readings are not appropriate for measuring TSS unless they are properly calibrated against TSS values at the same site. In a similar way, the acoustic backscatter from a Doppler Current Meter may be used to measure TSS (Land and Bray, 1998), but only if it is calibrated from TSS obtained from site-specific water samples. Correlations between turbidity and TSS are only valid for a specific site, sediment suspension, and time. What is most important from the point of view of environmental evaluation is the contribution (if any) of either turbidity or TSS to environmental impact.

When dredging resuspends sediments, the particles being put into suspension are initially of a wide range of sizes. Typically, the larger particles drop out of suspension relatively quickly near the dredging or placement site, leaving the finer particles in suspension. It is the fine particles that are usually seen as a turbid cloud or plume. These fine particles settle so slowly as they disperse from the dredging site that their effect on the bottom is almost imperceptible.

5.3.6 Interpreting the existing information

As the information gathered during the Initial Evaluation accumulates, the process of evaluating the information vis-à-vis potential environmental impacts begins. Note the exclusions under the London Convention in Section 5.2.2. Concern for chemically induced perturbations increase or decrease depending on a number of factors, as discussed below. The discussions of environmental risk assessment in Chapter 4 may be helpful in interpreting results:

Concern about chemically induced perturbations increases as the following increase:
- known/suspected historical or contemporary chemical inputs to the dredging site from point and non-point sources; and
- number, concentration, and toxicological importance of sediment-associated contaminants.

Concern about chemically induced perturbations decreases as the following increase:
- spatial or temporal isolation of the dredging operation from known/suspected chemical inputs;
- number of maintenance dredging operations since termination of inputs;
- mixing and dilution between sources of chemical input and the dredging site;
- deposition in the dredging area of sediment from uncontaminated sources;
- proportion of the dredged material consisting of sand or larger material; and
- lack of important biological resources in the dredging and placement areas.

These factors are complexly interrelated with many other factors. As a result, the acceptable level of each of the factors above depends on its interaction with other factors, so that what is acceptable for one project may or may not be acceptable for another project based on these interactions (Lee *et al.*, 1991).

At the end of the initial evaluation, the decision-maker must determine whether there is sufficient information to make a technically sound judgment regarding potential environmental impacts. This determination will be the basis for either expending additional resources for further pre-dredging investigations and characterisations to acquire additional information. Either choice may be challenged. Legal action may be pursued. Therefore, this decision is crucial and should be made carefully and with supporting documentation.

The two possible outcomes at the end of the Initial Evaluation, and their implications, are presented in the following text and illustrated in Figure 5.4. The initial evaluation provides a basis for determining how to proceed with pre-dredging investigations:
- The information is sufficient to make a regulatory decision. The proposed dredging project is either:
  - environmentally acceptable – no further environmental investigations are necessary, or
  - environmentally unacceptable – strong consideration should be given to terminating the project or revising the project plan based on the Initial Evaluation (revision might be to select an alternative placement site or institute management controls during dredging operations).
- The information is insufficient to make a regulatory decision regarding the environmental suitability of the proposed dredging project. In this case, the project manager

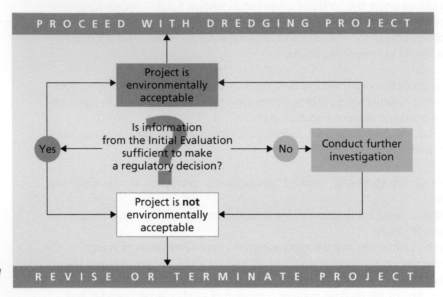

Figure 5.4 Initial
Evaluation.

has to take a decision, for instance, to conduct further pre-dredging investigations for materials characterisation or to terminate the project or to revise the project plan. Further investigations will focus on toxicity and bioaccumulation testing, field surveys and possible, further chemical characterisations. These investigations constitute Step 3, and are described in detail in the remainder of this chapter.

5.4 STEP 3, FIELD SURVEYS, SAMPLING AND LABORATORY TESTING

If the Initial Evaluation, Step 2, does not provide sufficient information, Step 3 can be initiated to collect more data in order to complete the environmental assessment for a proposed dredging project. These data are derived from laboratory and/or field studies. The laboratory studies focus on the physical, chemical, and biological characterisations of the sediment to be dredged. Field studies are aimed at developing a better understanding of the environs and biological resources at the dredging and placement sites.

Field studies are an essential element in the environmental assessment process. Field surveys at the dredging, and particularly at the placement sites, are playing an increasingly important role in the assessment process. As discussed in the Introduction (5.1), the emphasis here is on environmental aspects of maintenance dredging, that is, navigation dredging conducted to maintain existing waterways and the placement of the dredged material at aquatic sites. Pre-dredging studies to evaluate the potential environmental impacts of other types of dredging (e.g. new construction, contaminant clean-up) and other placement options (e.g. land placement, use of dredged material) are addressed in Section 5.6.8.

In addition to the environmental information discussed in this chapter, dredging projects may require data on geology, engineering, hydrodynamics and other subjects. Coordinating the field activities required to collect various types of information may be a cost-effective approach.

5.4.1 Iterative approach

An iterative approach to conducting the pre-dredging investigations outlined here is usually the best approach for some of the following reasons:

- The first chemical analyses for a project might be performed on a few samples for a comprehensive list of analytes, with the results used to select a more focused list of analytes to be run on a full set of samples.
- An initial field survey might consist of a general examination of many environmental issues, with the results used to design a detailed survey of the most critical issues.
- In countries where environmental evaluations of dredging operations are still developing, starting with very basic physical, chemical and biological evaluations is often best. More complex testing should be introduced only when the need warrants and experience allows. Unnecessary complexity is an indication, not of sophistication, but of naiveté.

As stated earlier, this chapter presents a comprehensive overview of pre-dredging investigations for materials characterisation, an approach which might be required for large projects or ones where environmental concerns are heightened. For such dredging projects, comprehensive planning documents may be required. For many other projects, however, potential environmental risks can be sufficiently characterised without conducting all the procedures presented here.

For most dredging projects, preparing at least brief planning documents for environmental evaluations is useful. These documents should be reviewed by and have the approval of authorities appropriate to the individual country's regulatory and administrative procedures.

Three types of planning documents are common:

- *Work Plan*: contains a detailed description of the approach, materials and methods taken to achieve project-specific objectives.
- *Health and Safety Plan*: outlines procedures and precautions for protecting project workers and others. Includes guidance to follow should an accident occur. The Health and Safety Plan can be project-specific or generic to a suite of dredging operations.
- *Quality Assurance/Quality Control (QA/QC) Plan*: This document outlines procedures and authorities for insuring data of known quality are produced.

The scope of pre-dredging investigations for materials characterisation will vary with each country's policy and regulatory requirements. Some general observations, however, can be made. For example, most countries assess the physical and chemical characteristics of sediments to be excavated. Some countries have established numerical standards to judge laboratory results (e.g. The Netherlands) while others consider each project on a case-by-case basis (e.g. UK, USA). A limited number of countries are beginning to conduct sediment bioassays, in addition to sediment physico-chemical analysis, to assess the potential biological impacts of dredging operations. These types of laboratory tests can be more costly but overcome many of the limitations of physico-chemical analysis.

Several issues, sometimes considered secondary, can be very important to some projects, but sometimes do not receive the consideration they deserve in dredging evaluations. Such issues, addressed elsewhere in this book, include:

- Noise and air quality evaluations should consider construction as well as resultant increases in traffic and such (EPA/USACE, 1992). Pre-dredging investigations should include collection of background data necessary to predict noise and air quality at key points coming from the project. (See Chapter 6.)
- Cultural resources, such as items facilities or locations of historical, archaeological, or cultural value (including shipwrecks and underwater sites) in the dredging and placement areas should be identified and evaluated as part of the pre-dredging investigations (Grosser, 1991). (See Chapter 4.)
- Aesthetics refers to that which is pleasing to the senses. In the context of dredging, insensitivity to aesthetics usually takes the form of public perceptions of impacting pleasing landscapes or views. This can result in bad public relations. Project planning evaluations should consider the aesthetic impacts of project location and design. (See Chapter 4.)
- Infringement on other public uses of the area, such as limiting access to the waterfront for fishing or potential effects of turbidity on a recreational diving site, should be considered as part of the early project concept and pre-dredging investigations. (See Chapters 4 and 6.)

5.4.2 Data collection

Environmental impact assessment requires the collection of data. Sometimes this entails only collection and compilation of existing information. More often, however, environmental assessments of dredging or dredged material placement also require collection of new field and/or laboratory data. This data collection is both time-consuming and expensive even for environmental assessments involving clean materials, where examining hydrodynamics, sedimentology, or complex biogeochemical processes may be useful. In cases where contaminated materials are to be dredged, the investigation process is certain to be far more expensive with additional costs for sampling and chemical and biological analyses.

The collection of representative data from marine sites is also complicated by the occurrence of extreme events such as storms. These events are part of the environmental background, and the local biota has developed the ability to cope with the environmental change associated with these naturally occurring events, at least of some degree and duration. Data collection often ignores these extreme events because of sampling difficulties. Yet the impact of a dredging project may be more fairly compared to that of an extreme event rather than to normal day-to-day background conditions.

Another complication may be funding and compliance. The London Convention requires that contracting countries evaluate dredging projects in jurisdictional waters in the spirit of the Dredged Material Assessment Framework (DMAF), as discussed above in Section 5.2.4. Since the collection of field and laboratory data is important and may be expensive, prioritising its collection thoughtfully to use the available funds most effectively is a necessity. Such an approach helps implement the spirit of the DMAF taking into account the practical constraints of available funds and exper-tise, especially in countries with limited resources or expertise in dredged material evaluations.

Stakeholders may play an essential role in discussing and planning necessary environ-mental data acquisition and assessment. Discussions should address topics such as:

- Given limited funds, which areas of investigation are the most important and for what reasons, social, scientific, aesthetic? (noting that this may vary according to the type of project and the site characteristics)
- Which types of investigation are most important in the priority areas selected?
- How will the DMAF be applied with the limited information available?

5.4.3 Field surveys

The purpose of pre-dredging field surveys is to characterise the biota, habitats and chemical conditions at the dredging and placement sites. This characterisation focuses on how the dredging operations may affect the environment. The reference or background site is also located during the field surveys. Field surveys and sample collection are conducted at the dredging site, the placement site, and perhaps other locations for background data.

At the dredging site and aquatic placement sites, the primarily biological resource of concern is benthic (bottom-dwelling) organisms. This is because environmental impacts of dredging, if they occur, are generally associated with the benthos. Benthic biota has limited mobility and thus cannot avoid the dredging and placement activities.

Impacts on biota that live in the water column (nektonic and pelagic organisms such as fish) are usually transitory if they occur at all. These organisms are generally highly mobile and can avoid dredging operations if necessary.

The planning and conduct of field activities is usually an iterative process. Guidance should be reviewed, and field activities carefully planned. Samples are then collected and field surveys conducted. After results are evaluated, re-visiting the field to fill data gaps may be necessary.

Field surveys of dredging sites
Field surveys may differ slightly at dredging sites depending on whether dredging has occurred previously. For previously dredged sites, expecting a "normal" benthic com-munity is unreasonable. Ship traffic, propeller wash, and non-normal hydraulics com-bine to make this a disturbed habitat. Some benthic organisms may recolonise the channel between dredging cycles, but they should be viewed as temporary. They will

be removed during the next dredging event. Some mobile species (e.g. sea turtles, commercial crabs) may use dredged channels as migratory pathways or over-wintering areas. If a potential concern, this should be evaluated during the field survey. It may be appropriate to consider scheduling dredging activities at times when such organisms are not present in the channels or to use deflecting or excluding devices to prevent such organisms from entering the dredging equipment.

The field survey should also evaluate biological resources adjacent to the dredging site. Particularly at risk are biota that are sensitive to suspended sediments or increased turbidity. They may include submerged aquatic vegetation, coral reefs, and some molluscs. Because turbidity is often a public and/or regulatory concern, monitoring the intensity, extent and duration of turbidity generated by dredging (and perhaps placement) activities should be carefully considered. The subject of turbidity caused by dredging is addressed in numerous publications including S. A. John *et al.* (2000) and in Proceedings of WODCON XVIII, Session 7C.

When dredging a new channel, the major concern is the lack of information. Valuable biological resources may exist at the future dredging site or biota may utilise the area seasonally. Also of concern is the altered hydrology at the proposed dredging site. Water salinity, temperature, dissolved oxygen, and circulation patterns may change in ways that adversely affect biota utilising the site. Existing information about the dredging site, if available, should have been assembled during the Initial Evaluation (Section 5.3). Additional information, if necessary, can be gathered during the field survey.

Field surveys at placement sites
Regulatory approval to use a placement site implies agreement to alter the physical, chemical, and biological characteristics of the site environs. The degree of alteration is subject to negotiation. Those negotiations may be based, in part, on information gathered during the pre-dredging field survey. Pre-dredging placement site surveys should address the following:
● conditions within the proposed placement site – characterise resources that will be affected to some degree by being covered with dredged material, and
● conditions around the site – characterise resources that may be affected via sediment transport away from the site (includes physical and hydrographic conditions as well as biota that would recolonise within the placement site).

For placement sites that have historically received dredged material, the major concern is for impacts adjacent to the site. However, because recolonisation of the site begins shortly after placement ceases, some degree of biological utilisation will have occurred. The potential value of this resource should be evaluated.

For new placement sites that have never received dredged material, "baseline" physical, chemical, and biological conditions should be characterised. Any valuable biological resources that could be adversely affected should be identified. As with dredging sites, the emphasis at aquatic placement sites should be on the benthic environment.

Post-dredging field surveys
Post-dredging survey or site monitoring is mentioned here because it forms the last link in what should be a continuum of evaluation of all aspects of the project. This site monitoring should provide the basis for feedback to improve the overall environmental assessment process. Monitoring should be structured around clearly stated objectives. These objectives should focus on determining the accuracy of the predictions about impacts which were made on the basis of the pre-dredging evaluations. Practical implementation guidance on the major aspects of pre- and post-dredging field surveys can be found in Chapter 8.

5.4.4 Sampling

The collection of field samples for pre-dredging laboratory investigations is similar for the different types of tests. Consequently, this section contains single sections for sampling plan, statistical design, sampling gear, and sample management, and ends with an overview of practical sampling considerations. IMO (2005) provides thorough guidance on dredged material sampling.

Sampling plan
The sampling plan is the foundation of all pre-dredging testing. However, owing to multi-variant factors no standard format exists. Consequently, each dredging project team must develop its own specific plan. All plans begin with a general statement describing the objectives of the dredging project. From this, a broad objective of the sampling plan is developed, for instance, "Obtain enough samples with the proper quality (e.g. proper sampling, preservation and storage) to adequately (i.e. enough properly located samples) and accurately (i.e. without bias) characterise the dredging, placement and reference or background sites".

Other major elements of the sampling plan should include the following:
● maps with project depth, over depth, side slopes, removal volumes and project segments;
● summary of existing information pertinent to sampling;
● anticipated methods of dredging (e.g. mechanical, pipeline) and placement;
● types of sampling gear and positioning technique(s);
● statistical designs;
● logistical considerations;
● provisions for field flexibility with follow-up documentation of deviations;
● anticipated costs; and
● anticipated sampling and analysis schedule.

Statistical design
All statistical designs vary between completely random and completely biased. The completely random design places no pre-conditions on where a sample is collected. Often, a grid map is placed over the site or segment and sample locations are determined via random number generation. In contrast, in a completely biased statistical design, the reasons for locating sampling stations have no statistical basis (e.g. it is by an outfall, or it is at the beginning of the project reach). Most statistical designs

for sediment sampling appropriately fall somewhere between completely random and completely biased (e.g. stratified random).

Statistical replication is an important consideration. Replicates are the smallest measure of independent observation made under identical experimental circumstances. More replicates decrease uncertainty associated with calculated statistics (e.g. means), but increase sampling, analytical, and data management costs. Taking multiple samples does not necessarily constitute true replication. For example, taking a grab sample in the field, and taking multiple sub-samples from that grab sample and treating them identically (e.g. chemical characterisation) would not make them true replicates and they are in fact called pseudo-replicates.

Likewise, in the laboratory, multiple sediment samples taken out of an aquarium at the end of a bioaccumulation bioassay and chemically analysed would be pseudo-replicates. Additional grabs and additional aquaria would have to be sampled to constitute true replicates. Field and lab replicates should also be distinguished. The former measure variation in the sample collection processes, while the latter measure variation in the analytical processes.

Frequently asked questions about statistical sampling designs
- "How many samples do I need to take?" A quantitative answer to this question that is practical and fully satisfactory to all interested parties is elusive. However, in general the number of samples will always be inversely proportional to the amount of known information and directly proportional to the level of confidence desired in the analytical results. The more samples that are collected, the higher the sampling and analytical costs will be. To maximise cost-effectiveness in the pre-dredging evaluation carefully balancing the amount of information (samples) collected with the costs required to collect them is essential. If the navigation project has a comprehensive, technically sound historical basis, the number of samples can be relaxed. If, on the other hand, the project is large and/or continuous, and the potential for environmental impacts is high, then the number of samples should be increased. Objective approaches to determine the number of samples to be collected are discussed by Bates (1981) and Csiti and Donze (1995). The subject is also addressed by the "FAST" computer model (POSW, undated).
- "Where should I take samples?" Sample location is a function of statistical design. For a completely random design, sample locations are selected at random from a grid that overlays the area to be dredged. This approach is generally applicable only for very large areas. A more common approach is to divide the project area into "representative" sections and sample each one of these. Another approach is to locate samples near known/suspected areas of contamination. Results of these samples most directly answer specific questions regarding environmental contamination; however, because they are not random, their statistical value is diminished.
- "Should I take samples of the sediment surface or of the entire depth of the sediment to be removed by dredging?" Surface samples are generally preferred for maintenance dredging projects where all the material is going to a single placement site. If this is new dredging or if the dredged material may have multiple placement/treatment fates, then samples taken at depth may also be required.

- "Can I composite samples?" Compositing refers to collecting several samples and combining them into one sample for analysis. This allows more locations to be included in the characterisation while minimising analytical costs. For dredging projects, the answer is a cautious "yes" if the area is relatively homogeneous and/or small but caution is advised when compositing samples from areas of vastly different grain size or known/suspected chemical "hot spots".
- "How much sediment do I have to collect?" Again, no quantitative answer is possible. It will be a function of how the sample will be characterised (physical, chemical, biological). Usually, the minimum volume required is a function of analytical chemistry detection limits. More volume means more chemical mass in the sediment and lower detection limits are possible. The maximum sample volume is usually determined by cost and what is practical to handle and transport.

Sampling gear

Sampling gear that is capable of providing samples with the required characteristics should be selected. This includes not only the volume of samples required for the planned analysis, but also the ability to provide an undisturbed sample if necessary. For example, many physical parameters require an undisturbed sediment sample, while many biological tests can be performed with sediments that have been disturbed or mixed.

Sampling gear for the collection of sediments can be broadly divided into two types: grabs and cores. Grabs collect surface sediments while corers collect a profile of sediment at depth. Selection of sampling gear is driven by why the sediments are being collected. For maintenance dredging, grabs are most often chosen because one is usually interested in the sediment that has deposited since the last dredging event. For characterising sediment at deeper, more undisturbed layers (e.g. for new construction dredging), a corer device may be more desirable.

Grabs fit the general description of a miniature clamshell dredge bucket. They come in a variety of sizes and actuating mechanisms, and range from small volume, lightweight devices to heavy, large volume grabs.

Coring devices also range in size and mechanisms. Hand corers and simple gravity corers are easy to manipulate, while a winch or crane is required to deploy long piston corers or large volume box corers. Vibratory corers (Figure 5.5) are designed to overcome the resistance of hard substrate or consolidated sediment. Sampling methods and gear (Figure 5.6) is discussed in detail in various sources (Bray, 2004; Murdock and MacKnight, 1991; EPA/USACE 1991, 1998; Pequegnat *et al.*, 1990).

The value of "visual samples" in addition to physical samples should not be overlooked. Photographs and videos can be very valuable for recording and conveying site conditions before, during, and after dredging or other construction. They are particularly helpful in communicating accurately with regulators, public groups and others who may not have visited the site or may not be familiar with dredging. General "tourist snapshots" are seldom very helpful; each scene should be thoughtfully selected to convey a particular message.

Figure 5.5 Standard vibrocore unit.

Sample management

Samples are subject to chemical, biological and physical changes as soon as they are collected. To minimise these changes and to ensure the sample collected is representative of actual conditions at the sampling site, proper sample handling is required. These procedures should be clearly outlined in a Sampling and Analysis Plan and documented in chain-of-custody forms.

Sample handling is driven by which analyses are to be performed on the sample for physical, chemical, and biological characterisations. The sample handling plan should specify type and number of collection containers, short-term preservation conditions in the field, transport methods/conditions from the collection site, and storage conditions in the laboratory prior to analysis. As a general rule of thumb, sediment samples should be placed in airtight containers with a minimum of air and water. They should be preserved and transported cold but not frozen (4°C) and kept in the dark. Analysis should begin as soon as practical. Long-term storage (i.e. for months) should generally be avoided.

Practical sampling considerations

Careful consideration to the practical requirements of the sampling operation can save much frustration and inefficiency. To minimise mobilisation costs, as many different

Figure 5.6 Remote ecological monitoring of the seafloor known as REMOTS.

types of samples for as many project purposes as possible should be collected by the same vessel and sampling crew. The vessel should be large enough to accommodate the necessary types of sampling gear, including lowering and raising the gear and removing samples. Storage space for empty sample containers and preservation of samples after collection are necessary. If cores must be collected, the vessel must be held motionless while the corer is lowered and raised: Multi-point anchoring or spuds are usually required.

The precision with which sampling stations must be located should be carefully considered and the appropriate navigation equipment must be available on the sampling vessel. To select precise sampling positions beforehand and try to position the vessel to collect samples from these exact locations is usually time-consuming, expensive and unnecessary. Identifying approximate sampling positions in advance, positioning the vessel as near to that as practical, and then recording the precise location from which the sample was actually taken is usually much better.

5.5 ANALYSES

Analyses of samples are conducted for physical and chemical sediment characterisation, including possible analytes, for biological characterisation including toxicity and bioaccumulation bioassays. In addition, Quality Assurance as well as Quality Control of the data collected is discussed.

5.5.1 Physical analysis

Physical analyses are important in the pre-dredging environmental evaluation because they help indicate how the sediment may behave during dredging operations, what type of organisms may be attracted to or repelled by the sediment, and the bioavailability of sediment-associated chemicals. These analyses may include grain size, percent water or solids, and density determinations.

Grain size is probably the most common physical analysis conducted and the most useful from an environmental assessment perspective. It tends to be a reasonably good surrogate for other physical characteristics. For example, sediments high in silts and clays tend to have higher water content and lower densities. Conversely, sandy sediments are usually denser. Categories of grain size, in order of decreasing diameter, are gravel, sand, silt, and clay. These categories and their associated particle size expressed on a micrometer or Phi scale are shown in Table 5.1 on the following page.

Grain size	Diameter (mm)	Phi (F) = −log (diameter)
Gravel	>2000	>−1
Sand	63 to 2000	4 to −1
Silt	2 to 63	9 to 4
Clay	<2	<9

Table 5.1
Grain Size Categories

Grain size can be determined via wet sieving or a settling velocity/hydrometer method (ASTM 1994). Results are typically expressed as percent sand or percent silt and clay by weight. Other methods for determining sediment grain size involve sample manipulations (e.g. drying, freezing, thawing) that compromise sample integrity and should generally be avoided.

Percent water or solids and sediment density are measured at the same time. Sediments with relatively high water content tend to resuspend more easily during the dredging process and can result in elevated suspended sediment concentrations at the dredging site. Sediments with high densities tend to settle rapidly once returned to water. Pre-dredging *in-situ* density and percent solids/water characteristics will be altered dramatically if hydraulic dredging occurs. The percent water or solids and sediment density analyses are also useful from an engineering standpoint if the material is to be placed upland, used for beach nourishment, or used to create wetlands.

Turbidity, the degree of water opaqueness, is usually determined *in situ* with a trans-missometer, Secchi disc, or similar instrumentation. For pre-dredging investigations, the potential to generate turbidity can be estimated from the historical records of similar dredging operations. Also a sediment-water slurry (or elutriate) can be pre-pared from the sediment sample, turbidity measured and the time-space distributions of *in situ* turbidity concentrations estimated with appropriate models (e.g. EPA/USACE 1991, 1998).

5.5.2 Chemical analysis

The potential environmental impact of sediment-associated chemicals is often a major environmental concern in dredging operations. Many pre-dredging investiga-tions include chemical analyses of sediments from the dredging site, placement site, and reference or background location. These laboratory procedures yield inventories of chemicals and the concentrations at which they are present.

Analyses of whole sediments are often called bulk sediment analyses. In contrast, analysis of a sediment-water slurry is called elutriate analysis, a laboratory procedure that approximates the release of suspended or dissolved contaminants in the water column from a dredged material discharge. Elutriate analyses may be useful in eval-uating potential of dissolved or suspended contaminants in the water column.

An elutriate is prepared by vigorously mixing sediment and water (often in a 1:3 or 1:4 ratio) for a relatively short period of time (30–90 minutes). The slurry is allowed to stand for a short period of time ($\approx$60 minutes). The overlying supernatant is then removed (siphoned or decanted) to produce the elutriate. The elutriate represents, in a very crude fashion, that fraction of dredged material that may become suspended during the dredging/placement process. Chemical analysis of the elutriate measures those contaminants associated with the suspended sediments. The elutriate can be filtered or centrifuged prior to analysis to determine what chemicals may be released to the water column.

The Detection Limit (DL) is an important concept in the pre-dredging chemical characterisation of samples. The DL is the lowest concentration of a particular ana-lyte that can be detected with accuracy. The DL is a function of many variables, including:
- the sample extraction/digestion process;
- instrument sensitivity;
- analytical method selected;
- interferences in the sample matrix; and
- mass of analyte in the sample.

The normal DL can be lowered but only at the expense of increased project cost and logistical difficulty (e.g. larger sample size). Therefore, the process of defining an appropriate DL is an iterative process. All DL should be specified *a priori* in the plan-ning documents. If numerical regulatory criteria are going to be used to interpret the chemical data, the DL must be below the criteria.

Chemicals that may be analysed include hydrocarbons, chlorinated compounds, pesticides, inorganics, and other miscellaneous analytes. Developing the analyte list of "constituents of potential concern" was discussed in Section 5.3.2. (An additional category of chemical, volatile organic compounds – VOCs – is rarely found in appreciable quantities in sediments because these low molecular weight compounds are generally very water-soluble and often volatilise rather than become associated with sediments.)

Possible analytes
IMO (2005) should be consulted for very helpful guidance on selection and analysis of chemical and physical parameters for dredged material assessments. Possible classes of analytes are described below.

Hydrocarbons
Hydrocarbons are naturally occurring compounds composed almost exclusively of carbon and hydrogen. In nature, hydrocarbons exist in very complex mixtures consisting of two basic molecular structures:
- Aliphatic hydrocarbons are linear (straight-chained) molecules that constitute the bulk (by mass) of most hydrocarbon mixtures. Aliphatic hydrocarbons generally represent low environmental risk because they are not very toxic and are very susceptible to degradation.
- Aromatic hydrocarbons are molecules consisting of multiple rings; hence the term Polycyclic Aromatic Hydrocarbons (PAHs). PAHs, in contrast to aliphatic hydrocarbons, are more resistant to degradation and are much more toxic to humans and ecological receptors. Common PAHs and their ring numbers that might be analysed in pre-dredging chemical characterisation include the following:
  - naphthalene – 2 rings,
  - anthracene – 3 rings,
  - chrysene – 4 rings, and
  - benzo(a)pyrene – 5 rings.

Hydrocarbons fall into three categories (which include both aliphatics and PAHs) based on source:
- petroleum hydrocarbons – from natural seeps, oil spills and surface water runoff,
- pyrogenic hydrocarbons – from combustion of coal, woods and petroleum, and
- biogenic hydrocarbons – from plants and animals.

Petroleum hydrocarbons range from heavy crude oil to highly refined petroleum products such as diesel or jet fuel. The relative proportion of PAHs generally increases with the degree of refinement. Pyrogenic hydrocarbons are composed mostly of PAHs. They enter the aquatic environment via atmospheric deposition and associate with sediments owing to their hydrophobicity. Biogenic hydrocarbons are synthesised by plants and animals and are largely aliphatic.

Chlorinated compounds
Chlorinated compounds are largely human-made chemicals in which one or more chlorine atoms have been inserted into a hydrocarbon molecule. Chlorinated compounds

are synthesised for a variety of industrial processes and commercial applications. Notable examples include polychlorinated biphenyls (PCBs), dichloro-diphenyl-trichloroethane (DDT) and 2,3,7,8-tetrachlorodibenzo(p)dioxin (2,3,7,8-TCDD). Chlorinated hydrocarbons are generally persistent in the environment and often very toxic to organisms. Because of their environmental persistence, chlorinated compounds are cosmopolitan and may occur in sediments far removed from any known or suspected source.

Pesticides

Pesticides are a diverse group of compounds synthesised for chemical combat of insects (insecticides), plants (herbicides), rodents (rodenticides), fungus (fungicides), and other nuisance species. These compounds vary greatly in their structure, composition, environmental persistence, and in their toxicity to non-target organisms. Pre-dredging chemical characterisation for pesticides will be very project-specific and largely driven by the presence of known or suspected local sources.

Organo-metallic compounds

Organo-metallic compounds, consisting of metals and an organic structure, have their own environmental characteristics. Amongst the environmentally most significant organo-metallic compounds in dredging investigations are the butyl tins, and especially Tri-Butyl Tin (TBT). This compound is highly persistent in the environment and chronic exposure to very low levels has been shown to cause adverse biological effects in many species, especially molluscs.

Inorganic constituents

Inorganic constituents include metals such as cadmium, chromium, copper, lead, mercury, and zinc; nutrients such as phosphates and nitrates; and perhaps compounds like cyanide. Many metals are natural constituents of clay minerals. Consequently, many sediments contain high concentrations of metals. These high concentrations of naturally occurring inorganics must be recognised and distinguished from metals from anthropogenic sources. Background or reference sediments are key to making this distinction.

Miscellaneous

In addition to the inorganic elements and organic compound described here, chemical characterisation of dredged material may include an analysis for Total Organic Carbon (TOC), Acid Volatile Sulphides (AVS), hydrogen sulphide, and ammonia. Results from the TOC and AVS analyses are used to assess the bioavailability of inorganic and hydrophobic compounds, respectively. Hydrogen sulphide and ammonia are two highly toxic, naturally occurring compounds that may seriously bias results of sediment toxicity bioassays. Analyses for these four miscellaneous analytes, therefore, are conducted to help reduce the uncertainty associated with biological characterisations (see page 113).

Chemical analysis of sediments usually begins with thorough mixing or homogenisation of the sample. The sample is then extracted or digested to remove the chemicals of interest from the sediment matrix. Organic compounds are usually extracted with one or more solvents. Inorganic elements are obtained by digesting the sample in strong acid. The sample extract or digest may then be subjected to a clean-up procedure to remove unwanted interferences prior to instrument analysis. Various analytical instruments are used to identify and quantify the target analytes. These include atomic spectrophotometers (primarily for inorganics), gas chromatography, mass spectrometry, and liquid chromatography. Sediments may also be analysed for total organic carbon. A sub-sample of sediment is dried and placed directly in a carbon analyser.

Tissues of biota may also be analysed chemically. The process for homogenisation, extraction, and instrument analysis is similar to that just described for whole sediments. Additionally, tissue samples may be analysed for percent lipid. To obtain percent lipid, a sub-sample of homogenised tissue is extracted with an organic solvent and the extract dried and weighed. Tissue samples may be derived from test species used in laboratory tests or field-collected animals.

5.5.3 Biological analysis

Pre-dredging laboratory testing for the biological characterisation of dredged material consists of toxicity and bioaccumulation bioassays. These tests provide insight regarding potential environmental impacts that cannot be gained through the physical and chemical characterisations just discussed. To conduct these tests successfully, however, requires expertise and laboratory capabilities that may not always be available or affordable. For example, selection and maintenance of appropriate test species is a major factor in the success or failure of sediment bioassays. Test procedures for toxicity and bioaccumulation bioassays are described in general terms below. Detailed guidance can be found in EPA/USACE (1991, 1998) and PIANC (2006).

Toxicity and bioaccumulation bioassays use living organisms to evaluate dredged material. Not surprisingly, the selection of the test species can influence the outcome of the bioassay. Test methods have been developed for a variety of fresh and salt water organisms. Selection will be a function of the receptor of concern, expected environmental exposures, regulatory requirements, public concerns and laboratory capabilities. Clearly species should be selected after careful consideration and agreement of all interested parties.

Guidance contained in EPA/USACE (1991, 1998) and PIANC (2006) may be helpful in selecting appropriate test species. For toxicity bioassays, most experts recommend using a suite of two or three species, because individual species vary in their chemical susceptibility. Benthic organisms are typically used to evaluate the toxicity of deposited sediments while necktonic and pelagic species are used in elutriate bioassays. For bioaccumulation bioassays, chemically tolerant species with large biomass (for chemical analysis) are desirable.

Toxicity
No analytical instrument can measure toxicity. All the chemical analyses described above measure the concentration of contaminants, but not their biological availability and consequently not toxicity. The primary purpose of toxicity bioassays is to provide a direct measure of the effects of all sediment constituents acting together on

appropriate and sensitive test species. Toxicity bioassays can evaluate potential effects of dredged material in the water column or the potential effects of deposited dredged material.

Deposited sediment bioassays
Deposited sediment bioassays may be relevant to many dredged material placement operations since most contaminants are associated with sediment, and organisms that live in and on the bottom may be exposed to deposited dredged material for extended times. In some circumstances the potential for exposure of water-column organisms to dissolved and suspended constituents from dredged material may exist, and elutriate bioassays may be used to evaluate such situations.

Deposited sediment bioassays are conducted by homogenising or thoroughly mixing a large volume of dredged material and placing it into replicate test containers. Test treatments also include reference and control materials. Test organisms are introduced into the containers to begin the bioassay. Exposure duration of 10 days is common. The most common test endpoint is percent survival. At test termination, the containers are emptied and surviving organisms counted.

The bioassay is judged valid if performance in the control treatment is satisfactory. Excessive control mortality (usually ≥10%) suggests a problem with test conditions, test execution and/or test organisms (see "Potential bias in sediment bioassays" below). The deposited sediment bioassay may be considered invalid unless excessive control mortality can be explained and demonstrated not to be a critical factor in test interpretation. For valid tests, response in the dredged material and reference treatments are compared statistically to determine the degree of toxicity of the dredged material.

Elutriate bioassays
Elutriate bioassays are designed primarily to determine the short-term, water column toxicity that may occur as a result of placing dredged material in the aquatic environment. The standard filtered elutriate is used to evaluate the combined effects of dissolved constituents and those associated with suspended sediments. To conduct this type of laboratory test, an elutriate is first prepared as described previously. A dilution series is prepared; 100, 50, 10 percent elutriate is typical. Test species are introduced into replicated containers including the control treatment. The test is terminated usually after 24 to 96 hours, a period stimulating dredged material suspension in the water column following disposal.

Results are usually presented as the LC50, or that concentration of elutriate which is lethal to 50 percent of the test organisms for the duration of the test.
● If the LC50 for the dredged material is equal to or significantly greater (less toxic) than the control treatment, then no impacts are expected.
● If the LC50 is significantly lower (more toxic), then mixing (dilution) in the water column is considered.

Elutriate bioassays generally utilise water column organisms as the test species. In contrast, benthic dwelling biota is used to evaluate the toxicity of deposited sediment.

Acute versus chronic

Acute vs. chronic tests refer to time of exposure in a sediment bioassay. Acute tests are conducted for relatively short periods of time (hours to days) while chronic tests may span weeks. The relatively rapid dispersion of constituents in the water column at typical dredged material placement sites limits the time to which organisms are exposed to dissolved or suspended constituents. Therefore, with only possible rare exceptions elutriate bioassays should be acute tests (see Table 5.2). A third possibility which is often recommended is the alternative design of a "partial chronic" sediment exposure that spans a significant portion of the test species life cycle or critical life stage of the test species.

	Near-field environmental effects (<1 kilometre)	Far-field environmental effects (>1 kilometre)
Short-term environmental effects (<1 week)	*Dredging Site* turbidity burial/removal of organisms reduced water quality	*Dredging Site* none generally expected
	Placement Site burial of organisms turbidity reduced water quality acute chemical toxicity	*Placement Site* offsite movement of chemicals via physical transport
Long-term environmental effects (>1 week)	*Dredging Site* disturbance by ship traffic removal of contaminated sediment	*Dredging Site* none generally expected
	Placement Site altered substrate type altered community structure chronic chemical toxicity bioaccumulation	*Placement Site* offsite movement of chemicals via physical transport and/or biota migration

Table 5.2 Spatial-Temporal Matrix of Potential Effects Associated with Dredging and Dredged Material Placement

Acute deposited sediment toxicity bioassays tests have been criticised because:
- exposures in the field are chronic, and
- historically, acute dredged material bioassays have typically demonstrated minimal toxicity.

Even so, at present chronic bioassays are not recommended for routine evaluations for practical reasons. The recommended alternative design is a "partial chronic" sediment exposure that spans a significant portion of the test species life cycle or critical life stage of the test species. Such a test should measure biologically important sub-lethal endpoints (e.g. growth and reproduction) that are more sensitive than survival. While these tests may provide a more accurate prediction of potential impact of dredging operations, few standard test procedures and species have been demonstrated to be suitable for deposited sediment toxicity bioassays. This type of test is also more costly and difficult to conduct.

Potential bias in sediment bioassays

Potential bias in sediment bioassays must be considered. The outcomes of both elutriate and deposited sediment bioassays can be unduly influenced by factors other than sediment-associated anthropogenic chemicals. Ammonia and hydrogen sulphide are two naturally occurring sediment constituents that are highly toxic to aquatic animals. Both can appear in high concentrations during dredged materials bioassays. This is especially true for high TOC sediments that have been recently homogenised for the bioassay. Sediment grain size can also introduce significant bias. Test species exhibit sensitivity to certain grain size fractions. Some cannot tolerate muddy sediments while others do poorly in sandy material. These potential biases can be addressed by having the appropriate experimental controls and by monitoring certain constituents during the bioassay.

Bioaccumulation bioassays

Bioaccumulation bioassays determine the potential for sediment-associated chemicals to move from the sediment matrix into biota. As such, they are not a measure of environmental impact *per se* as are toxicity bioassays.

Bioaccumulation bioassays are conducted in a manner similar to toxicity bioassays with deposited sediment. Animals are exposed to sediments in the laboratory usually for 10–28 days. Surviving organisms are collected at the end of the test and analysed for chemical constituents. If sediment-associated chemicals appear in biota tissues at environmentally significant concentrations (e.g. greater than the reference treatment), then the conclusion may be drawn that an environmentally significant potential for bioaccumulation may exist. Tissue samples are also collected on a subset of test animals used to initiate the bioassay. Elevated chemical concentrations are sometimes observed in this background sample. If these background data were not collected, a false conclusion might be reached that a significant potential for bioaccumulation exists when, in fact, it does not.

In some instances, bioaccumulation bioassays can be avoided by utilising simple bioavailability models (see Text Box on page 114). These models estimate the maximum bioaccumulation possible based on data generated during chemical characterisation. If these models suggest limited bioaccumulation potential, then the laboratory bioaccumulation bioassays can be avoided. Detailed guidance for bioaccumulation bioassays and bioavailability models can be found in EPA/USACE (1991, 1998) and PIANC (2006).

Bioavailability models may warrant consideration as screening tools in the initial tiers of environmental assessment. These models are highly conservative and provide upper boundary estimates of availability and trophic transfer potential, but should not be regarded as accurate predictors of environmental effects. For neutral non-polar organic compounds such as PCBs, sediment concentrations are normalised to percent organic carbon and maximum bioaccumulation potential is calculated based

---

Analytical specification: Bioavailability models

For neutral non-polar organic compounds such as PCBs, sediment concentrations are normalised to percent organic carbon and maximum bioaccumulation potential is calculated based on the percent lipid in a receptor of concern.

For inorganic elements, the ratio of acid volatile sulphides to simultaneously extracted metals (AVS:SEM) is calculated. If the ratio is less than one, theory predicts there will be enough sulphides to bind with the metals making them unavailable for uptake (and thus unable to exert toxic effects).

These models are highly conservative and provide upper-bound estimates of availability and trophic transfer potential. They are useful as screening tools in the initial tiers of environmental assessment, but should not be regarded as accurate predictors of environmental effects.

---

on the percent lipid in a receptor of concern. For inorganic elements, the ratio of acid volatile sulphides to simultaneously extracted metals (AVS:SEM) is calculated. If the ratio is less than one, theory predicts there will be enough sulphides to bind with the metals making them unavailable for uptake (and thus unable to exert toxic effects).

5.5.4 Quality Assurance/Quality Control (QA/QC)

In the broadest sense of the term, the quality of dredged material data is affected by six primary factors:
- the degree to which the samples reflect the important characteristics of the project;
- sample size, i.e. sufficient numbers of samples to characterise the project with appropriate statistical confidence;
- appropriate sampling technique to assure that the proper samples are collected;
- proper sample handling, transport and storage to prevent contamination and maintain sample integrity;
- analytical procedures appropriate to the analyte, matrix and detection limit requirements; and
- appropriate and careful data management.

Considerable discussion could be devoted to each of these factors. Clearly, the entire evaluation is no stronger than the weakest data point, and meticulous attention to every aspect of QA/QC at every step of the process is essential. QA/QC for dredged material evaluations is the subject of EPA (1995), and is discussed in EPA/USACE (1991, 1998), IMO (2005), Pequegnat *et al.* (1990) and PIANC (2006).

5.5.5 Projects other than aquatic placement of maintenance material

Much of the information in this chapter emphasises the aquatic placement of maintenance dredged material. The reasons for this were discussed in Section 5.1.

Nevertheless, much of what is presented is applicable to a wide variety of dredging situations. Additional perspectives for specific types of dredging are presented below.

New construction

New construction can also be subject to any of the procedures and evaluations discussed here although it is often less necessary. When sediments deposited prior to municipal or industrial development in the watersheds are dredged as in new construction, the potential for contamination is usually low. In such cases the pre-dredging investigations and materials characterisation focus primarily on potential physical impacts of the dredging and dredged material placement, and contaminant issues are much less important than they may be with many maintenance dredging projects.

Use of dredged material

Use of dredged material refers very broadly to any productive or beneficial purpose for which the dredged material is used instead of simply disposing of it. Historically, either clean or contaminated material might be considered for construction and development purposes, whereas only uncontaminated material was considered for ecological uses. Current practice recognises that under appropriate conditions somewhat contaminated material may be used in ecological projects if it is sufficiently isolated from interaction with the environment (e.g. by covering with clean materials).

The pre-dredging investigations for materials characterisation discussed in this chapter are appropriate for evaluating material for possible use. If the proposed uses of the dredged material would be in direct contact with the environment, emphasis will likely be on demonstrating a lack of toxicity or water quality impacts and little potential for bioaccumulation. If the proposed uses of the dredged material would be purposely isolated from contact with the environment, emphasis will likely be on the long-term physical and chemical integrity of the isolation. Uses of dredged material are discussed in detail in PIANC (1992) and PIANC WG 14 (2007).

Land placement

Land placement also requires evaluation and the potential for four general types of effects should be considered:
- physical effects such as long-term burial of the resources and habitat at the site;
- effects of contaminants and natural constituents associated with the sediments that will be placed at the site;
- effects such as suspended solids, low dissolved oxygen, elevated ammonia dissolved contaminants, caused by the effluent or return water during placement; and
- effects from the leachate and surface runoff after placement has been completed.

The kinds of considerations appropriate for each of the four categories are site-specific depending on the characteristics of the site and the material to be placed in it. All four should be considered; that some or all will prove to be of little concern at a particular site is entirely possible.

Salt can be a major issue when sediment dredged from brackish or salt water is placed at an upland site. As the sediment dries, salt is concentrated in the soil and can inhibit plant growth or limit the plant species that will develop on the site until it leaches out sufficiently. Less time is required for leaching in wet climates, if the original salt content is low or if the sediment and site characteristics allow good drainage. Leached salt can affect surface water and groundwater quality if it enters a freshwater system at a sufficient rate.

Grain size is also important in determining what plants will develop on wetland or upland placement sites. In general, very sandy sediments or soils support less dense and/or less diverse plant communities than do finer-grained sediments or soils.

When dredged material is removed from the water and placed at an upland site, changes occur in several geochemical conditions that can affect contaminant mobility and bioavailability. As the sediment gradually dries it is subject to local rainfaill that may be acid and sulphides oxidising to sulphates, both of which tend to lower the pH of the dredged material. To the extent that pH is lowered, the bioavailability and leaching potential of metals sorbed to the sediment increases, which can affect surface or groundwater quality or the toxicity and bioaccumulation of metals in terrestrial plants and animals on the site. The gradual oxidation of organic matter can progressively decrease the TOC available to sorb non-polar organics, and thus can increase the mobility and bioavailability of these compounds over time. In addition, the resources (e.g. groundwater, plants) potentially at risk at an upland site are different from those (e.g. fish) potentially at risk at an aquatic site.

Geo-physico-chemical conditions at upland sites differ greatly from those at the aquatic environment. Yet almost all sediment chemistry regulatory values have been set for the purpose of controlling potential impacts of sediments in the aquatic environment. Little or no consideration is given to the potential impacts if sediments exceeding the values were to be placed at an upland site. Still, many sediment chemistry regulatory values are not technically appropriate for land placement. The fact that a sediment exceeds some regulatory value(s) does not mean that it could not cause greater environmental impact if placed at an upland site than if left in the water. This should be kept in mind when evaluating chemistry data for dredged material for which upland placement may be an option (Peddicord, 1993).

Remedial dredging
Remedial dredging or clean-up dredging must focus heavily on potential contaminant effects, and rigorous chemical, toxicological and bioaccumulation evaluations should be conducted. The procedures discussed in this chapter are appropriate for such purposes, although in practice advanced tier tests may be used more often in remedial dredging than would be appropriate for most dredging projects.

5.6 STEP 4, INTERPRETATION OF RESULTS

Once the appropriate pre-dredging information is collected, the next step is to review the results and interpret their significance from an environmental assessment perspective (see Figure 5.1). An initial step in this review is statistical analysis. That analysis usually includes comparison to a reference and/or control treatment. These results are then interpreted from a spatial-temporal perspective; i.e. what short-term and long-term impacts may occur at or near the site of dredging operations. Section 5.7 concludes with a discussion of how to interpret the results of the individual suites of physical, chemical, biological, and field survey data.

5.6.1 Reference and background sediments

Once relevant procedures have been selected, a fundamental question concerns the benchmarks for the comparison for the dredged material results. Regulatory benchmarks for some contaminants of concern in dredged material have been established in some countries (see Annex A). Many countries, however, have not set quantitative chemical limits, and no country has limits for every contaminant that could be of concern in dredged material. Few quantitative regulatory benchmarks have been established for parameters other than sediment chemistry.

In the absence of quantitative benchmarks, results of tests on dredged material samples may be compared with the results of the same tests on reference or background sediments. Consequently, these sediments must be identified during the planning process so that the necessary data can be collected during the sampling and analysis programme. Comparison with reference or background sediments may be used in the evaluation of a wide variety of physical, chemical and biological data.

What defines a reference or background sediment? No answer to this question is generally accepted. The terms "reference" and "background" are used somewhat interchangeably by some people, while others assign specific meanings to each term. These meanings are not always consistent, and even all national regulatory programmes that use these terms do not attach the same meanings to them. In view of this situation, project planners should define how the terms reference and background are used in regulatory programmes applicable to the project under their consideration, and conduct evaluations accordingly.

5.6.2. Spatial-temporal considerations

Dredging and dredged material placement can be thought of as activities potentially affecting the site itself and activities potentially affecting the surrounding area. Some of the effects may be immediate; others might be manifested over a longer time period. Potential environmental effects should be considered in the context of a spatial-temporal matrix such as Table 5.2 (see also Text Box on page 118), illustrating the relative scales within which various environmental effects might be exerted and therefore the scales in which test results should be interpreted and evaluated. The contents of Table 5.2 are illustrative of effects that might occur under typical conditions, but are not intended to be either inclusive or exclusive under all conditions. For example, aesthetic impacts, noise and traffic congestion from equipment movements may be important considerations in addition to those shown in the matrix.

Temporal considerations also include the issue of seasonal restrictions on dredging activities owing to concerns about potential adverse impacts of dredging and/or

Definition of spatial-temporal terms

Near field: Phenomena occurring within the geographic bounds of the activity, or less than approximately 1 kilometre from the activity.

Far field: Phenomena occurring more than approximately 1 kilometre from the activity.

Short term: Phenomena occurring during the activity or within approximately 1 week of the cessation of the activity.

Long term: Phenomena extending more than approximately 1 week after cessation of the activity.

placement activities on sensitive life stages or activities of organisms. Situations could arise in which the timing, magnitude or nature of dredging-related alterations or the particular species of concern could warrant evaluation of the potential benefits of seasonal restrictions of the dredging and placement activities. A framework for evaluating the need for seasonal restrictions is provided by Transportation Research Board (2001). (See also PIANC ENVICOM-WG13, 2007).

However, hard information on the degree to which dredging-related activities actually affect the organisms of concern is often scarce. In many cases the magnitude of dredging-related alterations in the environment are similar to, and fall well within the range of, naturally occurring phenomena and probably impose little or no additional stress on organisms of concern.

After dredged material has been placed in the aquatic environment, it may mix and become diluted over space and time; this is particularly true for dissolved constituents and suspended sediments. Several mathematical models suitable for calculating the mixing zone associated with a dredged material discharge have been developed, including those provided and discussed in EPA/USACE (1991, 1998). These models predict the duration, size and shape of the area in which the end-point (e.g. water quality standard, toxic concentration) will be exceeded, and consider both aquatic placement sites and the water bodies receiving effluent from upland placement sites.

Several important concepts should be considered in determining the acceptability of the mixing zone predicted for any dredged material discharge. Concern over the potential to cause unacceptable impacts increases in direct relation to the following (Lee *et al.*, 1991):
- size and configuration of the mixing zone;
- proportion of the volume and cross-sectional area of the receiving water body occupied by the mixing zone;
- time required to achieve the desired dilution for each discrete discharge event;
- duration of the dredging and placement operation and management of the discharge;
- proximity to sources of drinking water;

- proximity to areas of high human water-contact activities at times of major use (e.g. swimming beaches in the summer);
- proximity to shellfish beds or other areas of recreational or commercial importance;
- proximity to major sport or commercial fishery areas at the time of major use;
- proximity to unique or concentrated fish or shellfish migration routes, spawning areas or nursery areas at the time of major use;
- proximity in space and time to other major dredged material operations, and
- proximity to major contaminant inputs to the environment from industrial, municipal or agricultural sources.

These and perhaps other factors are complexly interrelated, i.e. mixing zone conditions that are acceptable for one project may or may not be acceptable for another project.

5.6.3 Data analysis

Statistical analyses
Statistical analyses are techniques used to determine whether populations of numbers are similar or dissimilar to one another. Using these techniques, statistical differences or lack thereof are determined. In the collection of pre-dredging data, great efficiencies can be gained if statistical considerations are incorporated into the initial design of the sampling and analysis programmes. An experienced statistician may be helpful to the project team during the initial planning stages.

While a good statistical design is important for interpreting results, statistics should not be allowed to dominate the decision-making process. The tendency for decisions regarding potential environmental impacts associated with dredging operations to focus exclusively on statistical significance is unfortunate. Statistical significance, whether in toxicity or bioaccumulation bioassays, chemical data, field surveys, or other environmental data, does not imply environmental importance. Statistical significance (or lack thereof) is controlled by many factors (e.g. number of replicate samples) having nothing to do with potential environmental impacts. Statistical analyses are meant to support decision-making, and should never become the decision itself.

Pre-dredging field surveys
Pre-dredging field surveys may be conducted at the dredging site and/or the placement site. Results of field surveys at the dredging site are evaluated primarily for the presence or absence of valuable biological resources. If present, their spatial-temporal distribution should be determined.

Results of pre-dredging field surveys at the placement site are evaluated in two ways. One is a description of the site's physical characteristics such as depth and current conditions. The other evaluation is biological, i.e. the resources that exist at and adjacent to the site which merit special attention. The emphasis is usually directed to areas adjacent to the site, as this is where impacts are expected. These near-field effects (Table 5.2) are of greater concern. Pre-dredging field surveys at the placement site are most often qualitative unless a "baseline" is desired. In that case, results are quantified so that subsequent surveys (e.g. dredging and post-dredging monitoring) will have a quantitative basis for comparison.

Field survey results, both pre-dredging and post-dredging, should be evaluated in the context of natural variability and seasonal effects. Both natural variability and seasonal effects can have major impacts on the dredging and placement sites that are quantitatively more important than potential impacts of dredging operations. Field surveys are typically one-time events that provide a snapshot of the field at a particular moment and location. Rarely do field surveys collect enough data to document natural variability and seasonal effects.

Physical data
Physical data related to dredging and dredged material placement include turbidity, total suspended solids (TSS), and sediment deposition.

Turbidity and suspended sediments from dredging should be evaluated in relation to natural levels in the area and season(s) where dredging operations will occur. In many environments (e.g. estuaries), natural background turbidity and suspended sediment levels are generally high. Organisms in such environments are adapted to these conditions. They are able to withstand continuous exposure to high suspended sediment concentrations for much longer times than would be caused by most dredging and dredged material placement operations (Peddicord and McFarland, 1978; Pennekamp, 1996; Stern and Stickle, 1978).

Concern for turbidity and suspended sediment impacts increases if ambient water conditions are normally clear. Submerged aquatic vegetation, corals, and other species requiring clear water habitats may be especially vulnerable. These species may be adversely affected by changes in light penetration or by thin layers of fine sediment deposited from suspension. Turbidity and suspended sediments produced by dredging operations, however, are almost always a very near-field and short-term phenomena (Table 5.2). Field surveys should identify sensitive species that may be adjacent to the dredging and placement site.

Concerns have been raised about the possibility of elevated turbidity and/or TSS levels causing negative impacts on marine life by various means. While site-specific studies may be required, long experience has shown that in many cases these concerns are not as important as some have feared, if for no other reason than the generally short-term and near-field nature of the conditions. The concerns most frequently raised are:
- Depression of dissolved oxygen levels, reduction of photosynthesis, interference with respiration and feeding, impedance to mobility, for example. Much experience has shown that these conditions caused by dredging or placement are no more severe than those caused by natural conditions (e.g. storms, currents) or other human activities (e.g. trawling, ship traffic), and are very seldom ecologically important in practice (Pennekamp, 1996; Rhoads et al., 1978; Stern and Stickle, 1978).
- Release of adsorbed metals or organics from fine-grained suspended solids, irritation to tissue making infection or invasion by parasites more likely. Experience

has shown that such phenomena can occur, but that around dredging and place-ment operations they are much less common, and typically less intense and less widespread, than often anticipated.

- Deposited sediment changing the characteristics of the bottom or smothering benthic organisms near the dredging or placement site, suspended solids coating the surface of organisms. Much experience has shown that benthic organisms can be smothered at aquatic dredged material placement sites, but that most semi-mobile and mobile organisms have considerable capacity to accommodate such changes and actual adverse effects are generally much less than might be presumed. Experience has also shown that exceptions to this generalisation do exist, and that these exceptions are the most effects of turbidity and/or TSS to be realised in association with dredging and dredged material placement. Sessile benthic organisms (e.g. oysters) can be adversely affected by thin deposits of sediment. Mangroves and similar plants have little ability to accommodate rapid sediment deposition. Submerged aquatic vegetation (i.e. sea grasses) is often sensitive to thin deposits of fine sediment on leaves and stems. Corals are typically very sensitive to slight deposition of sediment from suspension. Measurement of suspended solids is particularly important for such very sensitive sites, and dispersion of solids into them may need to be prevented or severely controlled.

The typically short-term and/or near-field dispersion of turbidity and TSS will usually result in minimal negative impacts. Although turbidity and/or TSS are usually of little ecological consequence, turbidity is visible in areas frequented by people and this aesthetic impact causes much concern. The aesthetic impact is the real reason tur-bidity is often monitored and controlled by the regulatory authorities.

Turbidity and fine suspended solids can be assessed with predictive models. The behaviour of larger particles, which settle relatively quickly and may produce a layer on the water bed of considerable thickness, are less easy to monitor and are often not detected by the turbidity measuring devices. Consequently, for any site, deter-mining the location of the sensitive receptors in order to identify the important water quality parameters and to select the most appropriate methods of measurement should be emphasised.

Sediment resuspension and transport following placement are important environ-mental considerations. For example, the aim may be to place dredged material in a high energy environment to maximise dispersal and minimise potential navigational impediments. In this scenario, consider that fine-grained material is more susceptible to movement than sandy material. Thus, sandier dredged material will require a higher energy zone than clay sediments. Conversely, if the aim is to minimise disper-sion of dredged material, which is largely silt and clay, then a low energy or even depositional placement area must be selected.

Chemical fate and bioavailability
The chemical fate and bioavailability of sediment-associated chemicals are directly affected by grain size. In general, chemicals are found in higher concentrations in

fine-grained material, but are less available for bio-uptake than from sandier sediment owing to several factors:

- Fine-grained sediments have a greater surface area (relative to volume) than sandy sediments. This greater physical space accommodates a larger number of chemical associations.
- Clay minerals found in fine-grained sediment have small electrostatic charges and thus attract charged ions such as metals and some polar organic compounds.
- Fine-grained sediments tend to have a higher organic carbon content than sandy sediments, which bind hydrophobic chemicals (those that don't "like" to be in water) such as PCBs preferentially to fine-grained sediments from the water column.

In evaluating the environmental impacts of fine-grained dredged material, note that they usually have elevated but naturally occurring levels of inorganic elements, the mere presence of which cannot be assumed to have negative environmental consequences.

Fine-grained sediments have a large clay component. Clays are composed of hydrated minerals such as kaolinite, montmorillonite, and illite. These minerals, in turn, are composed largely of aluminum, silicon, magnesium, iron, sodium, calcium, potassium and barium, intermixed with many other metals. Thus, fine-grained sediments will contain these naturally occurring elements at higher concentrations than sandier sediments, exclusive of any anthropogenic (i.e. human-made, human-induced) inputs. This phenomenon must be kept in mind when evaluating sediment chemistry data and especially when the sediments under consideration vary in grain size.

Benthic (bottom-dwelling) community development
The benthic (bottom-dwelling) community development following the aquatic placement of dredged material is also significantly influenced by sediment grain size. When sediments are disturbed, benthic recolonisation begins almost immediately after the disturbance ceases. In temperate latitudes, benthic recolonisation proceeds through well-documented successional stages (Rhoads *et al.*, 1978). The progression is affected, to a large degree, by the preferences and tolerances benthic organisms have for different sediment grain sizes.

Generally speaking, the more similar the grain size of the dredged material is to the sediment at the placement site, the more closely the recolonising biota will resemble the surrounding benthic community.

Chemical data
Chemical data can be evaluated in three ways:
- *Comparison to regulatory values*: A common evaluation is to compare concentrations of individual chemical constitutes to a regulatory value. These numerical values may be standards (legally enforceable), criteria (generally accepted guidance), or some number that has some basis for application. One must carefully examine the derivation of these values to ensure they are technically sound, appropriate to the application being considered, and provide an adequate margin

of environmental safety. For example, soil criteria should not be applied to sediment and *vice versa*. Sediment criteria are generally inappropriate for dredged material placed at an upland site. Exceeding a regulatory value does not necessarily mean that unacceptable effects will occur, only that some threshold has been exceeded. International numerical regulations are discussed in Annex A.

- *Comparison to reference or background values*: A second approach for evaluating chemistry data is to compare concentrations of individual chemical constituents in the dredged material with a reference or background sediment. Sediment from a reference or background location is, by definition, considered "acceptable". Therefore, if constituents in the dredged material do not exceed the reference or background chemistry, the material may be judged environmentally acceptable.
- *Potential for individual constituents to persist and bioaccumulate*: Chemistry data alone cannot indicate the bioavailability and trophic transfer potential of persistent, bioaccumulative chemicals. Conservative mathematical screening models can use the TOC and AVS results to estimate bioavailability of certain non-polar organic compounds (e.g. DDT, PCB) and inorganics (e.g. cadmium, lead, zinc), respectively (see Text Box "Bioavailability models" on page 114).

In summary, when evaluating sediment chemistry data, any dredged material that exceeds applicable regulatory values must be dealt with as required by the regulation. For constituents that do not exceed regulatory values, concern over potential unacceptable environmental impacts increases in direct relation to the following (Lee *et al.*, 1991):

- number of chemicals present in concentrations statistically higher than the reference or background sediment;
- magnitude by which each chemical exceeds its concentration in the reference or background sediment;
- absolute magnitude of concentrations of chemicals in the dredged material;
- toxicological importance of chemicals present in concentrations statistically higher than in the reference or background sediment;
- tendency for chemicals present in concentrations statistically higher than in the reference or background sediment to bioaccumulate; and
- proportion of chemicals present in concentrations statistically higher than in the reference or background sediment.

Biological data
Biological data encompass toxicity bioassays and bioaccumulation bioassays.

Toxicity bioassays are used to evaluate the toxicity of sediment-associated contaminants which usually is a near-field phenomenon. It may be either short term (acute toxicity) or long term (chronic toxicity) (Table 5.2). Results of toxicity bioassays are usually evaluated by comparison with the reference or background sediment. For example, in acute toxicity tests, if survival in the dredged material equals or exceeds survival in the reference or background sediment, a conclusion can reasonably be reached that the potential for adverse effects is low. If survival is statistically less than that observed in the reference treatment, then the potential for adverse impacts associated with the dredging operations exists. For both deposited sediment and elutriate bioassays, the environmental risk represented by the test results increases in direct relation to the following (Lee *et al.*, 1991):

- exceedance of any guidance value that may exist for toxicity (e.g. <50% survival);
- number of species in which toxicity of the dredged material statistically exceeds toxicity of the reference or background sediment;
- magnitude of toxicity of the dredged material;
- magnitude by which toxicity of the dredged material exceeds toxicity of the reference or background sediment; and
- proportion of sediment samples from the dredging site with toxicity statistically higher than the reference or background sediment.

Bioaccumulation bioassays are not a measure of environmental impact *per se*. Rather, they provide empirical evidence regarding mobility of a chemical out of the sediment matrix and into biological tissue. Bioaccumulation of contaminants of concern by animals or plants recolonising an aquatic, wetland or upland placement site is often considered a long-term, near-field phenomenon, but may be a far-field phenomenon if it involves mobile organisms that move away from the placement site after the accumulation of contaminants in their tissues (Table 5.2).

Bioaccumulation should be evaluated in the context of potential impacts to both human health and ecology. Potential human health implications may be evaluated in relation to concentration limits for contaminants in food established by the World Health Organization, other international bodies, or national agencies. Such values have been established for a variety of potential contaminants of concern. Any dredged material that produces bioaccumulation of contaminants to concentrations that exceed applicable regulatory values for food must be dealt with as required by the regulations.

Limits developed to protect human health have no bearing on ecological impact, and there may be contaminants of concern for which applicable regulatory values for human health have not been established. When evaluating bioaccumulation results in such cases, concern over the potential to cause unacceptable adverse impacts increases in direct relation to the following (Lee *et al.*, 1991):

- number of species in which bioaccumulation from the dredged material statistically exceeds bioaccumulation from the reference material;
- number of contaminants for which bioaccumulation from the dredged material statistically exceeds bioaccumulation from the reference material;
- magnitude of bioaccumulation from the dredged material;
- magnitude by which bioaccumulation from the dredged material exceeds bioaccumulation from the reference material;
- toxicological importance of the contaminants for which bioaccumulation from the dredged material statistically exceeds bioaccumulation from the reference material;
- phylogenetic diversity of the species in which bioaccumulation from the dredged material statistically exceeds bioaccumulation from the reference material;

- tendency for the contaminants with statistically significant bioaccumulation to biomagnify within aquatic food webs;
- extent to which the dredged material is toxic as well as bioaccumulative (e.g. the magnitude of toxicity and number and phylogenetic diversity of the species exhibiting greater toxicity in the dredged material than in the reference material);
- magnitude by which contaminants whose bioaccumulation from dredged material exceeds that from the reference material also exceeds the concentrations found in tissues of comparable species living in the vicinity of the proposed placement site; and
- proportion of the dredged material producing bioaccumulation statistically higher than the reference material.

These and perhaps other factors are complexly interrelated; i.e. the acceptable level of each factor depends on its interaction with other factors, so that what is acceptable for one project may or may not be acceptable for another project.

5.7 ACCEPTABLE VERSUS UNACCEPTABLE IMPACTS

As was described in Chapter 3, a foundational issue for every environmental evaluation is "what constitutes unacceptable impacts?" (see Section 3.1.1). The decision to dredge is a decision to create a "permanent" transportation corridor that will change the environment at both the dredging and the placement sites. Dredging is fundamentally no different from converting a portion of a forest or grassland into a highway. The decision is ultimately made in the best overall interest of society after considering benefits accrued, potential environmental impacts, and costs (tangible and intangible) to construct and maintain.

The sites of navigation dredging and dredged material relocation may experience substantial environmental change, and cannot be expected to retain their pre-construction characteristics (although under some circumstances they may have a natural tendency to revert toward those characteristics over the years between dredging projects). The environmental changes might even be considered "adverse" but acceptable relative to the accrued societal benefits.

The critical question, therefore, is not whether there will be changes (there will be), or even whether the changes are adverse (they may be). The critical question is whether the proposed dredging and relocation of dredged material will be acceptable or unacceptable. Ultimately, what constitutes an unacceptable environmental impact is a socio-political and economic decision based, in part, on the information discussed throughout this book.

CHAPTER 6

Machines, Methods and Mitigation

6.1 INTRODUCTION

Dredging machines, how they carry out the dredging process, what effects this may have on the environment, and how these effects may be mitigated or eliminated is the focus of this chapter. To evaluate the environmental performances of different dredging equipment, techniques and procedures, a systematic approach using specific criteria is presented.

6.2 CRITERIA TO JUDGE THE ENVIRONMENTAL EFFECTS OF A DREDGING OPERATION

The environmental effects of an industrial activity are a complex amalgam of interacting processes in a wide range of different domains. Comparing different dredging equipment and projects with each other on their specific merits is a challenge. A framework is required to try to identify the most significant environmentally sensitive criteria which may be influenced by the dredging equipment and process. The following criteria have been identified to judge the environmental effects of a dredging operation:
- Safety of the people (on board and external)
- Accuracy of the excavated profile
- Generation of suspended sediment and turbidity generation
- Mixing of different soil layers
- Creation of loose (mobile) spill layers that remains on site after the dredging activity
- Dilution during the excavation and transport process
- Noise (both above and below the waterline)
- Normal output rate of the dredging equipment.

Each of these identified criteria is discussed in more detail below:
- *Safety of people*: Of primary concern during the execution of dredging projects is the safety and health of the crew on board the dredger, especially where contaminated sediments are being handled. Besides the crew, the safety of all other people in the vicinity must also be considered.
- *Accuracy of the excavated profile*: Taking into account the costs of treatment or relocation, the excavation of a pre-set profile as accurately as possible is essential in order to minimise the volume of dredged material for further transport, placement, treatment or storage. (In the case of contaminated sediments this factor can be critical.)

- *Suspended sediment*: The creation of excessive suspended sediments near the dredging or relocation sites may endanger sensitive local fauna and flora; where such sediments are contaminated, this increase can result in the spread of contaminants into the surroundings. Furthermore the suspension will increase the turbidity of the water which again will have its effects on the living conditions for the underwater fauna and flora at the site. As far as turbidity and suspended sediments are concerned, evaluating the effects of a dredging activity compared to the natural background variations should be established.
- *Mixing of different soil layers*: When layers of soil with different geotechnical or chemical characteristics are mixed together, material relocation, treatment or reuse can become complicated.
- *Creation of loose (mobile) spill layers*: Destruction of the cohesiveness of the upper soil layers, without complete removal, facilitates later natural erosion processes; these processes can present risks when close to sensitive areas (e.g. coral reefs, aquaculture ponds) or when contaminants adsorbed on the sediments can be dispersed.
- *Dilution*: Mixing fine-grained material with large quantities of water for hydraulic transport leads to considerable dewatering difficulties at the relocation or placement site. Moreover, it increases the volume of material to be treated and the costs of treatment in the case of contaminated sediments.
- *Noise generation*: Dredging equipment is often quite noisy. Where dredging is taking place in populated areas or near a nature reserve this noise can be disturbing. The effect diminishes considerably as the distance between the equipment and the potentially affected site increases. The effect of underwater noise on the aquatic life forms is another area of concern.
- *Normal output rate*: In general a reduction of the output rate will reduce the immediate effects of a dredging activity. However the output of the dredger has a significant bearing on the operator's ability to complete the work within the pre-set time limits. Note also that, for most types of dredger, operational costs increase if the output rate decreases.

The influence of each of these criteria depends greatly on the nature of each particular project. The task of the project sponsor is to identify the importance of each characteristic. These characteristics or parameters are of particular importance in the context of remedial dredging. Each parameter also helps in one way or another to identify, and eventually reduce, the potential environmental effects of more traditional dredging projects.

In cases where the dredging process is adapted or special equipment is developed in order to reduce the environmental effects of a certain type of dredger, the criteria defined above can also be helpful in evaluating the environmental efficiency.

6.3 PHASES OF A DREDGING PROCESS

The characteristics of the dredging cycle change considerably from one project to another. However, a number of different phases that are common to almost any dredging process regardless of the type of equipment used for its execution can be identified. These phases are:

- dislodging of the *in-situ* material;
- raising of the dredged material to the surface;
- horizontal transport; and
- placement or further treatment.

In the following section these phases are discussed briefly, with emphasis on the environmental effects which can result during each of them.

6.3.1 Dislodging of the *in-situ* material

The first phase of a dredging cycle is dislodging of the *in-situ* material. This dislodging process is essential as removing the whole volume *en masse* is impossible. The excavation process can be relatively easy in the case of soft sediments, but sometimes, where the removal of hard rock is concerned, it can be very difficult.

Dislodging is generally carried out by a cutting device such as cutterhead, draghead or the cutting edge of a bucket. Sometimes water jets are used for this purpose. The most significant environmental effects occurring during the disintegration process are:

- *Increase of suspended sediments*: During the dislodging process the cohesion of the *in-situ* material is broken and part of the material can be brought into suspension by the cutting movement, either rotating or straight. The quantity of material brought into suspension depends on the energy applied to the excavation and the way in which the material is raised to the water surface. When fine-grained materials are brought into suspension the particles will not settle again rapidly. This can increase the turbidity near the dredging site during an extended period.
- *Mixing of soil layers*: When using equipment designed to cut thick layers (combined horizontal and vertical cutting movement), mixing different layers may be unavoidable. When equipment with a horizontal cutting movement (trailing dredger, sweep dredger, environmental auger dredger, disc bottom dredger) is used, layer thickness can be controlled more precisely.
- *Dilution* (in case of hydraulic dredging): To facilitate transport of the dislodged material, water is mixed with the material during the cutting and suction process. The ratio of soil to water varies from one type of dredger to another. This variable is also influenced when the thickness of the layer to be removed is small compared with the minimum cutting layer thickness of the dredger.

The selection of suitable dredging equipment for a project is clearly one of the key factors in reducing the effects on some of those critical environmental criteria listed in Section 6.2. Suitable equipment must combine cost-effectiveness with environmental sensitivity.

6.3.2 Raising the material

During the second phase of a dredging cycle, the dislodged material is raised towards the water surface. This can be done either mechanically or hydraulically.

Using the mechanical alternative, the material is raised in a bucket (backhoe, dipper or bucket line dredgers). In the second case, hydraulic dredgers (cutter suction, trailing suction hopper, disc bottom, auger and sweep dredgers) use a suction pipe. The dislodged material is sucked into the suction mouth by means of a centrifugal pump. The material is then further raised through the suction pipe towards the pump and from there through the discharge line to the deck of the dredger.

The main environmental risks during this raising phase are:
- *Release of suspended sediments*: In the case of mechanically raising in an open bucket, the dredged material is in direct contact with the surrounding water which can result in dilution and an increase in suspended sediment content of the surrounding water layers. In addition, mechanically raising fine-grained material in an open bucket can increase the natural turbidity levels significantly because of a longer settlement period process.
- *Loose and mobile spill layers*: With hydraulic transport, potential problems are limited to the point at which the material enters the suction mouth. If the suction capacity is lower than the cutting capacity of the dredger, a residual spill layer of loose material remains on the seabed. The same effect is observed when dislodged material falls slowly through the water column, arriving at the suction depth after the suction mouth has moved away or when the material moves sidewards beyond the influence of the suction mouth during the dislodging process.
- *Density of material*: Problems can be encountered with hydraulic transport as this method requires the addition of water to create a mixture density suitable for pumping. The co-ordination of cutting capacity and pumping capacity is critical. When the pumping capacity is too high, the density of the mixture in the suction pipe becomes too low. This causes dewatering problems during the disposal or further treatment stage, which is even more of concern in the case of contaminated sediments.
- *Overflow during loading of hopper or barges*: The overflow of excess water inevitably brings sediment into the surrounding waters causing an increase in suspended sediments and turbidity at that location. This effect is greater when hydraulic loading is compared with mechanical dredging. Prohibition of overflow when dredging soft sediments will effectively prevent loss of sediment into the water column. This is especially important when contaminated sediments are being dredged.

Clearly, the proper selection of dredging equipment and the way it is used influences the environmental effects of this phase of a project, and handling contaminated sediments requires extra attention.

6.3.3 Horizontal transport of the material

The third phase of the dredging cycle is the horizontal transport of the excavated and raised material from the dredging area to the site for further treatment or final relocation. This can be achieved mainly by one of three methods:
- hydraulic pipeline transport;
- transport by hopper dredgers; or
- transport by barges.

Each method is linked primarily to a certain type of dredger. Barge transport is generally selected for mechanical excavation, while pipeline transport is used mainly with hydraulic dredgers with exception of the trailing suction hopper dredgers that are designed to load their hoppers hydraulically. Of course, other types of transport, such as truck or conveyor belt transport, exist but to date their application in the dredging industry has been limited. In the future these alternative transport techniques should be considered as possibilities, especially for remedial dredging projects.

The environmental effects of the horizontal transport phase can be summarised as follows:

- *Safety*: Using open barge transport, the crew may well come into direct contact with the dredged material. This is no cause for concern in normal dredging projects, but in the case of contaminated material a health risk can arise.
- *Dilution*: Dilution occurs mainly with the use of hydraulic dredgers where a maximum density is imposed to enable pipeline discharge with a centrifugal pump.
- *Spillage*: The main environmental risk here is the spread of material through leakage at pipeline joints or through spillage from barges during transport, especially in rough weather. Also leakage, mainly of fine material, through damaged or poorly closing bottom doors of hoppers and barges might occur. The impact of the transport process is slight compared with the other phases of the dredging cycle.
- *Noise and air pollution*: The potential effect on this variable is more significant in the case of barge transport compared with pipeline transport. However, should truck transport be used, the effects are greater.

6.3.4 Placement of the material

The final phase of a dredging project is the relocation of the excavated material to its final destination or to an intermediate site for further treatment. There are numerous options at this phase:

- reclamation of a site;
- beach nourishment;
- wetland creation;
- relocation on land;
- relocation in a pit;
- relocation at sea (underwater); and
- relocation to a contained site.

Some preliminary information on the physical effects of the relocation process is mentioned below, and some special low-impact disposal equipment is discussed in Section 6.7.2. For a detailed discussion on this subject, see Chapter 7, which is devoted to this topic.

The following effects of relocation can be noted:

- *Occupation of space and surfaces*: The major effect, especially for relocation on land, is the occupation of the ground and the alteration of the natural habitat at that location. The same occurs with underwater relocation, although the effect is less visible.

- *Dispersion of the deposited material*: At underwater relocation sites, depending on the water depth, the effects of natural wave and current conditions can result in the dispersion of the fine-grained material into the surroundings during or after relocation resulting in an increase of suspended sediment and turbidity at and near the relocation site. At land-based relocation sites, erosion and dispersion by wind or by rain run-off can occur after the relocated materials have dried. Also fine sediment can be re-dispersed by poorly decanted effluent from placement sites or settlement ponds.
- *Noise and air pollution*: The placement action, especially when the relocation site has to be formed with trucks and other earth-moving equipment, can generate noise and air pollution problems.
- *Groundwater quality*: If the selection of the relocation site is incorrect and the design or construction of protective measures (liners and such like) is poor, the groundwater can be affected by leakage.

6.4 DREDGING EQUIPMENT

Different types of dredging equipment are briefly described here, with special emphasis on those characteristics that influence environmental effects as defined by the criteria given at the end of Section 6.2. A more detailed description of the different types of dredging equipment can be found in Bray, Bates and Land (1997) and Bray (2004) or in other textbooks on dredging equipment.

At the end of this section, new technical developments introduced to avoid or mitigate the negative environmental effects of different types of dredging equipment are reviewed. In Section 6.6, new types of dredgers, especially developed for low-impact projects, are discussed. Other types of equipment, which are not included here, are in general not frequently used in the dredging industry.

6.4.1 Hydraulic dredgers

Hydraulic dredgers include all dredging equipment, which makes use of centrifugal pumps for (a part of), the transport process (raising or horizontal transport). Generally speaking, three main groups of hydraulic dredgers can be identified: Stationary Suction Dredgers (SD), Cutter Suction Dredgers (CSD) and Trailing Suction Hopper Dredgers (TSHD).

Suction Dredger
The Suction Dredger (SD) is the simplest form of hydraulic dredger. From the floating pontoon the suction pipe is lowered into the bottom and by the mere suction action of the dredge pump, often mounted on the suction ladder, bottom material is sucked up (Figure 6.1). Only relatively loosely packed, granular material or silt can be dredged with this equipment. The application of water jets to fluidise the soil layers near the suction inlet improves excavation capacities. After raising the material through the suction pipe the material is either hydraulically discharged through a floating pipeline to the shore, but more often loaded into barges.

Figure 6.1 A stationary suction dredger loading barges.

The SD is used mainly for sand winning purposes from very thick sand deposits in the sea, estuaries or enclosed areas. The excavated sand is generally placed into transport barges but sometimes direct pumping to a nearby processing plant is possible. This type of dredger is not used for environmentally sensitive projects.

Regarding the environmental effects of the SD, the following can be mentioned:
- *Safety of the crew*: The transport process occurs within a completely closed circuit. The crew has no direct contact with the material. Consequently, their safety, in relation to the dredging process, is guaranteed to a large extent, except when a blockage in the suction mouth or pump has to be removed.
- *Accuracy of the excavated profile*: Owing to the relatively uncontrolled excavation process by natural collapses, fluidisation and suction, normally an irregular pattern of pits is created. Accurate dredging is not possible. Only when the suction mouth is modified with a so-called "dustpan" suction head (a wide head with a low opening for thin layer dredging) can better precision can be obtained, in the order of 10 to 20 cm.
- *Increase of suspended sediments, turbidity*: Depending on the difference between jet flow and suction flow the SD has in principle a low tendency to resuspend sediments. During vertical and horizontal transport, increases in suspended sediments do not occur because the pipeline is closed. The effects on turbidity are similar.
- *Mixing of soil layers*: As for an effective use of the SD, the feed of material to the suction mouth has to flow or fall from a high bench, the SD is less suitable for selective dredging, unless equipped with a dustpan head.

- *Creation of loose spill layers*: The production process of a SD is based on a free, relatively uncontrolled flow of material to the suction mouth. Consequently considerable spill is to be expected.
- *Dilution*: Owing to the hydraulic character of the transport, water is added to the soil for transportation purposes. Depending upon the soil type and the attainable layer thickness, the amount of added water varies significantly. Taking into account the uncontrolled collapsing process it can be expected that the dilution is even higher compared to other hydraulic dredging methods which have better controlled cutting process.
- *Noise generation*: As only pump(s) and winches are to be powered for a quite constant process, a SD has relatively quiet engines. Noise disturbance is low, when engines are properly maintained. Underwater noise is moderate as there are no moving parts or motors in direct contact with the water layers. Transfer of the motor noise through the hull is the main source for such noise.
- *Output rate*: SD output rates vary widely from 50 to 5000 m³/hr depending upon the size of the SD and the soil characteristics. As transport is done mainly by barges the output of the dredger is also affected by the number and size of the barges, and the efficiency of the barge loading and transportation process.

Cutter Suction Dredger

The Cutter Suction Dredger (CSD) dislodges the material with a rotating cutter equipped with cutting teeth. The loosened material is sucked into the suction mouth located in the cutterhead by means of a centrifugal pump installed on the pontoon or ladder of the dredger. Further transport of the material to the relocation site is achieved by hydraulic transport through a discharge pipeline (partly floating, partly land based). Occasionally the material can be pumped into transport barges for further transport. The basic design of the CSD is shown in Figure 6.2A. A cutterhead being connected to the front end of a ladder is shown in Figure 6.2B.

The CSD is used mainly for capital dredging in harder soils, which have to be removed in thick layers. The transport distance to the relocation or reclamation site should preferably be limited (max. 5 to 10 km) to allow for an economical pipeline transport. In the case of environmentally sensitive projects, the dredging process must be controlled very carefully. The dislodging and hydraulic transport process must be carefully optimized such that dilution and spillage are minimized. To achieve this, the optimum setting should be found by carefully varying cutting face height, step length, cutter rotation speed, swing speed, pump engine power and pipeline resistance.

The CSD is usually rated according to either the diameter of the discharge pipe, which may range from 150 millimetres to 1000 millimetres, or by the power driving the cutterhead, which may range from 15 kilowatts to 6000 kilowatts. The total installed power can be as much as 35,000 kW or more.

Regarding the environmental effects of the CSD, the following can be mentioned:
- *Safety of the crew*: As with the SD, the transport process occurs within a completely closed circuit. The crew has no direct contact with the material.

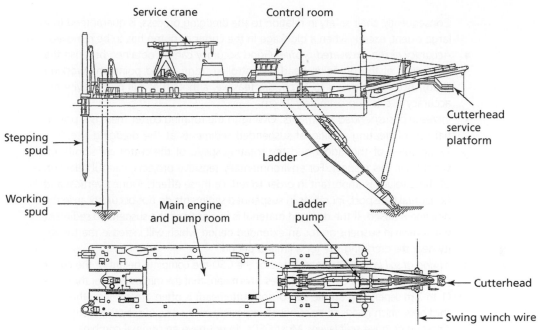

Service crane

Control room

Cutterhead
service
platform

Stepping
spud

Ladder

Working
spud

Main engine
and pump room

Ladder
pump

Cutterhead

Swing winch wire

Figure 6.2A Main features of a cutter suction dredger.

Figure 6.2B A cutterhead
being connected to the
front end of a ladder.

Consequently, their safety in relation to the dredging process, is guaranteed to a large extent, except when a blockage in the cutter or pump has to be removed.

- *Accuracy of the excavated profile*: Good accuracy can be obtained because the movement of the dredging head is controlled from a fixed point (the working spud). Accuracies down to 10 cm are feasible, although at full productivity the accuracy level is approximately 25 cm.
- *Increase of suspended sediments*: Owing to the rotating cutter there is a potential risk of creating additional suspended sediments at the dredging site. The swing speed of the ladder and the rotating speed of the cutter are significant variables in this respect. For environmentally sensitive projects, careful selection of these values is important in order to reduce these effects. During vertical and horizontal transport, increases in suspended sediments do not occur because the pipeline is closed. If the dredged material is fine grained the suspended sediment will remain in suspension for an extended period which will increase the turbidity near the dredging site for a limited period.
- *Mixing of soil layers*: For optimal use of the CSD the complete height of the cutter should be utilised for cutting purposes. This means that the minimum layer thickness (1 to 3 m depending on the size of the cutterhead) is often greater than the layer thickness which needs to be removed, especially in the case of selective dredging.
- *Creation of loose spill layers*: Most CSDs do not have an optimal combination of cutting capacity and suction capacity for all types of soil. In general the cutting capacity is over-dimensioned for softer soils; typically, therefore, a spill layer (25 to 45% of the cutting face, typically ranging between 0.25 m and 1 m) remains on the seabed after dredging if no special precautions are taken. An additional pass at the same dredging depth can remove most of this spill layer. The type of cutterhead may make a significant difference in the amount of spill left.
- *Dilution*: Owing to the hydraulic character of the transport, water is added to the soil for transportation purposes. Depending upon the soil type and the attainable layer thickness, the amount of added water varies significantly. It should be noted that dilution can be reduced by an under-dimensioned cutting power compared to the pumping power but this will increase the spill layer effect. An optimum has to be searched for each project.
- *Noise generation*: Generally the CSD has a powerful engine, which generates a high level of noise. Given that the CSD is a stationary vessel, which often works in populated areas, the dredger can be a continuous source of significant noise levels, reaching 100 to 115 dB in the immediate vicinity of the dredger. This noise level diminishes to acceptable levels (50–70 dB) a few hundred metres from the dredging site. Precautions, such as low-noise engines, noise-tempering covers and procedures to keep the engine room closed under any circumstance, are possible but are not implemented on a routine basis. Underwater noise caused by the cutting action and the presence of underwater engines on many of the larger cutter suction dredgers will be higher compared to the SD.
- *Output rate*: CSD output rates vary widely from 50 to 7000 m³/hr depending upon the size of the CSD and the soil characteristics. The challenge is to select the best size for a particular project. For a given soil type the cost per cubic metre of the dredging operation with a CSD generally decreases with an increase in the size of the dredger.

The most critical issue of the CSD in this respect is the creation of a spill layer. This is because the suction mouth is located inside the cutter approx 0.5 to 1 m above the actual cutting level. As such it is impossible to avoid the creation of a spill layer unless one accepts an important impact on the output of the dredger. This spill layer is easily erodable and will be a long-lasting source for an increased suspended sediment content or turbidity.

Trailing Suction Hopper Dredger
The Trailing Suction Hopper Dredger (TSHD) is a normal sea-going ship equipped with a suction ladder. At the end of the suction ladder is a draghead, which can be lowered onto the seabed while the TSHD navigates at a reduced speed. Most trailing suction dredgers have twin-screw propulsion and powerful bow thrusters, which provide a high degree of manoeuvrability. During the forward movement of the TSHD, the draghead agitates a thin layer of the seabed. The loosened material, together with some transport water, is sucked into the suction pipe by means of a centrifugal pump, which is installed in the vessel's hull or on the suction pipe. The basic design of the TSHD is given in Figure 6.3A. Figure 6.3B shows a modern TSHD at work.

The material is pumped into the ship's hopper until it is completely filled. Then the suction pipe with draghead is retrieved on board the ship. Even though the hopper is filled with a mixture of water and sand, the loading process can still continue. During this loading phase, excess water flows overboard together with some of the finer material, while the coarser (sand) fraction accumulates in the hopper, thereby increasing the quantity of sediment effectively loaded into the hopper during the dredging process. Modern dredgers have a "Light Mixture Over Board" (LMOB) arrangement. This diverts incoming pumped mixtures immediately overboard if the mixture density is less than a preset value. This device is principally used in fine-grained soils to optimise the load in the hopper and the overall productivity level. The overflowing process and the use of the LMOB system are critically important when assessing environmental effects: there has to be a balance between the increase in cost, which usually results if the overflow or LMOB system is restricted, and the increase in environmental effect if it is not.

Degassing equipment may be fitted within the dredge pipework to remove gas from the incoming pumped mixture before it reaches the dredge pump. Without such equipment, when dredging in bioorganic gaseous deposits, the pump performance may be adversely affected.

Horizontal transport is achieved by navigating the ship to the relocation site, which is often an underwater site. At this underwater relocation site, the bottom doors in the TSHD's hopper are opened and the sediment falls to the seabed. As an alternative to such direct disposal, many TSHDs are equipped with a system to use the suction pump to empty the hopper by pumping the material through a pipeline to a relocation site on land.

The TSHD is often used for maintenance dredging projects or for deepening existing channels. During such projects a limited thickness of softer material has to be removed, and relocation sites are at variable distances. This type of dredger is also used for winning good quality sand far out at sea for reclamation projects such as

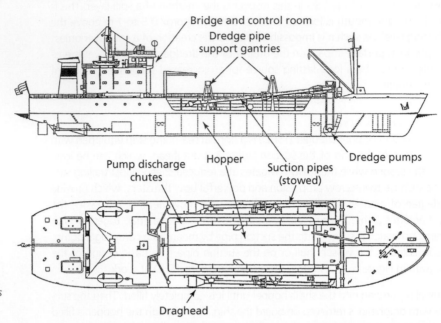

Bridge and control room

Dredge pipe support gantries

Pump discharge chutes

Hopper

Suction pipes (stowed)

Dredge pumps

Draghead

Figure 6.3A Main features of a trailing suction hopper dredger (TSHD).

Figure 6.3B One of the largest TSHDs in the world.

beach nourishment or the creation of artificial islands. Selection of the optimal duration of the suction process and limiting overflow losses during dredging are the major factors when trying to control the environmental effects of this type of equipment. For example, stopping the dredging process at an early stage will reduce the overflow of fine material from the hopper. However, this results in higher dredging costs per cubic metre of dredged material. To find an optimal solution, ecological and economic consequences should be evaluated together.

The TSHD is normally rated according to its maximum hopper capacity, which can nowadays be in the range 750 to 35,000 cubic metres. The size, length and number of suction pipes are other important elements of the TSHD.

Regarding the environmental effects of the TSHD, the following can be mentioned:
- *Safety of the crew*: Once the dredged material is pumped into the hopper, the crew may come into contact with it. In the case of, for instance, high methane or hydrogen sulphide gas content in the dredged material, sometimes arising from port maintenance dredging, the health of the crew can be in increased jeopardy. In some cases special precautions have to be taken. In exceptional cases this may even include the use of gasmasks on deck but normally adequate de-gassing procedures and clear instructions to the crew are sufficient to guarantee the safety of the crew.
- *Accuracy*: The accuracy of the dredging depth is low compared with the CSD, owing to the fact that the position of the suction pipe is flexible and more difficult to control. A vertical accuracy of approximately 15 to 25 cm can be obtained provided sophisticated monitoring and steering equipment is used. Normal accuracy is around 0.5 to 1 m vertically and 3 to 6 m horizontally.
- *Increase of suspended sediments and turbidity*: The actual cutting process creates less suspended sediments compared with a CSD as there is no rotating device in the draghead. However, when the LMOB system is in use or when loading continues with an overflow of the excess water and losses of fines, this causes a significant increase of suspended sediments throughout the water column at the dredging site and an increase in turbidity or reduction of the light penetrating trough the water column. As the sediments and turbidity is created in the upper water layers, the light attenuation effects will have an impact on the whole water column and the effect will last longer as the settling distance is relatively large. In the case of environmentally sensitive projects, such overflow can be limited or even prevented by stopping the dredging process earlier. This results in smaller loads in the hopper and higher dredging costs.
- *Mixing of soil layers*: The cutting process is strictly horizontal. As such, the mixing of soil layers can be controlled accurately. However, taking into account the lower accuracy level compared with the CSD, the TSHD is not specially suited to the removal of thin layers of (contaminated) material.
- *Creation of loose spill layers*: Because the cutting process is basically a scratching action, only limited amounts of soil are loosened. Consequently, most of the material is picked up by the suction process without leaving a residual spill layer. Larger spill layers can be generated by the settlement of large quantities of overflowing fine-grained material that settles in a fluid mud layer on the bottom after a certain time.

- *Dilution*: Significant amounts of water are added during the suction process. With modern monitoring and control equipment, this amount can be limited. Where the pump-ashore facility is used, an additional volume of water has to be mixed with the dredged material to facilitate pipeline transport.
- *Noise generation*: The TSHD is equipped with powerful engines generating significant noise levels. Sound levels of 100–110 dB in the immediate vicinity of the dredger can be expected. The noise level is reduced to acceptable levels (50–70 dB) at a distance of a few hundred metres. The machine room of the TSHD is often further away from the control room and slightly better insulated. Furthermore, the TSHD is generally at work in more distant areas; as such, noise generation is less critical. Underwater noise is generally less than from the suction dredger. The propeller tends to make more noise than the pump, even when mounted in the drag-arm, and the underwater noise is not significantly higher than a commercial vessel of similar size and power.
- *Output rate*: Output rates vary widely, ranging from 200 to 10,000 m³/hr depending on the size of the TSHD, the soil characteristics and the transport distance. For practically any type of project a suitable size of TSHD might be selected, as long as navigation is possible. The cost of a dredging project using a TSHD generally decreases with an increase in the size of the TSHD.

The most critical impact is generally the creation of a plume of fine-grained elements. This plume is generated by the overflow operation because of the extended loading cycle when the dredger needs to be used in the most economical way. This plume will increase the turbidity in the water around the dredger. The suspended sediments will slowly settle down once the dredging works stop, but depending on the soil characteristics, the time necessary for full settlement can take a few hours to one week. In case of long-lasting excessive use of the overflow system, a layer of fluid mud or loosely packed sand may be created on top of the natural sea-or riverbed bottom. This can have a negative impact on the benthic life in that area.

This type of impact can be reduced in different ways: green valves, recycling (part of) overflow water, overflow with bottom exit, reduced overflow or even forbidden overflow. The first three measures require some specific equipment to be installed on board but the impact on the productivity level of a dredger is almost non-existent. The last two measures do not require specific installations on board but do have an impact on the output level.

6.4.2 Mechanical dredgers

The second category of dredger is the mechanical type. This includes all plant which makes use of mechanical excavation equipment for cutting and raising material. Generally speaking, three sub-groups can be identified: Bucket Line Dredgers (BLD), Backhoe Dredgers (BHD) and Grab Dredgers (GD).

Bucket Ladder Dredger
The Bucket Ladder Dredger (BLD) was first employed in Europe and is the most traditional type of dredger. It consists of a large pontoon with a central well in which a ladder, equipped with an endless chain of buckets, is mounted. During dredging, the

endless chain rotates along the ladder. The lowest bucket digs into the bed material and the cut material falls into the bucket. It is then carried upwards as the bucket chain rotates. At the upper end of the ladder, the bucket turns upside-down and the soil falls into a chute which guides the material into a barge for further transport. The basic design and principal components of the BLD are given in Figure 6.4A and 6.4B.

The BLD is used mainly for accurate dredging such as for tunnel or pipeline trenches. However, taking into account the high density of the excavated material, the BLD is well suited to the excavation of fine-grained material when the addition of transport water can cause problems and if good geotechnical characteristics are required at the relocation site. The raising of the material in open buckets and the contact with the water column during this phase of dredging are drawbacks to using the BLD for remedial dredging projects.

The Bucket Ladder Dredger is normally rated according to the volume of one bucket, which can be in the range 100 to almost 1000 litres. The dredging depth is another important element of the BLD; it might range from 5 to 30 metres. As far as the environmental effect of the BLD is concerned, the following can be mentioned:

- *Safety of the crew*: The possibility of crew members coming into contact with excavated material is high compared with both the CSD and the TSHD. This danger exists throughout the whole process: when the material is being raised in open buckets, loaded into barges, and transported in open barges.
- *Accuracy*: The precision of the BLD is good because the cutting edge of the successive buckets passes at the same depth as long as the ladder remains in the same position with reference to the main pontoon. Accuracy to within 10 cm can be obtained. The BLD is, therefore, often used for dredging projects where precision is vital, such as the final cutting or cleaning of trenches.
- *Increase of suspended sediments, turbidity*: Some additional suspended sediments are released during the raising of the material in open buckets as they move at a relatively high speed through the water. During this raising movement some spillage can occur throughout the complete height of the water column. This effect can be limited by reducing the speed of the bucket line. However, this adversely affects output. In the case of fine-grained materials these sediments remain in suspension for a long period and the accumulation can increase the turbidity at the dredging site above the natural background levels.
- *Mixing of soil layers*: The BLD can easily cut relatively thin layers; consequently the mixing of different soil layers can be avoided. However, a minimal thickness is necessary for full productivity, otherwise the buckets become partially filled with water.
- *Creation of loose spill layers*: The cutting face of the BLD is the edge of each bucket. Almost all the soil loosened by the bucket is carried away by the rotating bucket chain leaving a clean surface. A minor risk of a spill layer remains if there is excessive spillage while the material is being raised.
- *Dilution*: As the material is raised mechanically, there is no need for transport water. Only when the buckets are not completely filled with soil, does the remaining space fill with water during the vertical movement of the bucket chain. These small quantities of water fall with the soil into the transport barges. The quantity of added water is nevertheless much less in comparison with hydraulic dredgers. However, when the relocation site is on land, the transport

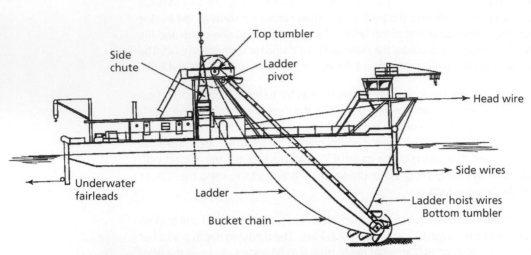

Figure 6.4A Main features of a bucket ladder dredger, the most traditional type of dredger.

*Figure 6.4B A bucket
ladder dredger at work.*

barges will be emptied using a barge-unloading dredger. In this case, water is added to the barge to enable suction and hydraulic transport to take place.

- *Noise generation*: Owing to the mechanical movement of large steel buckets over a steel framework, the BLD is the worst type of equipment with respect to noise. Taking into account the open character of the bucket line it is very difficult to implement mitigative measures. However, for smaller BLDs some trials have been carried out in which new types of material (instead of steel) have been used for the most critical parts. Noise levels of up to 115 dB can be expected in the immediate vicinity of the dredger. This decreases to acceptable levels (50–70 dB) at a distance of a few hundred metres. Also the underwater noise will be considerable as the mechanical movement of steel on steel occurs over the complete length of the ladder.
- *Output rate*: The output rate of the BLD is considerably lower than the CSD and the TSHD. Output ranges of 50 to 1100 m³/hr can be achieved.

Backhoe Dredger

The Backhoe Dredger (BHD) is basically a hydraulic excavator mounted on a pontoon equipped with a spud carriage system. Whereas the conventional land-based hydraulic excavator is normally mounted on a tracked or wheeled undercarriage, the dedicated dredging machine usually is mounted on a fabricated pedestal at one extremity of a spud-rigged pontoon. The basic design and principal components of a BHD are given in Figure 6.5A and 6.5B.

The backhoe dredger evolved with the introduction of larger backhoe excavators, which provided the greater digging depth and power required for many dredging applications. A secure, shock-absorbing pontoon mounting is important if the full digging potential of the machine is to be utilised. A rigid spud pontoon is usually necessary to provide a positive reaction to the hydraulic digging action, particularly when dredging in difficult ground (strong soils).

The backhoe is most efficient when working from behind the face, which means that the pontoon is located over the area to be dredged. If water depth is at any time less than the maximum draught of the pontoon, this method may not be practical.

The bed material is excavated by the crane's bucket, which is then raised above water by the movement of the crane arm. The backhoe is most commonly used to load a barge moored alongside the pontoon. Occasionally, such as in trench excavation, the dredged material may be side-casted alongside the trench, but outreach is limited and hence there is a strong possibility of the material re-entering the excavation.

Horizontal transport to the relocation site is generally carried out by transport barges. The material is either deposited through the bottom doors of the transport barge, pumped ashore using a barge-unloading dredger, or mechanically unloaded by a grab or hydraulic excavator.

The BHD is mainly used for the execution of relatively smaller dredging projects in the stronger soils as the mechanical cutting forces, which can be applied, are considerable. Until the 1990s this type of dredger was seldom used for environmentally sensitive projects because it lacked precision and dredged materials are raised in an open

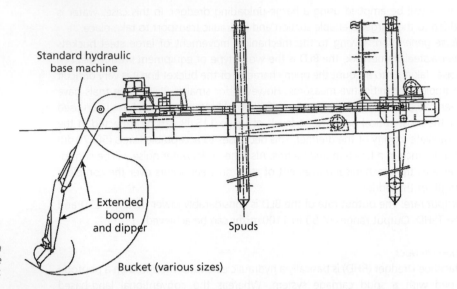

Figure 6.5A Main features of a backhoe dredger.

Standard hydraulic base machine

Extended boom and dipper

Spuds

Bucket (various sizes)

Figure 6.5B A backhoe at work.

bucket. More recent developments in sophisticated monitoring and control equipment have improved accuracy considerably, making this type of dredger attractive for more precise dredging projects in areas where debris is expected or where physical constraints prevent the use of more traditional equipment.

The backhoe dredger is normally rated according to the maximum size of digging bucket that the machine can handle. This may range from 1 to 40 cubic meters. The size of bucket employed will depend upon the nature of the material to be dredged and the maximum dredging depth. Bucket tear-out force decreases with increasing dredging depth, as does the total weight that can be handled. Thus, for dredging strong materials at maximum depth, a bucket smaller than the maximum should be used. The maximum dredging depth is another important element of the BHD; this dredging depth may vary from 4 to 32 metres.

Regarding the environmental effect of the backhoe dredger, the following can be mentioned:

- *Safety of the crew*: The risks encountered with a BHD are similar to those of a BLD in that the material is raised mechanically through the water column and the onward transport by barge is also identical.
- *Accuracy*: The accuracy of a traditional BHD is much lower than that of the BLD because the excavating bucket has to be repositioned at every cycle. Without sophisticated monitoring equipment, an accurate dredging depth is impossible. However, such monitoring systems exist and are now implemented on a routine basis for larger BHDs. Accuracy down to 10 cm is attainable, albeit with reduced productivity.
- *Increase of suspended sediments and turbidity*: The problem here is similar to that of the BLD. An additional problem with the BHD is the position of its bucket while being raised; the operator must give full attention to keeping the bucket in an optimal horizontal position in order to prevent excessive spillage. Closed buckets that limit spill are sometimes available.
- *Mixing of soil layers*: Thin layers can be excavated selectively provided a good monitoring and control system is available.
- *Creation of loose spill layers*: For the same reasons cited for the BLD, the BHD leaves a very thin spill layer.
- *Dilution*: Once again, the BHD is similar to the BLD. Where there is hydraulic unloading of the transport barges, there is a need for transport water.
- *Output rate*: The output rates of the BHD are limited; up to 1000 m^3/hr is achievable with the largest BHD.

Grab Dredgers

Grab Dredgers (GD), sometimes called clamshells, can exist in pontoon or self-propelled forms, the latter usually including a hopper within the vessel and therefore called grab hopper dredgers. The basic design and principal components of the GD are given in Figure 6.6A and 6.6B.

The GD is basically a conventional cable crane mounted on a pontoon. The bed material is excavated by the bucket of the crane and raised by the hoisting movement of the cable. Once above water the crane arm swings and the material is dumped into its own hopper for the grab hopper dredger or in a transport barge for the grab pontoon dredger. Horizontal transport is undertaken by navigating the grab hopper dredger or the transport barges from the dredging site to the relocation site.

At the relocation site the material is either discharged through the bottom doors or split hull of the dredger or transport barge, or pumped ashore by means of a barge-unloading dredger.

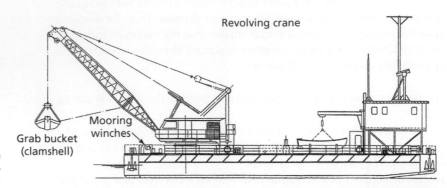

Revolving crane

Mooring
winches

Grab bucket
(clamshell)

*Figure 6.6A Main
features of a pontoon-
mounted grab dredger.*

*Figure 6.6B A clamshell
discharging into a barge.*

The GD is mainly used for the execution of relatively small dredging projects. Recent developments in sophisticated monitoring and control equipment and new types of buckets have improved the accuracy of this dredger considerably. This has made it also attractive for more precise dredging projects in areas where debris is expected or which are inaccessible to more traditional equipment. Furthermore, closed grabs, which prevent direct contact between the excavated material and the water during the raising movement, are available. Finally some grabs are equipped with a Remote Operating Vehicle (ROV) steering system to actively steer the grab to a pre-defined position. This is especially useful for very large water depths (above 100 m) where the grab dredger is often the only option for excavation activities.

The grab hopper dredger is normally rated according to its hopper volume, which may range from 50 to 2500 cubic metres. The grab pontoon dredger is normally rated by its grab bucket capacity. The capacity of grab buckets may range from 0.75 to 200 cubic metres, although buckets over 20 cubic metres are rare.

Regarding the environmental effects of the grab dredger, the following can be mentioned:

- *Safety of the crew*: The risks with a GD are similar to those with a BLD since the material is raised mechanically through the water column and onward transport (by barge) is also identical.
- *Accuracy*: The accuracy of the GD is limited because the excavating bucket has to be repositioned at every cycle. Without sophisticated monitoring equipment, precise positioning is impossible. Monitoring systems and special-purpose grabs allow vertical accuracies of around 0.35 to 0.50 m. Horizontal accuracy is poor, especially in deep waters and in water currents where a pendulum effect occurs except for the grabs equipped with an active ROV steering system.
- *Increase of suspended sediments, turbidity*: The problem here is similar to that of the BHD. Closed grabs, although not often used, can effectively reduce generation of suspended sediment during raising of the grab.
- *Mixing of soil layers*: With a traditional GD it is very difficult to achieve a horizontal cut as the excavation depth of each cycle cannot be kept under control. Therefore, mixing different layers cannot be avoided. Recently, new bucket types and monitoring and control systems have been developed with improved characteristics in this respect.
- *Creation of loose spill layers*: Because the GD uses a mechanical cutting process to scrape the material from the bed, only a very thin spill layer is left.
- *Dilution*: The situation here is again similar to that with the BHD. Where hydraulic unloading of the transport barges takes place, there is a need for transport water.
- *Output rate*: The output rates of the normal GD are limited to a few hundred cubic metres, heavily depending on the water depth at the dredging location. However, there are a few huge grab dredgers with considerably higher output rates (1000–2000 m^3/hr).

6.4.3 Hydrodynamic dredging

Hydrodynamic dredging is a form of dredging where material is lifted into and remains in the water and is transported away by the currents (Agitation Dredging) or under the influence of the natural gravity forces (Water Injection Dredging) or a mechanical push of the equipment (Underwater Plough or Sweepbar) as shown in Figure 6.7 (Van Raalte and Bray, 1999; Sullivan and Murray, 1999).

Hydrodynamic dredging may only be used with great care in environmentally sensitive projects as, in general, it is only possible to predict the movements of the suspended or fluid mud by means of a detailed analysis of the environmental conditions or by sophisticated mathematical modelling efforts.

The environmental desirability or acceptance of hydrodynamic dredging techniques should always be considered during the planning stage of a dredging project, so that

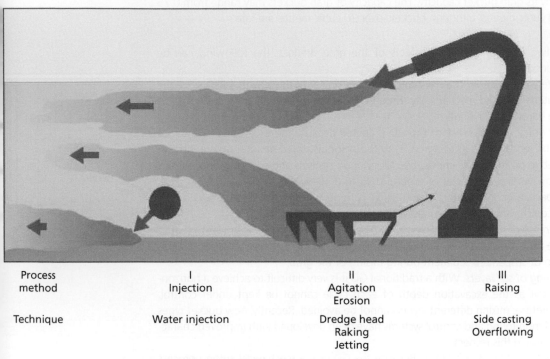

Process method	I Injection	II Agitation Erosion	III Raising
Technique	Water injection	Dredge head Raking Jetting	Side casting Overflowing

Figure 6.7 Hydrodynamic dredging processes.

the most appropriate technique can be selected and the necessary contractual conditions included to control the dredging activity in an appropriate way.

Hydrodynamic dredging is often cost effective and a major environmental benefit of hydrodynamic dredging is the fact that energy required to transport the materials is provided by the dynamics of the hydraulic environment and gravity. This makes hydrodynamic dredging the method with the lowest possible emission of CO_2 and other exhaust gasses.

Water Injection Dredger

The Water Injection Dredger (WID) consists of a fixed array of water jet nozzles that are lowered to penetrate the seabed and injects large quantities of water into the near surface seabed deposits. As a consequence, the density of these deposits decreases to a point were it becomes a liquid and the top level rises slightly. Because of the decreased density and the higher top level, the material starts flowing naturally until a new equilibrium is reached. In the case of (even small) gradients at the dredging site the transport distance can be significant. The material settles again at an adjacent site of lower elevation. The basic design and principal components of the WID are given in Figure 6.8A and 6.8B.

Unlike agitation dredging, the object is not to lift the individual sediment grains into the water column, although this can be achieved, if desirable, in strong current regimes using the same equipment.

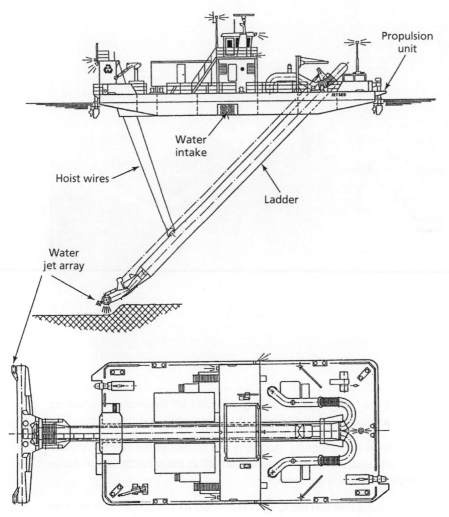

*Figure 6.8A The water
injection system.*

Although originally developed for soft sediments, purposely-built vessels and jet bars can also be equipped to cut and scrape harder materials like an underwater plough. Experiments are ongoing to equip WID equipment with oxygen-injecting systems to avoid oxygen depletion in the water caused by the fast removal of oxygen by, amongst others, hydrogen sulphide.

The WID is mainly used for the execution of dredging in tidal basins where significant quantities of natural sedimentation accumulate. The material flows either back into the main stream (e.g. the main channel alongside a basin) to reintroduce the material into the channel's natural transport process, or it is shifted from zones, where other dredging equipment has no access, to areas where it can be removed further by regular dredging. WID is also used to assist a traditional trailer dredger to remove local shallow spots. With WID localised shoals can be removed or irregular bottoms can be levelled to delay the need for a major campaign with a more traditional type

Figure 6.8B WID with artist's exploded underwater view.

of dredger. The jetting system allows for a small stand off distance to the bottom thus WID is also a suitable, safe dredging method near cables and pipelines.

Recent developments in monitoring, modelling and control of the injection process have improved the equipment's performance as well as the predictability of movements of the generated fluid mud layer.

As stated above, water injection dredging may be used for environmentally sensitive projects if used with care, even though the material is not physically removed from the local environment, but shifted to a nearby location in a predictable manner without significantly increasing the turbidity of the supernatant water. Although the WID is less suitable for remedial dredging projects, for other projects the technique can be an attractive alternative because of the low cost and the fact that the dredged material remains in the natural cycle of the river or estuary. (See also Netzband *et al.*, 1999; Verweij and Winterwerp, 1999.)

The Water Injection Dredger is normally rated according to its dredging depth or its pumping capacity. The dredging depth may range from 5 to 25 metres. The pumping capacity may range from 3,000 to 12,000 m^3/hr.

The following environmental effects can be mentioned:
- *Safety of the crew*: The transport process occurs almost exclusively at the riverbed. Therefore, contact between the crew and the material to be dredged is almost non-existent. Only when the water injection device is raised to the water surface is contact possible.

- *Accuracy*: The accuracy of a WID is reasonable during maintenance projects as the penetration of the water jets can be restricted to the layer of soft, recently deposited material. The original bed material will remain untouched. Sophisticated measurement and control systems make it possible to deliver a flat bed surface with an accuracy of 0.10 metres. However, during capital dredging works the accuracy is more difficult to control.
- *Increase of suspended sediments and turbidity*: During the injection process a layer of fluid mud is created at the dredging site which is more sensitive to the natural erosion forces by the water currents above this layer. Most of the material, however, remains close to the riverbed and is therefore less subject to spreading over the full water column. The effect on turbidity of the upper water layers is limited.
- *Mixing of soil layers*: Soft recent deposits will not mix with underlying sand layers or firm materials. Only during removal of a soft layer overlying another soft layer some mixing of soils will occur.
- *Creation of loose spill layers*: The actual dredging process involves the creation of a fluid mud layer that moves under natural hydrodynamic forces. The remaining seabed contains a thin layer of weakened soil.
- *Dilution*: Dilution is inherent to the WID process.
- *Output rate*: The output rates of the WID vary according to location and soil type. In favourable conditions the output rate can be higher than that produced by a TSHD. In other circumstances the output rate can be far less. Output rates of up to 4000 m³/hr have been recorded.

Agitation Dredger

The Agitation Dredger (AD) will include any method of dredging that aims to lift the excavated material towards the overlying water layers. This can be done either by a side-casting TSHD or by a stationary hydraulic dredger that discharges its water-sediment mixture in near surroundings with a spreader pontoon or an open pressure pipe end. In fact, any method that raises the individual particles of sediment into the water column, such that they behave as individual particles rather than as a mass, could be termed agitation dredging. The main aim is to bring the material into suspension for further transport by the natural currents at the site.

Agitation dredging is mainly used for maintenance dredging projects in areas with a strong (tidal) current and high background turbidities. When currents are variable, agitation dredging can be restricted to periods when currents are high.

Because control of the transport and relocation process is impossible, and because the material is not removed from the natural system, this dredging method is less suitable for environmentally sensitive projects. For other projects, however, the technique can be an attractive alternative because of the low cost and the fact that the dredged material remains in the natural sedimentary system.

The following environmental effects can be mentioned:
- *Safety of the crew*: The excavated material is lifted and spread into the water column, including occasionally the water surface. Therefore contact between the crew and the material to be dredged may occur.

- *Accuracy*: The dredging accuracy of AD is similar to the accuracy of the type of dredger (TSHD, CSD or other). However, owing to the local discharge and the fact that it is very difficult to predict the settlement zones, which can be partly inside the dredging area, this accuracy may become lower.
- *Increase of suspended sediments*: This is the main aim of AD. Therefore the increase will be considerable.
- *Mixing of soil layers*: The mixing characteristics depend on the base equipment used for AD. However, AD is certainly not suitable for an accurate control of the mixing of soil layers. Furthermore, selective dredging is not an option as differentiation between the transport and relocation site is almost impossible.
- *Creation of loose spill layers*: Especially when using a CSD the spill layer will be considerable. Furthermore it can be expected that the settlement of the suspended material will increase the spill effect, especially in the relocation area where the material may settle as a fluid mud layer.
- *Dilution*: Significant amounts of water are added during the suction process. With modern monitoring and control equipment, this amount can be limited. It should be recognised that a higher dilution will result in a better suspension and a further natural transport of the dredged material.
- *Output rate*: The output rates of the AD is determined by the base dredger used for the agitation action. Very high productivities can be reached in case of large TSHD (above 10,000 m³/hour).

Underwater Plough or Sweepbeam

The Underwater Plough (UWP) or Sweepbeam (SB) can be described as a frame that is pulled over the riverbed by a tugboat. The frame is equipped with a cutting blade that scrapes over the riverbed cutting the bottom layers. The cut material remains in front of the cutting blade and is pushed forward until the volume in front of the blade is full. At that moment the transport process is either stopped or the material falls off the blade and goes partially into suspension. The basic design and principal components of an UWP are given in Figure 6.9.

The UWP is mainly used for the execution of relatively small dredging projects, often to assist a traditional trailer dredger to remove local shallow spots or ridges. In addition, this type of dredger is now also used as a stand-alone piece of equipment to remove mud from restricted locations such as harbour basins or mooring sites or to level a soft irregular bottom and to delay the need for a major campaign with a more traditional type of dredger.

Experiments are ongoing to equip this type of dredger with a water jet system to extend the applicability of the equipment towards the more compacted soil types such as hard clay and compacted sand. Recent developments in sophisticated monitoring and control of the equipment and natural transport process have improved the performance of such equipment.

The UWP is mainly used for maintenance dredging in tidal basins where significant quantities of natural sedimentation accumulate. The material is either pushed back into the main stream (e.g. the main channel alongside a basin) to reintroduce the

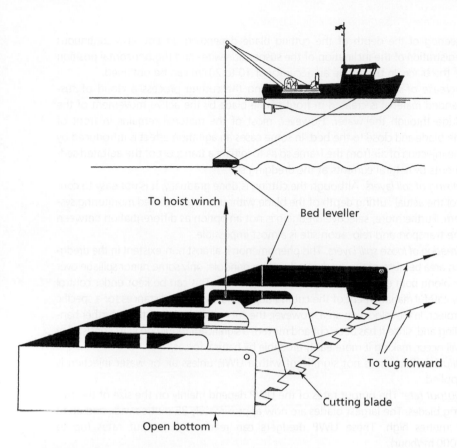

To hoist winch

Bed leveller

To tug forward

Cutting blade

Open bottom

Figure 6.9 Main features of the underwater plough.

material into the channel's natural transport process, or it is shifted from zones, where other dredging equipment has no access, to areas where it can be removed further by regular dredging. This dredging method is less suitable for environmentally sensitive projects because perfect control of the transport and relocation process is impossible. Because the material is not removed from the natural system, the method is also not suitable if the material is contaminated. For other projects, however, the technique can be an attractive alternative because of the low cost and the fact that the dredged material remains in the natural cycle of the estuary.

The UWP is normally rated according to its width or plough surface. The width can vary from 3 to 35 m and the surface from 1.5 to 50 m².

The following environmental effects can be mentioned:
- *Safety of the crew*: The transport process takes place almost exclusively at the riverbed. Therefore contact between the crew and the material to be dredged is almost non-existent. Only when the cutting blade is raised to the water surface is contact possible.
- *Accuracy*: The accuracy of UWP is low, as it is difficult to control effectively the penetration depth of the cutting blade in the bed. This depends largely on the characteristics of the material to be dredged. Recent developments include an active

steering of the depth of the cutting blade depending on tide and continuous registration of the inclination of the suspension wires and the horizontal position of the blade. In this mode an accuracy of 10 to 20 cm can be obtained.

- *Increase of suspended sediments*: During the cutting process a cloud of suspended material is created in front of the blade by the active movement of the blade through the water. However, most of the material remains in front of the blade and close to the bed. In some cases an agitation effect is introduced by the injection of air from the frame so as to initiate a transport of the agitated sediments by natural currents at the dredging site.
- *Mixing of soil layers*: Although the cutting is done gradually, it is not easy to control the actual cutting depth of the blade without a sophisticated monitoring system. Furthermore, selective dredging is not an option as differentiation between the transport and relocation site is almost impossible.
- *Creation of loose spill layers*: This phenomenon is almost non-existent in the dredging area because of the mechanical cutting principle; only some minor spillage over or along both edges of the blade can be possible. This can be kept under control by careful management of the cutting depth and sweeping distances for a specific project. In the relocation area, however, the cut material is left without further handling and, since it has been cut and moved, a significant reduction in its consistency will occur, making it more easily erodable by natural forces.
- *Dilution*: Dilution is not significant with a UWP, unless air or water injection is applied.
- *Output rate*: The output rates of the UWP depend mainly on the size of the cutting blades. The largest blades are now approximately 35 metres wide and up to 2 metres high. These UWP dredgers can reach high output rates (up to 2000 m³/hour).

6.4.4 Evaluation of environmental performances

Table 6.1 gives a comparative overview of the environmental performance of the dredgers described above. A plus indicates a better than average environmental acceptability; a minus denotes a less than average environmental acceptability; and a square indicates an about average performance. Note this table is very general and indicative only, and should not be used for the direct selection of equipment. Case-specific assessments regarding equipment should always be made.

	Safety	Accuracy	Turbidity	Mixing	Spill	Dilution	Noise
SD	+	−	+	−	−	□	+
CSD	+	+	□/+	□/+	□	□	+
TSHD	+/□	−	−/□	−	□	−	+
BLD	−	+	−/□	□/+	+	+	−
BHD	−	+	−/□	+	+	+	+
GD	−	−	−/□	□	+	+	+
HyD	+	−	−	−	−	−	+

Table 6.1 Comparison of dredgers' environmental performance

6.5 INNOVATIONS IN EXISTING DREDGING EQUIPMENT

Besides the development of almost completely new equipment, especially designed for remedial dredging (described further in Section 6.6), a significant effort has been put into the further development and optimisation of existing dredging equipment.

During the last decades, considerable change has taken place with respect to the constraints imposed on dredging activities, particularly in response to environmental considerations and limitations. This change has led to a technical and conceptual rethinking about dredging as a whole. It has also led to continuous adaptation of traditional dredging equipment to improve the plant's ability to cope with new constraints.

The main developments that have guided this adaptation of existing dredging equipment are:
- increase in dredging accuracy in order to decrease dredging tolerances and, as a consequence, decrease dredged volumes;
- decrease in the generation of suspended sediments at the dredging and relocation sites, in order to avoid problems arising from dredging slightly or heavily contaminated material as well as for projects in or near environmentally sensitive areas;
- improvement of monitoring and control on board dredgers in order to enable the crew to recognise and correct negative tendencies and effects at an early stage;
- development of layer-by-layer excavation methods to permit selective dredging of layers with different contaminant contents and removal of surface layers;
- reduction of the spill layer which is easily eroded; in the case of contaminated material this can prevent unacceptable dispersion problems;
- increase in the density of the dredged material during transport (in the pipeline for hydraulic dredgers and in the barges for ship transport) to decrease the volume for further treatment and/or to reduce the time needed for full consolidation or ripening in onshore placement areas, and
- increase in automation over manual control, to guarantee a continuous process and to allow the crew to concentrate on the main controls, leaving the repetitive tasks to the automatic control system.

These developments have led to a large number of new, innovative systems and equipment being installed on board different types of dredgers. Several of these improvements are described and discussed below, especially those which are designed to reduce the environmental impact of dredging equipment.

6.5.1 Developments for all types of dredgers

Dredging Information Systems (DIS)
During recent years there has been a spectacular improvement in the monitoring of dredging activities, both on board vessels using a continuous data logging system as well as in the office using "DIS" (see Figure 6.10). These systems continuously record a large number of variables such as dredging depth, position, volumes, concentration, vacuum, pressure, swing speed, velocity, power consumption, and so on. The data are interpreted automatically and supplied to either a control and steering

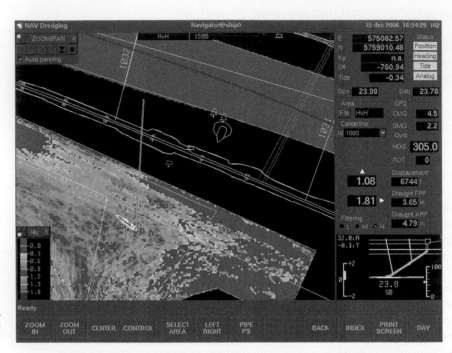

Figure 6.10 Computer screen showing the DIS system.

system that is virtually automatic, or, at least, to a real time feedback system for use by the work supervisor and dredge master. The development of silent inspectors, profile monitors, depth indicators, dynamic tracking systems and multi-page monitors are a few visible results of this trend.

Automatic control systems

The further development of information and monitoring systems has led to an increase in the use of automatic control systems on board dredgers. Clearly a computer is best for repetitive tasks within the dredging process. This allows the dredge master to concentrate on the creative side of the job, namely the further improvement and optimisation of the dredging cycle. An additional consequence of this development is the permanent need for accurate and precise measuring systems for all process variables. This has resulted in better quality of information being made available to the dredge master.

Improved feedback of the operational dredging parameters to the dredge master, or even the use of fully automatic control systems, results in a smoother dredging process in which higher densities can be achieved, less mixing with water for transport purposes is necessary and a higher cutting accuracy is obtained.

6.5.2 Developments for hydraulic dredgers

Degassing system

One factor, which limits the density of the transport mixture pumped by centrifugal pumps, is the gas content of the bed material. Cavitation (imploding of the gas bubbles inside the dredge pump) can occur as a result of the rapid pressure increase inside the pump in the sediment/water mixture with an excessive gas content (e.g. when silty beds with a high organic content have to be dredged). In order to overcome this limitation the gas content of the material before the sediment/water mixture enters the suction pump must be reduced.

In recent years significant progress has been made in this respect by various dredging contractors and specialised shipyards. The different systems developed aim to extract (a part of) the gas content just in front of the suction pump. This results in significant increases in density in the transport pipeline or in the TSHD's hopper. This in turn results in reduced relocation volumes and lower dilution factors.

Environmentally improved ("Green") TSHD
As far as the TSHD is concerned, one of its major environmental constraints is the suspended sediments generated by the overflow of excessive transport water with a high content of fines. One obvious way to overcome this problem is to stop the dredging action when the hopper is full. This unfortunately results in a rather uneconomical dredging cycle. In response to this problem new technologies have been developed, such as:
- low-density trailers that have a relatively large hopper well, with better features to accelerate and improve the settlement process of dredged material in the hopper;
- controlled overflow to improve the retention capacity of the dredger;
- trailers with an overflow with an outflow to the open water close to the keel of the ship. This results in a lower exit point of the water-sediment mixture that will favour the rapid settlement and reduce the turbidity of the upper water layers;
- the "green valve system" aims to reduce the amount of turbidity from the dredging process. The valve is located in the overflow system of the dredge. To reduce turbidity from the overflow, the green valve is adjusted to stop air being entrained in the overflow mixture. Entrapped air carries the sediment particles towards the surface. But, when the green valve is used, the suspended sediment starts to settle towards the seabed immediately, staying closer to sea-or riverbed with a settlement close to the dredging area and a reduced amount of turbidity (as shown in Figure 6.11);

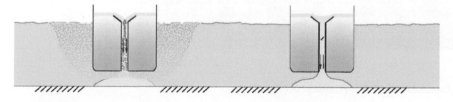

Figure 6.11 Effects of using the "green valve" system; left, without, right, with the green valve.

- controlled overflow, using a special guidance system along the suction pipe, of the excess water to the lower water layers. This system directs the overflow water to the riverbed or seabed and reduces the addition of suspended sediment to the upper water layers. As such, the spread of suspended sediments into the surroundings can be reduced considerably and the turbidity effect in the upper water layers is reduced drastically;

- the reuse or recirculation of overflow water in the jets that are installed in the draghead of the dredger. This most recent development drastically reduces the excess water that is discharged freely overboard during an overflow stage of the dredging cycle. Its major mitigating effect is a reduction of the dispersion of fines at the dredging site, and
- the use of submerged pumps on the suction pipes of trailer suction dredgers, which permits a higher density of material to be dredged and which therefore reduces the need to overflow by achieving a higher hopper load.

The turtle deflecting device, developed by the US Army Corps of Engineers, is another specific development to reduce the environmental effect of dredging activities (see Figure 6.12). Mounted on a TSHD draghead, it is used to avoid the entrapment of sea turtles (an endangered species) through the suction pipe of a TSHD during maintenance dredging in Florida, USA. A detailed description can be found in Nelson and Shafer (1996). This is a typical example of an engineering solution for a project-specific environmental problem arising during normal maintenance dredging activities.

Figure 6.12 Turtle deflecting device.

6.5.3 Developments for mechanical dredgers

Monitoring and control systems
The most significant development to improve the accuracy of mechanical dredging equipment, especially the BHD and the GD, is the continuous computer-based monitoring of the position of the bucket underwater.

Over the past 15 years or so, the traditional mechanical monitors have gradually been replaced by electronic monitors which give the dredge master accurate information (up to 5 cm) on the actual position of the cutting edge of the bucket.

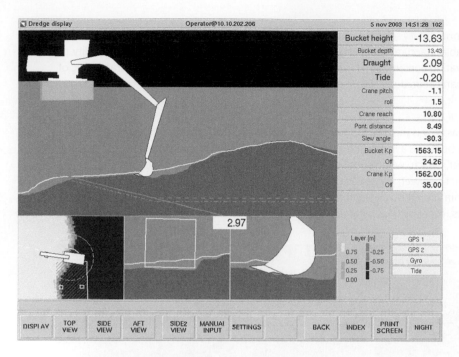

Figure 6.13 Computer screen of backhoe monitoring system.

The more sophisticated systems memorise the dredging history, thus helping the dredge master avoid dredging the same place unnecessarily (Figure 6.13) and producing half-filled buckets that will reduce the density in the transport barges.

For the GD additional control systems exist to pre-define automatically the excavation depth (cable length), in order to avoid differential depths from one cycle to another. A new development is the use of an ROV to guide accurately the positioning of the bucket at great water depths.

For the BLD, systems have been developed to control automatically the movement of the dredging pontoon along a pre-defined curve. These devices allow the dredge master to improve the accuracy of this type of equipment.

Positioning systems
A major aid to dredging accuracy is the development of significantly more accurate positioning systems and the application of some of them to measure continuously, within centimetres, the actual cutting depth and position. This is achieved by using laser equipment and real time kinematic DGPS systems which are, for the most demanding cases, installed on the ladder or excavating arm of the dredger. As such, the instantaneous measurement of the movements of the cutting edge and the correction of the position either automatically or manually becomes possible.

Regarding environmental effects, these systems allow for much reduced tolerances. As a consequence, mechanical dredgers can now be used for a new range of environmentally sensitive projects, which was previously not the case owing to unacceptable inaccuracies of the dredged profile. This has introduced the possibility of excavating contaminated material at almost *in-situ* densities, a basic characteristic of mechanical dredgers, without unacceptable over-dredging.

Special buckets

Noteworthy is the development of the closed bucket. Raising dredged material in an open bucket is a major source of suspended sediments. This is mitigated by use of a bucket, which, when shut, completely encloses the material picked up. Thus direct contact between the excavated sediments and the surrounding water is reduced to a minimum. This development opens up new possibilities for the removal of sediments, especially contaminated sediments (Figure 6.14).

Figure 6.14 Visor dredging grab.

Encapsulated bucket line

The major environmental constraint of the BLD is the use of open buckets while excavating and raising the material. The surrounding water comes into free contact with the buckets and their contents, which results in turbulence during the raising movement and, consequently, the generation of suspended sediments around the dredger which are then spread by the natural movement of the water. Of course, this is unacceptable for environmentally sensitive projects. In response, a completely encapsulated bucket line ladder has been developed (Figure 6.15). With this new BLD the bucket line is raised in still water (no currents around the buckets) and, should some spillage occur, this spill remains in the encapsulated area and is guided in a natural way down to the river- or seabed close to the excavating buckets.

Figure 6.15 Encapsulated bucket line.

This development reduces the turbidity created by a BLD during the raising of the excavated material. As a result, it allows the use of the BLD for environmentally sensitive projects while maintaining the BLD's traditional advantages of high accuracy and high density of the excavated material.

6.6 MODIFICATIONS AND INNOVATIONS IN LOW-IMPACT DREDGING EQUIPMENT

The worldwide increase in environmental consciousness in recent decades is also evident in the dredging industry. Awareness has grown that in some locations sediment to be removed was polluted to a greater or lesser degree with contaminants, mainly originating from the pollution of the surrounding waters. Concerns were also raised about potential risks (e.g. the spreading of adsorbed contaminants, leakage from the disposal areas and direct health risks facing people working with dredged material). Because of the contaminant content, finding suitable reuse or disposal options for the material became more and more difficult.

Along with growing awareness and concern, the dredging world (port authorities, dredging companies and consultants) started working towards solutions, not only for the removal of the siltation, but also for the careful handling, disposal and/or treatment of contaminated dredged material.

This has led to new concepts in which dredging techniques are used for remediation of riverbeds and seabeds. In order to realise such new concepts, the dredging industry has had to develop new types of dredgers, adapted to the requirements of these new remedial tasks. Simultaneously it has had to find techniques for the neutralisation of contaminated sediments or at least for their safe disposal within economically and socially acceptable limits.

6.6.1 Objectives and constraints of the dredging industry

Listening to the general concerns of the public, the responsible authorities or sponsors and the dredging industry concluded, in some cases jointly, that normal practices and procedures had to be changed, especially where contaminated sediments are involved. This understanding forced the dredging world to rethink its role. It was deemed necessary to:

- reconsider maintenance dredging in industrialised ports, taking into account the possible contaminant content of the sediments;
- develop measures to overcome negative environmental effects other than by ceasing the dredging operation;
- develop a policy concerning the utilisation of dredging equipment for the removal of heavily contaminated sediments (remedial dredging);
- adapt existing dredging equipment in order to reduce their environmental features (e.g. resuspension of sediments, accuracy, and so on);
- initiate monitoring and control procedures to measure in real time the environmental effects of dredging activities;
- initiate studies to gain a better understanding of the contamination characteristics of the sediments and the potential risks linked to these contaminants;
- initiate a series of laboratory and prototype tests to evaluate new dredging, treatment and disposal techniques for contaminated sediments, and
- reconsider the selection of the most appropriate dredging equipment for certain types of dredging projects, taking into account the presence of contaminated sediments or other environmental conditions.

6.6.2 Direction of new developments for remediation

In the context of remedial dredging, fairly spectacular innovations have taken place during the last 15 years. In this period a completely new range of dredging equipment has been developed. The design of this new equipment is based on existing technology, but takes into account the overall goal of improving the performance of the equipment with regard to criteria (see Section 6.2) that are critical to an ecologically sensitive task, namely removal of contaminated sediments.

Various characteristics of standard dredgers were thoroughly investigated, and newly developed improvements were incorporated, where possible, within one or more of the new dredgers for remedial work. The following improvements were made:

- the accuracy required to realise a pre-set excavation profile was increased in order to reduce as much as possible the volume of contaminated sediments dredged and to facilitate selective dredging of layers with different contamination characteristics;
- the automation of the monitoring and control equipment on board the dredger, especially for positioning the excavating device, was improved enormously in order to cope with a variable geometry of contaminated layers;

- the equipment was adapted to increase the density of the mixture in the pipeline (minimise dilution by advanced pump automation systems), in order to limit the volume of material requiring further treatment and/or relocation;
- the generation of suspended sediments during the dredging and disposal processes was reduced so as to avoid dispersion of contaminated sediments into the surroundings;
- refinements were made to avoid, or at least reduce, the creation of spill layers;
- the basic attitude of going for the highest production rate (and lowest economic cost) was softened, especially for remedial dredging projects, to take into account ecological costs, and
- to guarantee a continuously high level of quality during an entire dredging project, Quality Control and Assurance procedures were adopted by the dredging industry.

In the following pages some of the specially developed dredgers are described in more detail. These descriptions illustrate how quickly special-purpose dredging equipment is evolving. The innovations described here have been tested in depth on full-scale environmental dredging projects. In the future new developments will surely emerge, providing even further improved equipment and output rates as research and development in the dredging industry is a constant and continuing process.

It should, however, be borne in mind that all developments that aim at accurate and precise operations are highly vulnerable to debris and other obstacles that might be present in the material to be dredged. Processes have to be interrupted when such elements are encountered, immediately resulting in spill and turbidity, with associated environmental effects. Efforts have been made to detect debris and separately remove it at an early stage, but so far no suitable and environmentally acceptable processes or tools have been developed.

6.6.3 Specially developed equipment

Environmental Disc Bottom Cutter
The Environmental Disc Bottom Cutter (EDBC) is a classic stationary dredger equipped with a cylindrical-shaped cutter with a flat, closed bottom and a vertical rotation axis. The suction mouth for removal of the cut material is situated inside the cutter to avoid spillage. A shield over the full cutter height, at places where no soil is encountered, and an adjustable shield above the soil to be cut, prevent both the cut material from entering the surroundings, as well as the intake of excessive water volumes (see Figure 6.16).

An automatic steering and control unit systematically adjusts the cutting depth, the height of the shield to the height of the material to be cut in front of the cutter, and the flow rate through the pump in relation to the amount of soil being cut. A degassing system is installed on board the dredger to avoid cavitation problems and to improve the density of the mixture during the pipeline discharge process.

The disc bottom dredger is a powerful tool for the execution of environmentally sensitive, remedial dredging projects where accuracy within acceptable budgets is required. The technique has been in use for approximately 15 years. As such this equipment has

Figure 6.16
Environmental disc
bottom cutter.

passed the experimental stage, although continuous upgrading takes place based on project experience. Where the material to be dredged contains debris (such as plastic bags, chains and stones), the cutting device can frequently become blocked because of the relatively small suction openings of the disc bottom cutterhead.

The disc bottom dredger features a number of environmental improvements when compared with more traditional dredgers:

- *Safety of the crew*: The basic arrangement is a closed-circuit hydraulic pipeline system. The health and safety risks for the crew are therefore minimal.
- *Accuracy*: The accuracy of the disc bottom dredger is basically the same as that of a stationary dredger which can position its cutting edge within centimetres of the target depth. Combined with the specially designed cutting device, an accuracy of less than 5 cm can be achieved.
- *Suspended sediments and turbidity*: The completely closed shield around the cutter of the disc bottom cutting device is designed to avoid the spread of the cut material. The creation of suspended sediments is therefore minimal and limited

to the near surroundings of the cutting device. As a consequence the generation of turbidity is very small and certainly limited to the waterlayers near the riverbed.

- *Mixing of soil layers*: The automatic steering device makes it possible to excavate different layers selectively and the cutting device allows a clear cut at a well defined depth.

- *Creation of loose spill layers*: The closed shield around the cutter prevents the material cut by the dredger from escaping through the suction mouth. Furthermore the horizontal cut is done by the bottom plate with a clear separation in between the cut material and the virgin bottom. Consequently, the residual spill layer is minimal or non-existent.

- *Dilution*: The disc bottom dredger remains a hydraulic dredger which requires a minimal quantity of transport water. The automatic steering system is programmed to realise pipeline transport with minimal quantities of water. However, during start-up, slow down, and spud and anchor changes, some free water passes through the pipeline. To reduce the quantity of undesired process water during these unproductive stages, the disc bottom cutter dredger is equipped with a system that automatically slows down the flow in the discharge pipeline to the critical transport velocity.

- *Noise generation*: Noise generation, both above and below the waterline, is similar to that of other CSDs.

- *Output rate*: The output rate is somewhat restricted compared with a traditional CSD, because most effort is not put into optimising the output level, but into reducing negative environmental effects, such as the creation of additional suspended sediment content, low accuracy, and mixture density. Output rates of up to 500 m^3/hr can be achieved with the existing equipment.

Sweep Dredger

The Sweep Dredger and Low Turbidity Dredger (see Figure 6.17A and 6.17B), jointly termed Sweep Dredger, are based on a classic stationary dredger equipped with a sweephead, which is a modified draghead of a TSHD or a special designed suction box. A movable visor or valve makes it possible to operate this type of suction head in two opposite swing directions. The cutting height can be adjusted in order to accurately excavate layers with variable thickness. During the dredging process numerous variables (*in-situ* levels, water content, mixture density, suspended sediment, pump power, accuracy and output) are monitored and controlled by the highly sophisticated steering and control system.

The sweephead and the low turbidity draghead, equipped with a visor and using the lower cutting edge, shave the designated soil layer as defined during the preparatory survey. The upper visor adjusts to the bottom profile in order to prevent the inflow of excess water. The swing speed is controlled automatically according to the pre-set hourly production rate, while the pump speed is fixed to permit hydraulic transport of the cut soil with a minimum volume of additional water.

The dredgers are also equipped with a sophisticated degassing system, which prevents unacceptable cavitation in the suction pump, even when sediments to be dredged have a high, variable gas content.

Figure 6.17A Close-up of a sweephead with a movable visor.

Figure 6.17B Close-up of a low turbidity dredge head.

Sweep Dredgers are powerful tools for the execution of environmentally sensitive, remedial dredging projects where accuracy within an acceptable budget is sought. The technology is new but based on well-known concepts. The efficiency of the Sweep Dredger can decrease rapidly when the consistency of the material increases, since no active cutting device is provided to reduce lumps in the suction mouth.

Compared with more traditional dredgers, the sweep dredgers introduce a series of improvements:

- *Safety of the crew*: The sweep dredger is basically a hydraulic dredger with a completely closed circuit. The health and safety risks for the crew are consequently minimal.
- *Accuracy*: Given that the sweep dredger is based on a traditional hydraulic stationary dredger, accurate steering of the cutting edge is possible. This, combined with the sweephead design, means that an accuracy of less than 5 cm can be achieved if the dredger is used carefully.
- *Suspended sediment and turbidity*: The sweepheads contain no rotating devices, which could generate resuspension of sediments around the sweephead. The creation of suspended sediments is therefore minimal and limited to the near surroundings of the cutting device. As a consequence the generation of turbidity is very small and certainly limited to the waterlayers near the riverbed.
- *Mixing of soil layers*: The automatic steering device allows the accurate excavation of different pre-defined layers selectively. For optimal steering, the information on the layers to be cut must be provided in a GIS database.
- *Creation of loose spill layers*: The cutting edge continuously shaves the material to be excavated, guiding the material into the sweephead and suction mouth. This, combined with optimal control of the suction pump, prevents the creation of a spill layer.
- *Dilution*: The sweep dredger requires the addition of water for hydraulic transport through the pipeline, although the automatic steering system is programmed to keep the quantity of additional water to a minimum. During start up, slow down, and spud and anchor changes, some free water passes through the pipeline. During these unproductive phases, however, the sweep dredger is equipped with a system that automatically slows down the velocity in the pipeline to the critical velocity for entrainment. The application of the concept of an intermediate (small) buffer basin in the horizontal pipeline close to the dredger can reduce the need for additional water to a bare minimum.
- *Noise generation*: Noise generation, both above and below the waterline, is similar to that of other CSDs.
- *Output rate*: The output rate is probably the highest of the special-purpose dredgers used for remedial dredging. Up to 1200 m^3/hr can be reached for a full layer thickness. This is close to the output rate of a traditional CSD. That said, the environmental factors (suspended sediments, accuracy and mixture density) were considered of equal importance during the design of this new dredger.

Auger Dredgers

Auger Dredgers, especially smaller types, have been utilised for many years, mainly in lake clean-up projects. Combining their working principle with some specific environmental features, the environmental Auger Dredger (see Figure 6.18) is specially designed for the removal of thin layers of contaminated sediments. The dredger is a normal stationary dredger equipped with an auger that cuts the material in layers with a thickness ranging from a few centimetres to one metre. The thickness of the layers being cut can be maintained continuously within this range. The screws in the auger transport the material to the centre, where a dredging pump sucks away

Figure 6.18 Close-up of an auger dredger.

the material through a suction mouth. The suction force and a screen around the auger prevent dispersion of the material into the surrounding water.

The width of the cut depends on the width of the auger, which can vary from 2 to 14 metres. A sophisticated monitoring and control system is installed to optimise environmental efficiency. A degassing system is fitted to the dredger to avoid cavitation problems and to improve the density of the mixture during the pipeline discharge process.

The environmental auger dredger is a powerful tool for remedial dredging projects where accuracy and environmental effects are of prime concern. The equipment has been in use for approximately 10 years and has passed the experimental stage, although ongoing upgrading is necessary to deal with continuously changing demands.

The large dimensions of the cutting device mean that the auger is less suitable where very high accuracies are required in areas with high three-dimensional variability. The environmental Auger Dredger is specially suited to dredging accurately and producing smooth surfaces, because of the large dimensions of its cutting device.

Compared with more traditional dredgers, with respect to environmental criteria, the environmental Auger Dredger offers a number of improvements:

- *Safety of the crew*: The cutting and pumping system of the environmental Auger Dredger is completely enclosed. Thus, the risks of direct contact between the crew and the transported material are minimal.
- *Accuracy*: The environmental auger dredger is based on a stationary dredger with a spud system. This allows, in principle, precise positioning of the cutting head which, combined with the automatic steering and control system, allows the environmental Auger Dredger to work to tolerances of less than 5 cm. The layout of the auger allows the equipment to produce a very smooth surface. However in case of a three-dimensionally variable target depth the auger is less suitable.
- *Suspended sediments and turbidity*: The auger is completely closed off from the environment by a skirt, which covers the cutting opening. Combined with the rather slow rotating movement of the auger, this results in little suspension of sediments around the cutting device. As a consequence the generation of turbidity is very small and certainly limited to the waterlayers near the riverbed.
- *Mixing of soil layers*: The sophisticated automatic steering system allows for accurate layer-by-layer excavation. Only when there is a rapid three-dimensional variation of the inter-layer boundary, does the length of the auger cause some difficulties in following small-scale variations in both directions at the same time. However, such small-scale variations are not often encountered in nature and it is very difficult to measure them during the pre-dredging survey.
- *Creation of loose spill layers*: The auger cuts and conveys the material towards the suction mouth of the dredger, which is located at the centre of the auger. This feature, combined with good control of the pumping process, eliminates the spill layer.
- *Dilution*: The environmental auger dredger is a hydraulic system, which requires the addition of transport water. It is very similar to the sweep dredger and the disc bottom dredger, which means that with a good monitoring and steering system, the amount of water needed can be reduced to a minimum.
- *Noise generation*: Noise generation, both above and below the waterline, is similar to that of other CSDs.
- *Output rate*: The output rate is determined by the size of the auger and is generally lower compared with a traditional CSD, as most of the effort is aimed at reducing the environmental effects of the equipment. Output rates up to 500 m^3/hour can be achieved.

Environmental Grab
The Environmental Grab is a specially designed grab with the following features:

- during the opening and closing of the bucket, the cutting edge remains on the same horizontal plane;
- the opening and closing of the grab is undertaken hydraulically with a built-in hydrogroup, or mechanically with a special rigging of cables;
- when the grab is closed all openings are sealed to minimise spill, and

- the crane is equipped with a positioning system on top of the crane boom to accurately measure the position of the turning point of the cable. An encoder is used to measure the paid-out cable length to identify precisely the depth of the grab during the excavation process.

The environmental grab can be installed either on a traditional grab dredger (cable crane), where it is suspended from cables, or on a backhoe dredger (hydraulic excavator). The latter permits better positioning and guidance of the cutting edge during excavation, as the pendulum effect can be avoided (horizontal profiling grab) (Figure 6.19).

The environmental grab dredger is a mechanical device, which has significant advantages in respect of density of the excavated material. For other environmental criteria, the grab dredger is less advantageous compared with the other systems described. It should, therefore, generally be used in combination with other protective measures such as silt curtains. It is suitable for the removal of small quantities at sites that are difficult to reach and for projects where the costs of dewatering or treating the excavated material are high in proportion to the volume of material.

Compared with the more traditional dredgers in relation to environmental criteria, the environmental grab dredger offers a number of improvements:
- *Safety of the crew*: Excavation and barge loading are done mechanically. Although efforts have been made to optimise the process, considerable possibilities remain for direct contact between the crew and the dredged material at different stages of the dredging cycle (e.g. a hard object does not permit perfect closure of the bucket and some mud can stick to the outside of the bucket, only to be washed away during the raising movement). Detailed discharge procedures are necessary to avoid careless loading of the barges and the consequential risk of significant splashing.

Figure 6.19A ECO grab dredger suspended from cables.

Figure 6.19B Horizontal profiling grab installed on a backhoe dredger.

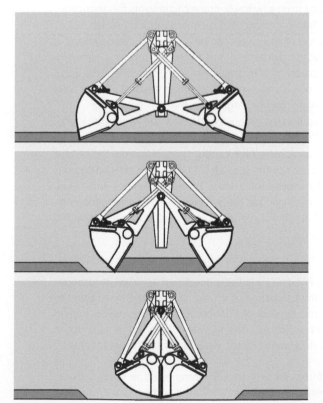

Figure 6.19C Schematic drawing of the closing pattern of a horizontal profiling grab.

- *Accuracy*: The accuracy of the cable crane version is reasonable, provided that the most modern technologies are applied. However, some disturbances occur when the dredging actions take place in currents where the depth measurements are not completely precise. Angle sensors on the cable can partially correct this error. Accuracy levels of approximately 10 cm can be achieved vertically. Horizontal accuracy is less as a result of the free suspension of the grab, which causes a pendulum effect, especially in deep waters and in tidal currents. When the environmental grab is mounted on a hydraulic crane, horizontal accuracy improves drastically, achieving levels similar to those of a backhoe dredger.
- *Suspended sediments and turbidity*: The generation of additional suspended sediments is low compared with normal grab dredgers. However, similar problems can occur as described above under "safety of the crew". The suspended sediments will be mainly generated by the "washing" effect while the bucket is raised. This means that the sediments will be present over the whole water column with an influence on the turbidity of all water layers. However, this effect is significantly reduced, as the hydraulic grab buckets are watertight and are operated in such a way that minimal sediment is collected on the outside and on top of the grab.
- *Mixing of soil layers*: Clearly, an advanced monitoring and steering system is necessary in order to control the lowering and cutting movement of the grab and the penetration depth, and thus achieve the required accuracy at each intersection of the different layers. The limited accuracy in the horizontal plane is a drawback in this respect. Again, deployment from a hydraulic crane circumvents this restriction.
- *Creation of loose spill layers*: Taking into account the mechanical characteristics of the cutting process, the creation of spill layers can be avoided to a large extent.
- *Dilution*: The dredging process is mechanical and further transport is generally undertaken by barges. Compared with hydraulic dredgers, this offers considerable advantages as far as dilution is concerned. Only when the bucket is partially loaded will some additional water be discharged into the transport barge.
- *Noise generation*: Noise generation (both above and below the water line) is similar to that of a traditional GD or BHD.
- *Output rate*: Grab dredgers can achieve high output rates. However, the output is basically determined by the size of the bucket, and it is more difficult to control the opening and cutting process of a large bucket than of a small one. Similar observations can be made about the selective dredging requirements. Therefore, the output rate of an environmental grab is limited to a few hundred cubic metres per hour.

Pneumatic Dredgers

Pneumatic Dredgers such as the Pneuma have been developed at different locations throughout the world for more than a decade. It should also be mentioned that the system has been developed for the removal of siltation in fairly deep waters behind hydraulic dams. The efficiency with which the material enters the cylinders depends on the pressure difference, which is determined by the water depth at the dredging site.

The Pneuma system consists of three cylinders, shovels, compressed air supply/exhaust pipes, a compressor and a delivery unit. The system is based on hydrostatic pumping principles. A differential pressure is induced in a cylinder by a vacuum and the external hydraulic head. This creates an influx of the soft sediments into the cylinder. When the cylinder is filled, the inlet valve is closed and compressed air is pumped into the system to force the sediments through an outlet valve into the delivery pipeline. When the cylinder is almost empty, the pressure in the cylinder is released, the vacuum is applied again and the entrance valve is opened ready for a new cycle.

Pneumatic dredgers could be attractive for remedial dredging projects because of their high density pumping method, the closed circuit principle, and the fact that no moving elements are in direct contact with the material to be removed (reduced suspended sediment generation). The system is, however, vulnerable to debris in the dredging area (a common problem during removal of contaminated harbour sediments), and the automatic steering and control system is far less developed compared with other dredging systems. In port projects including the removal of contaminated sediments, the water depths are generally rather limited, which puts a constraint on this type of dredging equipment for remedial projects because of a lower productivity owing to small pressure differences.

6.7 TRANSPORT AND PLACEMENT; EQUIPMENT AND TECHNIQUES

How dredged sediments are transported and where they are placed is of great environmental concern. Many different methods are available and special attention is paid here to those techniques and equipment that can be used to mitigate the environmental effects of transport and placement activities. For each of the transport modes, new developments and possibilities to improve the environmental characteristics are discussed.

6.7.1 Transport equipment

Pipeline transport
Pipeline transport is basically an environmentally friendly transport method. This generally applies to the transport of dredged material whether it is contaminated or not (Figure 6.20). The only major disadvantage is the requirement to mix the excavated material with transport water. This increases the volume for storage and/or further treatment, which in the case of contaminated fine-grained sediments can be a serious issue.

Pipeline transport is also a safe, clean method because it takes place in a closed circuit system. The major contact points with the outside environment are the entrance point at the suction mouth and the outlet point at the relocation or destination site. In principle, between the entry and outlet points the transported material cannot be in contact with the outside world unless a pipeline failure or leakage occurs (which is unlikely provided proper maintenance procedures are followed). Complete failure (breakage) of a pipeline is virtually unheard of and leakage control can be instigated when the transported material is heavily contaminated. In most cases, however, the impact of minor leakage can be ignored and, in any event, it is far less significant than the potential risks of release from most other methods of transporting dredged material.

Figure 6.20 Pipeline transport is one of the safest and most often utilised means of transporting dredged material.

To improve the pipeline transport system, the following developments are cited together with suggestions for further improvements:

- The automatic control systems that have been developed include a pump monitoring and steering system for a smoother discharge process with fewer high and low peak values of the mixture density. Given that physical laws limit the upper peaks, this development results in an increase of the average density in the pipeline and a reduction of the dilution effect.
- Regular interruptions of the dredging process at the end of each swing movement and during anchor and spud carriage moving activities are a major source of additional water in the transport pipeline of the dredgers. The introduction of a buffer stock between the suction and discharge processes enables the operator to stop the suction process during unproductive periods (spud changes and slow downs). This offers great possibilities for avoiding frequent interruptions of the discharge process during such periods. This facility can be compared with the LMOB system on the TSHD.
- The use of high-density pumps reduces considerably the need for additional transport water. The types of high-density pumps available, however, have rather limited output capacities. This is the main drawback to their application in the dredging industry.
- In cases where the pipeline transport involves the use of a barge unloading dredger, returning the transport water from the relocation site to the dredger and recycling the transport water for the unloading process should be considered. In the case of contaminated sediments, such a procedure is environmentally and economically beneficial.

Hopper or barge transport
The second means of transport frequently used in the dredging industry is hopper or barge transport. In this case the dredged material is loaded onto a ship either

hydraulically (TSHD) or mechanically (BLD, BHD, GD). Horizontal transport between the dredging site and the relocation or treatment site is done by navigation of the barge or hopper. Barge and hopper transport is relatively environmentally friendly with limited environmental impact as regards noise generation, the emission of exhaust gases and the creation of road blockages.

The main advantage of this means of transport, compared to the pipeline method, is the elimination of the need for transport water. Excavated material can be transported at almost its original density and consistency, provided this density can be maintained during the disintegration process. The exception here is the TSHD, which, with the use of hydraulic pumps, raises the material from the riverbed or seabed into its hopper.

A limitation of barge transport is the prerequisite that the dredging site and the relocation site be linked by a navigable channel with sufficient water depth. If this is not the case, transport must be executed in another way, which is most often pipeline transport. In that event, the dredged material has to be rehandled by a hydraulic dredger (barge unloading equipment) with consequent increases in dilution and volume, in addition to cost effects (see Figure 6.21).

Figure 6.21 In a combined transport cycle, barges are used together with conveyor belts.

The fact that most barges are open making the risk of spillage slightly greater compared with pipeline transport is a minor disadvantage of this method. Furthermore, contact between the crew and the dredged material is a continuous risk, which can be a problem in the case of heavily contaminated sediments or sediments with a highly volatile content.

An important element in control is ensuring the regular inspection and maintenance of barges and hoppers, to make certain that barge and hopper bottom doors close properly. Also the use of automatic monitoring devices to allow total surveillance of vessels during transport and disposal improves environmental performance.

New developments and options for improving the characteristics of this means of transport focus primarily on the limitations previously mentioned:
- *Spillage of material from the barges during transport.* This can be avoided either by placing a cover over the hopper during transport (something as simple as a canvas cover) or by continuously reminding those involved to leave sufficient freeboard (50 cm minimum) in the hopper above the loaded sediments. A third possibility is to allow the sediment to settle for a period of time after finishing the loading operation. Subsequently, the water on top can be pumped out, either overboard or to a purification plant, before further transport takes place. Until now neither of these options for removal of excess water have been applied frequently as both have adverse financial implications.
- *Unloading procedure.* Either the barge is unloaded via its bottom doors, discharging at an underwater relocation site (a rather uncontrolled procedure not requiring the addition of transport water), or the barge is unloaded using an hydraulic barge unloading dredger resulting in significant dilution during the suction process phase. To avoid this problem, new systems to unload barges mechanically are being used. The main advantage of such a mechanical unloading system is the high density of the material during and after the unloading process. This advantage can only be maintained, however, if further transport, between the unloading site and the placement or relocation site, is changed from pipeline transport to an alternative high-density transport mode.

Road transport
Although pipeline and barge transport are used for almost every dredging project, alternative transport modes should be considered, especially if reduction of the overall environmental effects of the dredging cycle is an objective, in particular when there is a need for further transport after unloading the barges (i.e. when the destination is not located in the immediate vicinity of a waterway).

The first alternative transport mode is road transport by means of trucks. The main advantages of this method are:
- trucks can be loaded mechanically at any density, and
- choice of destination is flexible, which is a major advantage when different qualities of material have to be transported to different relocation or treatment sites.

The main disadvantages of this method are:
- tipper trucks, which are difficult to make spill-free against leaking fluid, are normally used;
- considering the normal output of a dredging project, the number of trucks necessary for transport of dredged material is high; and
- the environmental effects of road transport are greater compared with pipeline transport (e.g. noise generation, exhaust gases, road usage, spillage on public roads, and so on).

Considering the above, road transport is obviously only acceptable for the transport of dredged material in certain specific cases with the following characteristics:
- where there is highly contaminated material, which would result in excessive costs for treating the transport water;
- where excavation is realised mechanically; and
- where dredged material with different qualities will have different destinations (on-line measurement of the soil quality and decision-making about the destinations are not simple).

The opportunities for the use of road transport in the dredging industry are limited. Still given the trend to dewater fine-grained dredged material (either mechanically or naturally) before final (beneficial) use, new demands have been placed on the transport process and this method can certainly not be ignored.

Conveyor belt transport
A fourth possibility for transporting dredged material on a large scale is the conveyor belt. The system is not commonly used in the dredging industry, as the installation costs of such a system are high. Furthermore, the basic characteristics of a wet dredging process are not compatible with the mechanical characteristics of the conveyor belt. However, as the boundary conditions of dredging projects continue to change drastically, all possible transport processes and their applicability to certain types of dredging projects, especially for remedial works, need to be examined.

The application of conveyor belt transport to dredging offers a number of advantages to the industry:
- dredged material can be loaded mechanically on a conveyor belt system with no need to add transport water;
- the conveyor belt is a continuous transport system capable of conveying large volumes of material;
- transport costs are reasonable provided that a large volume is to be transported over the same route; and
- the environmental effects during transport (e.g. noise, exhaust gases and such) are relatively low, provided special precautions are taken to reduce noise and avoid dust.

Disadvantages of the conveyor belt transport are:
- alignment of the conveyor belt transport is fixed; changes to this alignment during the transport process are difficult and costly;

- special precautions have to be taken to avoid material losses during transport of dredged material with a normal water content; and
- logistics problems can arise should the transported material have to be spread at the destination site.

From the above it is evident that application of this transport method in the dredging industry is limited to specific cases with the following characteristics:
- where dredging is carried out mechanically or where a dewatering process has taken place;
- where there are fixed loading and destination sites; and
- where the work is of sufficiently long duration to allow for depreciation of the installation costs.

Conveyor belt transport is not commonly used in the dredging industry. However, it has potential advantages worthy of consideration for use in the future, especially when there are environmental concerns. As with road transport, the tendency to consider and apply dewatering techniques prior to reuse or disposal, offers new possibilities for utilising this transport method in cases where origin and destination are fixed for a long period. Used in combination with barges for transport between unloading quay and reuse/relocation site, conveyor belts can be an attractive alternative, considering the high density, which can be transported with minimal environmental risks.

Combined transport cycles

Another aspect of dredged material transport is the increasing complexity of the process, which regularly leads to bi- or even tri-modal transport procedures where two or three different transport modes are used to reach the final destination site (see Figure 6.21).

The following are some of the combinations possible:
- barge transport combined with a barge unloading dredger that pumps the unloaded material towards the final destination site;
- barge transport with mechanical unloading and conveyor belt or truck transport to the final destination site;
- trailing suction dredger with hopper transport and pump ashore facilities (pipeline discharge); and
- pipeline discharge to an intermediate treatment installation and further truck, barge or conveyor belt transport.

Selection of the optimal transport cycle has to be planned with great care, as a bi-modal transport process not only combines the advantages of both transport modes but also the disadvantages. As such, it has to be taken into account that each transfer between two transport modes creates the risk of material losses or other environmental risks. Therefore, during the environmental effect analysis of a project, it is necessary to consider the different phases of dredging, transport and treatment or relocation in an integrated way.

6.7.2 Placement techniques

Placement of the dredged material at the placement site is another major phase in the dredging process which potentially can have significant environmental effects. Selecting the most appropriate placement site and considering the infrastructure at that destination site are of major importance. In addition, equipment and techniques used for the placement of dredged material also have an influence on the overall environmental effects of dredging. A discussion on this and on potential effects of the various placement options follows. (For more details See Chapter 7.)

Land placement
One option for the placement of dredged material is on land within a confined area surrounded by dikes. This is generally applied when the use of dredged material is required on land, or in case contaminated material has to be stored on land.

Discharge pumps on board dredgers are used to pump the dredged material through a pipeline, which ends in the confined destination area (see Figure 6.22). The most

Figure 6.22 Dredged material placed on land in a confined area surrounded by dikes.

significant environmental effects of this procedure are:
- burial of (environmentally sensitive) surfaces;
- change of the topography; and
- leakage of (contaminated) transport water into subsoil layers.

Potential environmental risks during the actual placement action are overflow of the material and rupture of the surrounding dikes. Both can result in serious damage and spreading of deposited material into unwanted areas. Proper design of dikes and regular monitoring of the water level at the relocation area can help avoid these problems.

Another critical item is the removal of the excess transport water from the confined placement area. Given the large volumes of water to be removed, this can create serious environmental damage if not properly managed. The removed water still carries a small part of the fine material that is being deposited in the confined relocation area. If this material is contaminated, it can result in environmental problems. Moreover, the quantity of fines that pass through the outlet can generate secondary difficulties such as blockage of the smaller waterways used to carry away the transport water.

Proper design of the placement area, to ensure maximum opportunity for the material to settle within the area, reduces this effect considerably. Installation of decanter-type basins can be an additional safety measure where strict limitations are imposed on suspended sediment content of the removed water.

Careful management of the placement procedure can also reduce the risk of losing fines through the outlet. For example, closing the outlet during actual placement operations and thus interrupting placement each time the confined area is filled to its maximum might be feasible. The confined area is then left for a few days to settle, after which the outlet is opened carefully in order to remove the overlying water layers, containing a much reduced suspended sediment content. To implement this procedure without interrupting the dredging project, several confined areas, which can be used simultaneously, must be available.

The salt content of the transport water can differ from natural groundwater characteristics at the relocation site. Again, groundwater quality problems can arise if the site is not managed properly.

Detailed studies of the natural conditions at and around the planned relocation areas are needed in order to evaluate potential risks and to take necessary precautions such as relocation of the destination site to another less vulnerable place or installation of protective liners such as HDPE or natural materials (e.g. clay and peat).

Underwater placement

Another option is to place dredged material in open waters. This is mainly done with clean material and with slightly contaminated material, if suitable open water sites are available. Underwater placement generally follows hopper or barge transport, where the barge or hopper (TSHD) sails directly from the dredging area to the placement site. At that site the vessel's bottom doors are opened and the material falls from the hopper onto the seabed or riverbed. This results in a large cloud of heavily laden water that falls through the entire water column onto the seabed at the placement site. Clearly part of the material will remain in suspension for a certain period during which both the suspended sediment content as well as the turbidity will be significantly higher compared to the background conditions at the placement site. Whether such an increase will be acceptable or not will largely depend on the characteristics of the prevailing ecosystem at that location (e.g. in a mudflat area the effect will be far lower compared to a coral reef environment).

A second, potentially important effect is the burial of the natural river bottom including the local fauna and flora at that location by the rather massive volumes of sediment-water mixtures that fall down during such placement operations.

Similar conditions occur when hydraulic dredgers (TSHD, CSD or SD) pump their mixture directly to an underwater placement site. The main difference compared to bottom door placement is the longer duration of the disposal operation. This will result in lower peak values for suspended sediment content and turbidity; but, on the other hand, the time of the impact will be considerably longer.

Again, the choice of site has a major effect on the overall environmental consequences of the project. A site with large tidal or other currents generates a greater risk of erosion, resuspension and further dispersion of the materials into the surroundings. Indeed, some sites are deliberately chosen to be dispersive, thus making use of natural tidal currents. This matter of placement site selection is discussed in more detail in the next chapter.

Equipment and techniques used for placement can, however, be adapted to reduce environmental effects of dredging. Instead of directly opening bottom doors, the hopper can be emptied by means of a pump linked to a vertical pipeline reaching down to the river- or seabed. In this way the material is guided to its final placement depth without intermediate contact with overlying water layers. Losses of fines and dispersion during the fall of the dredged material from surface to river- or seabed are reduced considerably. This will limit the effect of the placement operation mainly to the lower water layers and turbidity increase in the main water body will be significantly lower. The effect on swimming species will, therefore, be limited but the burial effect on the species living on and in the river bottom will not be very different.

To further optimise this procedure, an underwater diffuser can be fitted to the lower end of the discharge outlet. The diffuser has two main purposes:
- to change the flow from a vertical downward movement in the last section of the discharge pipeline towards a horizontal flow just above the seabed. This change considerably reduces the impact of the outflow on the previously placed soft layers. As a consequence, the resuspension of the placed material is minimised, and
- to reduce the outflow velocity of mixture from 4 to 5 m/sec in the normal discharge pipeline to less than 0.5 m/sec in the outlet of the diffuser. Again, erosive forces on the previously placed material are further reduced. The design of the actual diffuser is crucial to efficiently reducing the velocity, as it is critical that flow occurs over the full section of the diffuser opening.

This procedure will further reduce the resuspension effect of the disposal operation. The turbidity in the water layers above the diffuser head will be very small and the location of the final settlement of the material can be predicted fairly accurate. This leads to a better control of the potential burial effects provided that an in-depth knowledge is available of the local ecosystem and a management system is put in place to control the disposal operations properly.

When dredging with larger TSHDs equipped with two suction pipes, a recent development to improve placement action is to adapt one of the suction pipes by equipping it with a type of diffuser head, using it in reverse flow direction as a kind of fallpipe.

Other means for reducing environmental effects of the placement process can be realised by restricting the period during which placement is allowed. For instance, avoid underwater placement when maximum tidal currents occur or during seasons when there is intensive biological activity in the area. The cost for a seasonal restriction is rather limited when this restriction is already included in the proper planning of the placement works, but can be very high in case the dredging and placement operations have to be stopped by the client during the actual execution.

The construction of underwater bunds or use of underwater pits are other ways of reducing the dispersion of material from the relocation site after actual placement has occurred.

Capping techniques
In case of underwater placement of contaminated material, isolating the material from the environment by means of a capping layer may be necessary. This is a layer of clean material installed on top of the (underwater) placement site of contaminated material; the type of material has to be selected depending on the isolation requirements imposed for such capping layer. This layer protects the contaminated material from erosion by natural water currents and prevents, or at least reduces, the uptake of contaminating elements by the aquatic life, as well as the migration of those elements to the overlying water layers.

Capping is defined as the controlled, accurate placement of contaminated dredged material at an open water site, followed by a cover or cap of clean isolating material (Palermo, 1994; Whiteside et al., 1996). Chapter 7 gives more details about requirements for such capping systems. The discussion here focuses on equipment and techniques used for placement of such layers, as this is a critical factor in guaranteeing the expected performance of the capping layer.

Placement of contaminated materials can be undertaken in several ways:
● by conventional discharge with barges or TSHDs;
● by discharging material through a pipeline from the water surface, and
● by guiding the material close to the underwater bed by means of an underwater diffuser.

The choice of the best placement method is dependent on the actual level of contamination. Because it obviously brings the material as close as possible to the underwater bed, the last method provides the best possibility to control the placement and limit the dispersion or spreading of the material during the placement operation.

Methods for placement of the capping layer depend primarily on the material used for capping purposes (sand, clay lumps, gravel, rock or geotextile containers are all possible) (see Figure 6.23). Furthermore, the water depth and the characteristics of the contaminated material will have a major impact on the selection process for the best technique. The following options are possible:

- *Surface discharge with barges or TSHDs*: The main advantage of this technique is its low cost per m³. It has to be determined whether additional dispersion of fines in overlying water layers is acceptable for the project. Control of thickness of the capping layer is difficult; therefore the volume necessary to realise a minimum capping thickness over the whole surface will be high.
- *Spreading by barge movements*: The material can be spread gradually by controlled opening of the barge, while tugboats or a Christmas tree anchoring system move the barge laterally. In this case the impact of the falling sediments will

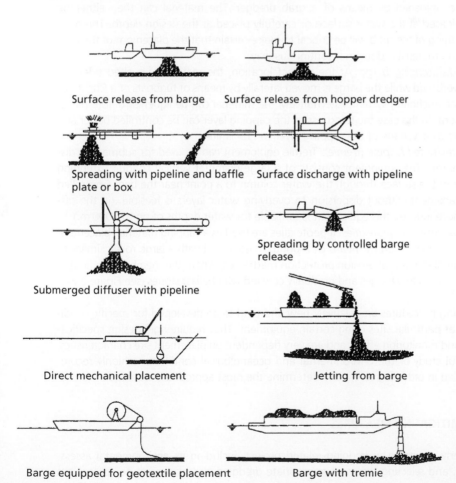

Surface release from barge   Surface release from hopper dredger

Spreading with pipeline and baffle plate or box   Surface discharge with pipeline

Submerged diffuser with pipeline   Spreading by controlled barge release

Direct mechanical placement   Jetting from barge

Barge equipped for geotextile placement   Barge with tremie

Figure 6.23 Conceptual illustrations of various equipment used for capping.

be lower and the contaminated material is less disturbed by such impact and covered more evenly. The total volume of the capping layer can be smaller compared to surface discharge.

- *Surface pipeline discharge*: The material can be pumped through a pipeline located on the water surface. This will result in a thin layer, which can be built up gradually. Additional dispersion of suspended sediments may be significant and it has to be determined if this is acceptable for the project. A baffle plate or sandbox at the end of the pipeline can be installed for better control of the actual placement location.
- *Submerged discharge/diffuser*: To reduce the amount of additional dispersed sediment and the turbidity in the overlying water layers, the pipeline outlet can be located nearer to the lower water layers. To reduce the impact and outflow velocity above the contaminated material, an underwater diffuser can be used. In such a way the capping layer can be placed quite accurately and the thickness of this layer can be kept close to the theoretical minimum requirements.
- *Mechanical placement*: Capping material is brought to the site by barges, which are unloaded by means of a grab dredger. The material can then either be released at the water surface or carefully placed at the design depth. The positioning of the grab will be critical to make certain that the placement of the capping material is done in a continuous way.
- *Side-dumping barge*: At the project location, the material is pushed sideways overboard while the barge is moved laterally by means of tugboats or a Christmas tree anchoring system. Alternatively, the material can be liquefied before placement. In this case the thickness of the capping layer can be controlled better and the total volume of the capping layer can be limited.
- *Gravity-fed fallpipe (tremie)*: Tremie equipment can be used for submerged discharge. The equipment consists of a large-diameter vertical conduit extending from the surface through the water column to a point near the bed. Controlled placement without dispersion to overlying water layers is feasible. (In the offshore industry this equipment is available for water depths of up to 1000 m.)
- *Placement of geotextiles*: If geotextiles are used as a capping material, placement can be executed by means of a pontoon equipped with a large roller, similar to the placement of erosion protection mattresses, where the membrane is floated between two barges and gradually covered with ballast stones (see Figure 6.24).

Capping procedures are relatively new and tend to be developed for specific conditions at particular sites using certain equipment. They require rigid design specification and monitoring effort, and are very dependent on placement site characteristics. Careful study of the environmental and oceanological conditions is highly recommended in order to successfully determine the most appropriate technique.

6.8 MITIGATING MEASURES

Properly designing a dredging-related project, including its environmental assessment, and selecting the most appropriate dredging and placement equipment are undoubtedly the major actions required for reducing its overall environmental effects.

Figure 6.24 The placement of geotextiles from a pontoon equipped with a large roller and a sand spreader.

The need for such measures is not only limited to projects for remedial dredging. Even when dredging and relocating clean sediments the evaluation of potential environmental effects is necessary. This should be followed up by taking measures to reduce such effects to a value acceptable to all stakeholders. For specific purposes well-defined actions can be taken to reduce environmental effects.

These actions may comprise physical measures on or around the dredger to avoid the spilling or spreading of suspended sediments or to reduce turbidity around the dredger. Also organisational measures can be taken in planning dredging works, for instance, the adoption of working restrictions (e.g. tidal dredging, seasonal restrictions) to reduce disturbance of the ecological system at and around the dredging or relocation site.

6.8.1 Mitigating measures on board the dredgers

There are numerous measures that can be implemented on board a dredger. Firstly, a dredger can be equipped with special equipment (as discussed in detail in Section 6.5), such as:

- specially designed cutterheads to reduce spillage and the creation of suspended sediments at the dredging site;
- dragheads to improve suction efficiency (thus reducing the dilution effect) for silty materials which contain significant quantities of organic material and natural gases;

- green valves to reduce the turbidity when using the overflow system of a TSHD;
- degassing systems to avoid irregular cavitation within the pump;
- specially designed grabs to limit losses during the raising movement (thus reducing generation of additional suspended sediments in the surroundings);
- monitoring and automation systems to improve the crew's information regarding the various dredging parameters (which improves dredging accuracy and efficiency); and
- control and monitoring systems to alert the crew early on to leaks or any other potential risks.

Secondly operational measures can be taken such as:
- limit speed (revolution and swing speed) of the cutter and ladder of the CSD respectively, in order to reduce generation of suspended sediments and turbidity;
- carefully control pump speed to maximise the concentration where density of the transport mixture is critical (dilution);
- carefully navigate in shallow water to avoid additional turbulence (which generates suspended sediments);
- limit overflow quantities by good management of the LMOB system (which reduces generation of suspended sediments);
- limit hoisting speed of grab and backhoe dredgers to avoid spillage; and
- reduce navigation speed of laden barges and hoppers during bad weather to avoid excessive spillage.

Implementation of a quality control system that takes into account potential environmental effects, as well as different options for mitigating actions, can help reduce negative effects to a large extent. The results of this part of the QA/QC system should be implemented in an "Environmental Protection Plan" and an "Environmental Protection Manual" which describe the various critical characteristics of the dredging and relocation sites and the possible actions to be taken during the planning and execution of the dredging project.

6.8.2 Mitigating measures at the dredging site

Mitigating measures can also be taken at the dredging site itself. Apart from careful planning and control of the dredging actions, implementation of physical barriers to prevent the spread of suspended sediments is an important option. This can be achieved by the installation of silt screens (also known as curtains) or turbidity barriers at or near the dredging site providing there are relatively slow current conditions (see Figure 6.25). The following options exist:
- *complete enclosure of the dredging equipment with a silt screen*: this can only be done with stationary dredgers using pipeline discharge methods; in other cases the surface to be enclosed is too large or the curtain has to be opened too frequently in order to allow barges to enter and leave the protected area;
- *complete enclosure of the dredging zone*: this can be done around the dredging area of grab or backhoe operations, enclosing the dislodging and raising operations, but allowing barges free access to lie alongside and to be changed without hindrance;

Figure 6.25 Underwater artist's rendering of a silt screen.

- *protecting a sensitive area nearby the dredging site*: in this case the dredger operates freely, unhindered by the curtain. The curtain is installed, during the whole dredging operation or only during the sensitive season, in such a way that the suspended sediments cannot pass through the curtain towards the sensitive area, or
- a combination of these options.

Silt screens/curtains are silt-impervious, vertical barriers that extend from the water surface to a specified water depth. The flexible, polyester reinforced vinyl fabric forming the barrier is maintained in a vertical position by flotation material at the top and a ballast chain along the bottom. A tension cable is often built into the curtain immediately above or just below the flotation segments (top tension) to absorb stresses imposed by currents and other hydrodynamic forces. The screens are usually manufactured in sections that can be joined together at a particular site to provide a screen of specified length. Anchored lines hold the screen in a deployed configuration that is usually U-shaped or circular.

The silt curtain does not indefinitely contain turbid water but instead controls the dispersion of turbid water by diverting the flow under the curtain, thereby minimising the turbidity in upper layers of the water column outside the silt curtain.

A silt screen's effectiveness, defined as the degree of turbidity reduction outside the screen relative to the turbidity levels inside the screen enclosure, depends on several factors such as the nature of the operation; the quantity and type of materials in suspension within or upstream of the screen; the characteristics, construction, and condition of the silt screen, as well as the area and configuration of the screen enclosure; the method of mooring; and the hydrodynamic conditions (i.e. currents, tides, waves, etc.) present at the site. Because of the high degree of variability in these factors, the effectiveness of different silt curtain operations is highly variable. The efficiency of a silt screen depends strongly on the local oceanographic conditions. In the perfect case (no currents, no waves, no tide) the retention can be as high as 80 to 90%. Any deviation from these "perfect conditions" will decrease the efficiency (Jin, J-Y. 2003). Currents above 0.5 m/s, waves above 1 m and high tidal range (above 3 m) will result in retention factors of 25 to 40% only and conditions under which it will be almost impossible to maintain the silt screen in a working condition.

The installation of a silt screen is often a difficult operation and demands great skill and experience on the part of the dredging contractor in order to avoid problems of leakage through the curtain. The following should be considered during the planning and installation of such a curtain:
- surface of the site to be surrounded;
- water depth to be protected;
- currents to be expected (both speed and direction are of importance);
- wave climate at the site to be protected;
- size of the sediments to be retained;
- water depth where the additional suspended sediments are created; and
- movements of the dredging equipment during the project.

Based on the physical conditions of the site and the environmental restrictions at the location, the type of silt screen, method of deployment and anchoring system can be selected. The use of a silt screen, however, clearly limits the output level of the dredger, lengthens the execution period, and increases the costs of the project. Yet, under certain conditions, as mentioned above, a silt screen might effectively be used in the vicinity of an environmentally very sensitive area.

As an alternative to a silt screen, a bubble curtain is sometimes considered. A bubble curtain is formed by laying a perforated pipe on the sea- or riverbed and pumping air continuously through it. The upwelling of tiny bubbles from the pipe to the surface of the water has the effect of preventing fine sediments from passing across the line of the pipe. This is accomplished because the vertical current formed by the bubbles catches the suspended sediment on the turbid side of the "curtain", moving the sediment to the surface and deflecting it back towards its point of origin in the part of the surface current that moves in the direction of the turbid water.

Bubble curtains are effective in benign conditions where water velocities are very low (see Figure 6.26). They use a large amount of air, which leads to high energy consumption.

Figure 6.26 A bubble curtain.

6.8.3 Measures at the relocation site

Finally, mitigating measures at the placement site can be implemented. Again, both planning and physical measures are possible. The following physical measures can be considered:

- installation of a silt screen around the underwater relocation site or around the outlet of a confined placement area to reduce the turbidity mainly in the upper water layers;

- utilisation of underwater diffusers to reduce the suspended sediment content and the turbidity at the placement site; and
- application of settlement ponds at the outlet of a confined placement area, in order to reduce suspended sediment content in the excess transport water that is returned to natural water courses and will increase the turbidity levels of these waters.

In the operational field, the following measures can be considered:
- seasonal restrictions on placement at certain locations;
- tidal restrictions for underwater placement to avoid or at least reduce natural transport of suspended sediments to sensitive areas nearby the dredging or relocation site; and
- use of absorbent or impermeable liners at the bottom of confined placement areas.

The choice of the most appropriate measures depends largely on the actual conditions at the relocation site. A careful analysis of these conditions, including an environmental impact analysis of the planned dredging and relocation operations, is a prerequisite to defining the correct infrastructure and procedures for an optimal project both in terms of economics and environment.

CHAPTER 7

Reuse, Recycle or Relocate

7.1 INTRODUCTION

Every year millions of cubic metres of sediment arising from dredging projects are transported to another location – in water or on land – and placed there.

More than 90% of the sediment from navigation dredging is relatively uncontaminated, natural, undisturbed sediment, and is considered acceptable for a wide range of uses or placement alternatives. The remainder is more or less contaminated as a result of industrial, municipal and agricultural activities. The presence of contaminants may cause some environmental problems regarding placement and sometimes evokes strong reactions, which may prevent essential waterway maintenance and construction. Some of the contaminated dredged materials may also be suitable for use, but will usually require treatment to remove or stabilise the contaminants. Failure to find solutions could have serious consequences for local, regional or even national economies.

Sediments from remediation dredging, by definition, contain harmful substances at levels that pose a threat to the survival of bottom-dwelling organisms and their consumers, and impair the quality of the surrounding water. Handling and placement of these sediments require special control measures that do not simply transfer the problems to another place.

For environmental and socio-economic reasons the potential environmental effects of each dredged material relocation project must be assessed and addressed in a technically and scientifically sound manner, within the framework of relevant international/national regulations.

The major environmental concerns relate to potential direct physical impacts at the placement site and its vicinity, and to potential contaminant-associated impacts on the biota of the receiving environments, including humans.

This chapter provides guidance on the available options when faced with the question, "What should we do with removed sediment?". The discussion of alternatives includes related environmental considerations as well as potential control measures to reduce or eliminate unacceptable impacts. The reader should note that no "cookbook" solution exist to sediment management; the relevant factors and conditions vary from site to site. It is recommended that expert advice always be sought.

Dredged sediments are often called dredged material and these two terms are used interchangeably.

7.2 OVERVIEW OF MANAGEMENT ALTERNATIVES

Before considering the various management alternatives available for dredged sediments, it is important to realise that to achieve sustainable sediment management, further reduction of emissions of contaminants (source control) is necessary both from industrial point sources and from non-point sources (e.g. traffic, agriculture). Investments at the source very often are more beneficial both in economic and environmental terms than investing in costly "end-of-pipe" solutions downstream.

A great variety of management alternatives exist for dredged sediments with use and placement being the two main categories. Treatment can be used in combination with any management option to render the sediments suitable for a specific use or placement, either in terms of quality or quantity. The most common placement options are shown in Figure 7.1.

Although some inconsistency in the international literature exists in the terminology used to identify particular management alternatives, an effort has been made for uniformity with PIANC. In agreement with the PIANC working group reports in preparation (PIANC EnviCom 2008a and 2008b), the following definitions are now introduced and used here:

Use: This is any use that regards dredged sediments as a resource. In this book the term "use" is preferred to the commonly used term "beneficial use", which implies that the use of the material will be of benefit to the community. Most people do not define what exactly is meant by "beneficial", nor do they define who benefits from it (see Section 9.3.4).

Confined placement: This is placement of sediments in an enclosed area to prevent the horizontal spread of the sediment particles after placement. Confined placement is possible in water or on land. Natural or artificial depressions, pits, built berms or dams may be used to provide confinement. Placement types b–i in Figure 7.1 are semi-confined and confined placements.

Confined Disposal (Placement) Facility (CDF): This is a special confined area that provides a complete enclosure (horizontal and vertical) of sediments from the surrounding waters and soils not only after but also during placement. CDFs can be constructed in water or on land. CDFs are a distinctive subgroup within confined placement. Placement types d–i in Figure 7.1 are CDFs.

Containment measures/features: These are control measures used with (semi)confined placement options to hold the contaminants within the confinement area. All placement types shown in Figure 7.1 can be constructed with containment features. A CDF may be contained in terms of contaminant releases to water, but not contained in terms of releases to air.

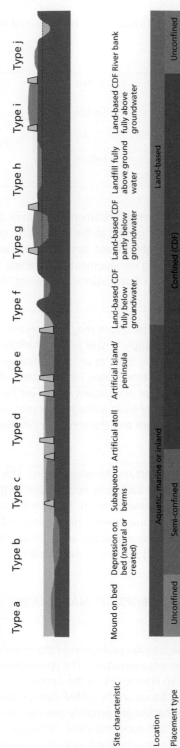

Dredged sediment placement options

	Type a	Type b	Type c	Type d	Type e	Type f	Type g	Type h	Type i	Type j
Site characteristic	Mound on bed	Depression on bed (natural or created)	Subaqueous berms	Artificial atoll	Artificial island/ peninsula	Land-based CDF fully below groundwater	Land-based CDF partly below groundwater	Landfill fully above ground water	Land-based CDF fully above groundwater	Land-based CDF River bank
Location	Aquatic, marine or inland							Land-based		
Placement type	Unconfined	Semi-confined				Confined (CDF)				Unconfined
Physico-chem environment	Water saturated, anoxic, neutral in pH						Mixed		Dry, oxic, acidic	
Suitable for … material	Mainly clean or mildly contaminated					Mainly contaminated				
Description in this book	Section 7.5					Section 7.6		Section 7.6	See 1 to 3	Section 7.4

Note: Containment measures are possible for each placement type.

Figure 7.1 Dredged sediment placement options.

The discussion of management options in this chapter has been divided into four sections as illustrated in Table 7.1.

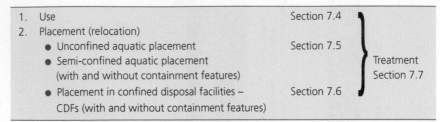

1. Use   Section 7.4
2. Placement (relocation)
   • Unconfined aquatic placement   Section 7.5
   • Semi-confined aquatic placement
     (with and without containment features)   } Treatment
                                                                  Section 7.7
   • Placement in confined disposal facilities –   Section 7.6
     CDFs (with and without containment features)

Table 7.1
Dredged material
management options

Sediments are an integral part of aquatic ecosystems. If sediments have to be removed from the waterbed at certain locations, returning them to the same water system at other designated locations is the most straightforward destination/management alternative for them. Therefore, when assessing management alternatives, the first two options to be considered should be "Unconfined aquatic placement" together with "Use" options (SedNet, 2007).

If the material is not suitable for either of these, treatment may be considered to alter the sediment characteristic as required. Increasingly, a new way of thinking is gaining ground: The traditional local and short-term focus in sediment management is being replaced with larger spatial and longer time scales. Many experts share the view that assessment of sediment management alternatives on a river basin scale and integrating sediment management into river basin management plans offer significantly more benefits both to the environment and the economy (SedNet, 2007; Vogt *et al.*, 2007).

Each management alternative will have an effect on the environment. This may be positive or negative, short term or long term. Each may occur either in the close vicinity of the placement site or at distances far away from it. Environmental considerations for each of the above three main management alternatives, and control measures available to eliminate potential adverse impacts or to reduce them to acceptable levels, are reviewed in the remaining sections of this chapter.

7.3 FRAMEWORK TO SELECT THE MOST APPROPRIATE MANAGEMENT ALTERNATIVE

This section describes a systematic approach to the evaluation of the environmental acceptability of dredged material management alternatives. The approach is consistent with the Dredged Material Assessment Framework of the London Convention (DMAF) and other international guidelines based on the DMAF (see Chapter 5 and Annex A). These seek to ensure that alternatives are properly assessed in terms of environmental acceptability and technical and socio-economic feasibility. Here the focus is on that part of the decision-making process which evaluates environmental acceptability and identifies control measures to reduce potential adverse impacts to acceptable levels. The description is based on LC-72 (1996), OSPAR (1998), PIANC (1997), DGE (2002), USEPA/USACE (2004), and SedNet (2007).

The decision-making process is applicable to all types of dredged material, from clean to highly contaminated. It reflects the state-of-the-art technical and scientific knowledge and experience. The more advanced evaluation procedures are expensive, require skill and sophisticated equipment.

A variety of techniques such as multi-criteria analysis, risk assessment, cost-benefit analysis, and decision-making analysis are available to facilitate decision making in complex cases. Some of these techniques are described in Chapters 3 and 4.

The main steps in the decision-making process are (see flowchart in Figure 7.2):

1. Establishing the need for dredging

2. Characterisation of dredged material

3. Assessment of use options

4. Preliminary screening of potential placement alternatives

5. Detailed assessment of placement alternatives

6. Selection for final design and implementation

7. Permit application and processing

8. Monitoring programme design

Step 1. Establishing the need for dredging

In some circumstances dredging might be avoided or the volume of the sediment which needs to be dredged might be reduced by careful consideration. With remediation of contaminated water beds, *in-situ* isolation of contaminants by capping is in some cases a viable alternative to dredging.

Step 2. Characterisation of dredged material

Physical, chemical, biological and engineering data on the sediment to be dredged are collected and evaluated. The amount of detail and type of analysis required depend on the potential management alternatives and will vary from case to case. In this phase contaminants of concern, if present, should be identified. Following a tiered approach (Chapter 5), beginning with physical characterisation, to avoid unnecessary testing is advisable. This step may need to be revisited during the detailed analysis of management alternatives for more detailed specific information.

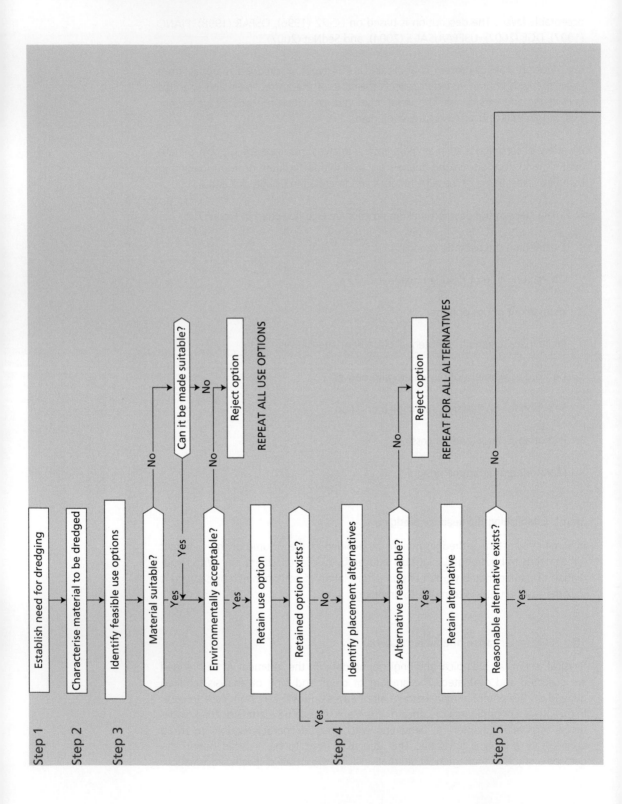

Step 1 — Establish need for dredging

Step 2 — Characterise material to be dredged

Step 3 — Identify feasible use options

Material suitable? — No → Can it be made suitable? — No → Reject option

Yes / Yes

Environmentally acceptable? — No → Reject option

Yes

Retain use option

Retained option exists? — No

Yes

REPEAT ALL USE OPTIONS

Step 4 — Identify placement alternatives

Alternative reasonable? — No → Reject option

Yes

Retain alternative

REPEAT FOR ALL ALTERNATIVES

Step 5 — Reasonable alternative exists? — No

Yes

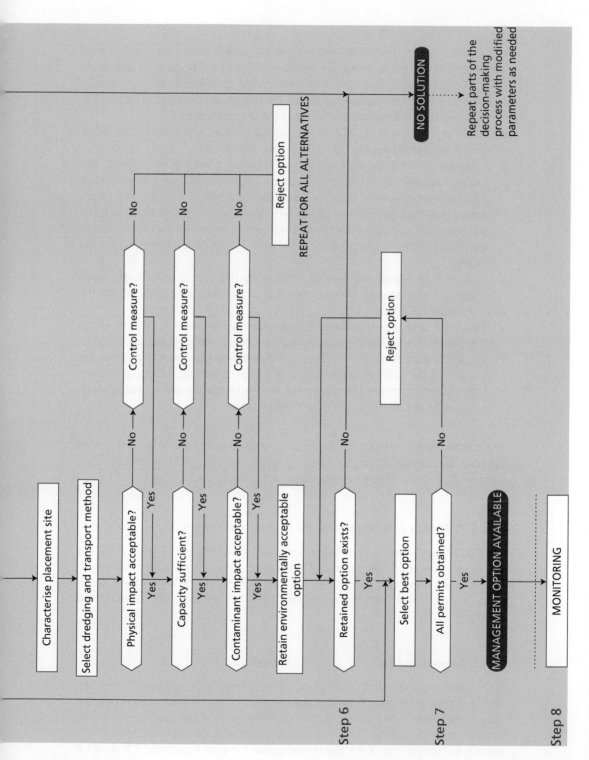

Figure 7.2 Flow chart for selecting the most appropriate management alternative.

Step 3. Assessment of use options

The DMAF of LC-72, the OSPAR Dredged Material Guidelines and the environmental legislation of many countries require that possibilities of use be properly considered before a marine disposal licence will be granted. The conventions, however, give no real guidance on what those uses might be. The options are described in Section 7.4. The assessment of use options involves the following activities:

● Assess whether the quality of dredged material complies with standards for relocation of dredged material in marine or fresh water systems.
● Evaluation of physical/engineering and chemical suitability of material: The physical/engineering suitability of material for various purposes varies widely. Chemical suitability may also vary. Generally, the physical, chemical and engineering characterisation of the material provides sufficient basis to decide whether the material is suitable or can be made suitable by treatment for the considered use. Additional parameters specific to the intended use may, however, have to be tested.
● Evaluation of operational feasibility: Feasilibty depends upon having suitable material in the required amounts available at the same time and place as the need arises for its use.
● Evaluation of environmental acceptability: Sensitive resources in the area and in the vicinity of the intended use should be carefully mapped, and all potential impacts should be assessed. With construction material production schemes, impacts of the process chains also have to be considered. Most uses involve open-water or confined placement. For those schemes, the same considerations apply as for the respective placement alternatives.

The DMAF allows cost factors to be taken into consideration. Attempts should be made to consider intangible benefits, which are not easily quantifiable (see Chapter 9), such as improved environment and aesthetic improvement.

Step 4. Preliminary screening of potential placement alternatives

Once potential placement alternatives have been identified, a preliminary screening can be done to reduce the number of alternatives subjected to detailed evaluation. The screening criteria are project specific. They may include: geohydrological conditions, environmental importance, distance from the dredging site, capacity and costs.

Step 5. Detailed assessment of placement alternatives

Reasonable placement alternatives are scrutinised. In particular, the nature, temporal and spatial scales of the anticipated impacts should be determined based on reasonably conservative assumptions. Control measures to manage specific anticipated impacts may be required. This step should result in a concise statement of the expected consequences of the placement option ("impact hypothesis"). If an EIA is required, this statement will be in the form of an EIS. Most EISs require that the rationale behind rejecting alternatives be stated. This step involves the following activities:

- Detailed characterisation of considered placement sites (see Chapter 5).
- Selection of compatible dredging and transport method for each potential site: Dredging and transport methods affect the characteristics of the material and consequently its behaviour upon placement. Decisions should be based on the results of placement site and material characterisations.
- Assessment of potential direct physical impacts and site capacity: Considerations for assessing physical impacts and site capacity are reviewed in Sections 7.5 and 7.6.
- Assessment of potential contaminant impacts: All contaminant pathways of concern, identified on the basis of sediment and site characteristics, have to be addressed. The pathways have to be analysed for the target contaminants identified in Step 2. Site control measures are reviewed in the Sections 7.5 and 7.6.

Step 6. Selection for final design and implementation

The assessment may result in a number of environmentally acceptable options. The final selection is based on weighing and balancing a broader set of case-specific factors which may include: additional environmental aspects (e.g. potential for noise and air pollution, potential damage to aesthetics and cultural resources, and impacts on the health of the operating crew); implementability and availability; site operation and management feasibility; costs; safety, and public acceptance. Law (EIA procedure) may require involvement of the public and concerned agencies.

Step 7. Permit application and processing

All required permits for implementation must be obtained. Permit requests may be published and made available for public review. An early involvement of all interested parties in the evaluation process may generously pay back at this stage (see Chapter 3). If all the necessary permits cannot be obtained, project requirements and decision criteria may have to be re-evaluated and part of the decision-making process may have to be repeated until a solution is found.

Step 8. Monitoring programme design

Monitoring of the placement operation and of its long-term impacts is an integral element in the decision-making process. It is often a condition of the licence granted. Monitoring is carried out before, during and after completion of the placement operation. The objectives of pre-placement monitoring are:

- to provide information for site selection; and
- to establish baseline conditions.

The objectives of monitoring during and after placement are:

- to evaluate the physical/engineering integrity of the placement site;
- to evaluate environmental compliance and adequacy of the impact hypothesis; and
- with some use projects (e.g. habitat development), to establish project success rate.

Monitoring is addressed in detail in Chapter 8.

7.4 POTENTIAL USES

7.4.1 General

Guidance on potential uses, based on numerous case studies, is described in the PIANC EnviCom. There are two broad categories of potential use:

- Environmental Enhancement, including habitat creation and enhancement, maintenance of sediment supply, aquaculture and recreation, and
- Engineering Uses, such as construction materials for reclamation and flood defence.

Although some countries already make extensive use of dredged material – in Japan more than 60% of dredged material is used – this is not yet common practice. In many countries a number of constraints, such as complex and inconsistent legislation and regulation, the difficulty of finding suitable schemes for using the material or markets for treated products and, not least, a negative public perception, have prevented more extensive use. Higher costs than traditional placement may also be a constraint even though sometimes higher costs may be offset by the value of the use.

Some recommendations for the success of a scheme to use dredged material are:

- Communicate with all stakeholders to promote better understanding that dredged material is a valuable resource, not a waste;
- Take into account the saving on the costs for primary resources and make a proper evaluation of the costs and benefits for society;
- Work to ensure that national policies do not classify dredged material as "waste" in their legislation intentionally or unintentionally. The real barrier to use of dredged material is often inadequate legislation, and a change of legislation or its interpretation is required before the most effective use of dredged material can be realised worldwide;
- Promote regional strategies to bring together supply and demand. Appropriate planning, and site-specific solutions are needed; and
- Make an appropriate evaluation of environmental improvements. An adequate risk assessment requires a good understanding of environmental processes to minimise uncertainty about the interaction between dredged material and its receiving environment.

Publications providing guidance on use include: PIANC (1992), HR Wallingford (1996), Great Lakes Commission (2004), Rijkswaterstaat (2004), PIANC EnviCom (2008a).

The key issue for use of dredged material is matching the available material with an appropriate use taking into account:

- physical/chemical suitability of material;
- operational feasibility;
- environmental acceptability; and
- costs and benefits.

Physical, engineering and chemical suitability of material
The engineering properties of the material are of primary importance and may limit the range of potential use options. For instance, sandy material which can be used as fill material or in dike construction may not be suitable for vegetation establishment because of its low water-retaining capacity. Chemical properties may also be important. For instance, the aforementioned sandy material may also be unsuitable for vegetation establishment because of its low nutrient content. The relevant properties are described in Chapter 5. Descriptions of tests to determine the suitability for use can also be found in Great Lakes Commission (2004). If the material is not suitable, the possibility of treatment to make it suitable can be considered.

Operational feasibility
Uses of dredged material are generally driven by the need to "do something" with the material rather than by a demand for it. A better idea would be to consider demand while developing a dredging project, for example, by planning to use dredged material instead of using primary resources for road construction. In most cases a dredging operation will produce material at a faster rate than it can be placed and steps have to be taken to either adapt the rates or provide storage.

Environmental acceptability
While use of dredged material has to conform with legislation, in many cases this is dependent on regulators allowing flexibility instead of a strict interpretation of waste legislation. In reality, formal guidelines specifically covering the environmental assessment of uses of dredged material do not exist. This means that any use may have negative effects as well as the intended benefits. For example with sediment cell maintenance and coastal protection schemes, the movement of newly placed material could potentially disturb well-established marine resources. Likewise, in schemes that use contaminated dredged material, dispersion of contaminants during transport, rehandling and product manufacturing may raise concern.

For each use scheme, the potential impacts need to be identified and assessed using appropriate tools, for instance, laboratory testing and computer modelling. As far as possible impacts should be presented as comparative solutions with standards or using an appropriate level of risk analysis.

Some use schemes have failed, not because the material is unsuitable or the scheme badly designed but rather because of a lack of proper control on the handling or placement of the material. For this reason, most schemes demand that the work, besides being well designed, should be well supervised and controlled.

Support from stakeholders is to be encouraged and can be crucial for the success of a use scheme.

Costs and benefits
In the majority of cases (sediment cell maintenance being a significant exception) extra costs will be incurred when implementing a use programme compared to mere

placement. One of the main costs will be handling and transport. These costs should be offset against the benefits of the use as well as the savings on placement capacity and primary resources, although who bears the financial burden of these bene-fits must be stipulated. The following sections list and briefly describe possible uses.

7.4.2 Sustainable relocation

A number of situations exist where regular removal of sediment by dredging may cause physical problems such as erosion and unacceptable impacts to the environ-ment: In these cases, recycling the sediment within the natural sediment transport system may provide a better solution.

Many tidal estuaries are "in regime", i.e. a net balance exists between the amount of material deposited and the amount eroded. This is a dynamic, self-regulating process. If excess erosion takes place (e.g. during a period of high-velocity flow), the fact that the riverbed becomes deeper reduces the speed of flow, which allows deposition to occur. If excess accretion occurs, the flow is forced through a smaller cross-section, speeds up and becomes capable of re-eroding the accreted material.

Such a balance can be disturbed when an estuary is dredged. Continuous removal may eventually lead to erosion of intertidal banks and salt marshes. Projects are cur-rently carried out in the United Kingdom to "trickle charge" such estuaries with dredged sediment to mitigate long-term damage. Schemes to avoid such damage in the first place comprise "in-estuary" placement.

7.4.3 Flood and coastal protection

There are several ways that dredged material can be used in flood and coastal pro-tection schemes. These include:
- the direct replacement of eroded beach material (beach nourishment);
- encouragement of the development of new, unenforced coastlines (managed retreat);
- adjustment of intertidal mud profiles;
- the formation of offshore berms designed to modify the wave climate. These berms may be hard – designed to withstand wave energy; or soft – designed to absorb wave energy;
- Placement on banks of waterways; and
- Dike, dam or dune construction.

Beach nourishment
Beach nourishment is now an accepted form of "soft" coastal defence. In the past the specification has usually required that the beach is nourished with similar mater-ial to that which has eroded. This usually means that the responsible authority (with or without assistance of a contractor) prospects the immediate area for a source of sufficient size. The real challenge is to use maintenance or capital dredged material. To do this may require more research into stability criteria. Although the historically

pertaining beach profile may not be achieved, a more stable beach with a different profile may be created.

Managed retreat

One approach to the problem of the erosion of muddy coasts backed by low-value land is "managed retreat". The objective is to create a buffer zone by setting back the defence works and breaching the existing wall, behind which dredged material has been placed, in order to create new wetlands. Experiments to evaluate the concept are in progress in several areas of the world. A pilot project where dredged material is used is Wallasea (see Figure 7.3) in the UK.

Figure 7.3 Breaching of sea defence to create new wetlands at Wallasea (UK).

Offshore berms

Three types of berm are considered:
- feeder berms;
- hard berms; and
- soft berms (see Text Box on page 204).

The latter two are offshore berms that can be used to reduce the force and vary the direction of waves striking the shore, thereby reducing shore erosion. The first is sacrificial: it supplies the sediment intended to move onshore.

Soft berms

An offshore mud berm to absorb wave energy constructed using maintenance dredged material was built in the Gulf of Mexico off Mobile Bay (USACE 1992), the aim being to reduce wave erosion of the coast. The design concept necessitated the berm being placed in waters shallow enough to absorb wave energy via penetration of wave orbits into the mud bottom, but also deep enough such that wave-induced shear stresses never exceeded the bottom shear strength of the berm. The design parameters have been studied by Mehta and Jiang (1993) using computer modelling (see Figure 7.4). The model was used to calculate wave attenuation over a non-sacrificial mud berm designed by the US Army Corps of Engineers in the Gulf of Mexico, off Dauphin Island in Alabama. Fine-grained material for the berm was derived from dredging the ship navigation channel into Mobile Bay. The berm has been effective in reducing wave energy in the sheltered area. In the two cases studied the measured reduction was 29% and 46%. Considering the nature of the wave field and the water depth over the berm crest at the site, the high degree of wave damping is believed to be mainly the result of wave energy absorption by the berm. The model gave reasonably good correlation between predicted and measured wave spectra in the sheltered area.

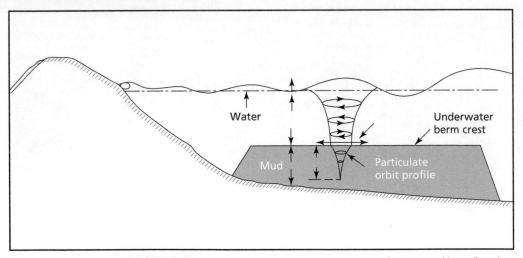

Figure 7.4 Schematic drawing showing wave propagation over an underwater mud berm (based on Mehta and Jiang 1993).

Placement on banks of waterways

Placement of dredged material on the banks of rivers and ditches and in polders is a very old practice in, for example, The Netherlands, Belgium and the USA. The dredged material is beneficial for agriculture and raises the land level. Clean and slightly contaminated dredged material can be placed on the banks sometimes after dewatering (see Section 7.7) under certain conditions. Temporary placement sites are made where the dredged material is dewatered in a natural way. The ripening process takes about one year, after which the ripened dredged material can be spread on the land.

Dike, dam or dune construction
Some of the dikes along the coasts and inland in countries like The Netherlands and the USA are made of clay from dredged material. Dunes may also be made of dredged sand.

7.4.4 Construction

In the construction industry, dredged material is used for:

- construction material and reclamation;
- the production of aggregates and mudcrete (mud stabilised with cement);
- filling geotextile containers; and
- the manufacture of synthetic building materials.

Construction material and reclamation (direct use)
Dredged material such as sand, clay and gravel can be used as an alternative for some primary resources used in the construction industry. These materials may be used as construction material for a large number of applications such as land reclamation, replacement fill, road foundations and noise barriers.

To use dredged material on land, dewatering, and sometimes desalination or some type of separation process may be needed. Sands and gravels are suitable for reclamation schemes, especially if the land is to be used for building purposes. Silts and clays, however, consolidate over a longer period and so are only suitable for foundations after a period of some time. Most dredged material from maintenance projects is muddy, but material from capital dredging schemes may be suitable. Certain schemes, such as for parkland, can be accommodated settlement without concern. If material is supplied from maintenance dredging it should be noted that the rate of supply is dictated by the siltation/maintenance-dredging rate. Techniques exist for speeding the consolidation process.

In Hamburg, Germany dredged material was used to backfill harbour basins that were no longer in use. Both dewatered and natural dredged material was used. In both cases time is needed for consolidation of the material. Future uses of the site have to take this into account.

Dewatered fine dredged material can be used to isolate contaminated material in both upland or aquatic facilities. Examples are covering of (household) disposal sites or landfills and rehabilitation of contaminated and abandoned brownfield sites. A very good example is the Fasiver site near Ghent in Belgium, which is now used again for industrial activities.

Another example of the use of dewatered fine dredged material is as a liner to seal confined placement facilities for contaminated dredged material.

Clean dredged material can be used for capping contaminated sediments to mitigate negative ecological effects. Usually this would be sandy material but dredged clay has been used as a capping material in Hong Kong.

Aggregate
The sand and gravel component of dredged material may be used in the making of concrete. However, much maintenance-dredged material contains a high proportion

of silt which makes it unsuitable for aggregate. Material arising from capital dredging and maintenance of certain sites, where the predominant material is sand or (rarely) gravel, is the most useful. Logistics and economics then become the biggest issues.

The fine cohesive fraction can be removed. Separation methods include hydrocyclones and sedimentation basins (see Section 7.7). Salt is not acceptable in aggregate to be used for reinforced or other high specification concrete. Therefore, if the source of the material is marine, it will require washing before use. If the salinity is not too high, the sandy fraction can be used as backfill material or in the production of bituminous mixtures or mortar. Once the material is in the form of clean sand and/or gravel, it may require screening to achieve the desired grading for a specific purpose.

Stabilising with cement
Adding cement (or other additives) may transform muddy dredged material into a construction material. "Mudcrete" is marine mud stabilised with ordinary Portland cement. This gives it the advantage of greater strength while minimising the leaching of contaminants. The process produces a material of very low permeability.

Mudcrete can potentially be used on large structures. With suitably selected design criteria, such large structures would be lighter than conventional quay structures. The use of cement to stabilise and immobilise dredged material is common practice in the USA (see Figure 7.5) and Western Europe.

Figure 7.5 Adding cement to contaminated dredged material from New York/New Jersey in barges to create material suitable for structural fill.

Clay
Clay can be obtained from capital dredging or from natural dewatering of dredged material in ripening fields or lagoons. Clay produced from ripening fields near the CDF De Slufter (see Figure 7.6) is used for construction alternated with sand layers to provide drainage.

Figure 7.6 New sedimentation basins near CDF De Slufter (The Netherlands).

Fine-grained dredged material can be a very efficient sealing material after dewatering. Examples are known in Germany where such a material is used for dike construction or in muncipal waste disposal facilities. Permeability coefficients can be as low as for natural clay. Dredged material, therefore can be used to substitute natural resources.

Filled geotextile containers
Synthetic fabrics have been used for the past 30 years for various types of containers such as sandbags, geotextile tubes and geotextile containers. Geotextile containers filled with fine dredged material are now being used in construction (Fowler *et al.*, 1996).

Geotextile containers filled with granular dredged material have been successfully used in constructing groynes (Harris, 1994). The material is placed in bags, tubes or large containers either *in situ* or in split-hull barges. The tubes can be filled hydraulically with dredged material straight from the delivery pipe of a cutter suction dredger or other slurry pump system. They can be placed using a cradle bucket on a barge-mounted crane or they can be installed using a continuous position-and-fill procedure. When using the hopper barge technique a geotextile sheet is placed over the whole hopper and the dredged material is loaded in the normal way (taking care not to damage the fabric). The two long sides are then drawn together along the centre line of the barge and joined using portable industrial stitching equipment. The tube is then allowed to free-fall through splitting of the split-hopper barge.

Guidance on fabric design is available in the UTF Geosynthetics Manual (Rankilor, 1994).

Existing and possible uses for geotextile containers include:
● river training works;
● estuary dike construction (e.g. for wetland schemes);
● beach nourishment schemes (e.g. for groynes, breakwaters or sills);
● offshore breakwaters;
● sand dune stabilisation; and
● wetland stabilisation.

Synthetic materials
Several methods have been investigated to develop technology to reuse dredged sediments for manufacturing artificial building materials, such as bricks and artificial gravel.

Bricks
Various attempts have been made to produce bricks from dredged material. Collins (1979) reported that small test bricks were extruded from combined samples of silt representative of three ports. The main problems were that the high moisture content caused expansion when the bricks were fired or shrinkage if they were dried slowly before firing.

In Hamburg, silt separated from unpolluted sand in the METHA processing plant (see Section 7.7, "Pre-treatment, Hydrocyclones") had replaced much of the clay in producing bricks during a 4 year test period (Detzner et al., 1997, Detzner and Knies, 2004). The moisture content problem was overcome by using a dry forming process. Before pressing, the material was dried in a closed system. The bricks were baked at about 1100°C. Cleaning of exhaust gases, treatment of water from the drying process and the closed drying system ensured that the process did not pollute the environment.

The demonstration site had a production capacity of 5 million bricks per year. The construction of full-scale plant was investigated but the costs for acquiring METHA-dredged material in an industrial brickyard remain considerably above those of the environmentally safe placement as a result of which a large-scale project has not come into operation.

Synthetic aggregate
Smits and Sas (1997) reported on a promising technology developed in Belgium for the reuse of heavily contaminated silt by producing pellets (artificial gravel) to replace

natural gravel in concrete. The pellets formed from dewatered silt are dried, heated and sintered to a temperature of 1150°C. All organic pollutants are burned out and the heavy metals are encapsulated in the ceramic pellet structure. The technology has been tested at full scale. Both the engineering and leaching properties make the material suitable for production of concrete blocks as well as mass concrete that can be used in housing, road and dike construction works. The relatively high costs and public wariness of the final product may limit the current applicability to large infrastructure projects and industrial buildings.

Detzner and Knies (2004) reported that a comparatively small amount of METHA-silt has been used to make lightweight pellets. As an ongoing procedure a total of 8,000 tonnes/year of pre-treated silt has replaced 10–20% of the natural clay in the process without the need for adaptation of the process technology. Following the mixing of natural clay, dredged material and other additives, blown-clay pellets are produced in a rotary kiln at 1250°C with a density of 0.3 to 0.6 tonnes/m^3. The pellets are sold and used as gravel substitute, geological filling material or as a supplement for lightweight concrete.

7.4.5 Agriculture, horticulture and forestry

Dredged material, mainly from rivers and inland waterways, has been used in each of these industries. The main problems are possible contaminants, including salt (in the case of marine sediments), and the need to dewater.

Some placement sites, especially in river systems, have provided livestock pastures, allowing natural grass colonisation or by planting. Other uses include amending marginal soils for agriculture, horticulture or forestry. Marginal soils feature poor drainage, unsuitable grain size and poor physical and chemical conditions. They may also be of low productivity because of high water tables or frequent flooding. Dredged material can be used to enhance their value.

Soil factory
A full-scale soil factory was set up on a former quay on the River Clyde in Scotland which was capable of producing 2000 tonnes/week of topsoil. The material was supplied free by the Port Authority who had the benefit of reduced costs through not having to transport the material to the licensed placement site. It was used to regenerate redundant dock land for the Glasgow Garden Festival.

Agriculture
When dredged material is free of nuisance weeds and has the proper balance of nutrients, it is no different from productive agricultural soils. Dredged material can alter the physical and chemical characteristics so that water and nutrients become more available for crop growth. In some cases, raising the elevation of the soil surface may improve surface drainage, reduce flooding and thereby lengthen the growing season. With saline dredged material, a conflict of interest arises in that drying is essential to successful handling, whereas washing with fresh water is required to reduce salinity.

In the USA and in Belgium, bio-remediated dredged material is mixed with compost and treated municipal sewage sludge to produce "manufactured soil" for reuse in sanitation and landscaping projects. In Germany treated, dewatered dredged material is used for agricultural purposes in orchards. It has to fulfil special requirements to improve the soil and to meet specific quality standards.

Horticulture

Horticulture crops are vegetable, fruit, nut and ornamental varieties of commercially grown plants. Dredged material applications on soils for vegetable production, orchards and nurseries are similar to those for agriculture. All commercially grown vegetable crops can be produced on dredged material amended soils. The best types are sandy silts or silty dredged material that can be incorporated into an existing sandy site. Clays are too heavy for good vegetable production but could be improved by the addition of sand.

Urban and suburban areas require large quantities of readily available grass for such uses as residential lawns, parks, golf courses and so on. Marginal soils near urban areas may be brought into turf production through applications of dredged material. Since grass is less exacting in its growth requirements than most food crops, the type of dredged material used is not as critical. The material should be a loamy or silty sand substrate to ensure the best growth.

Forestry

The improvement of marginal timberland by the application of dredged material shows promise. Several rapid growing pulpwood species can be grown in dredged material. The same physical and chemical material properties discussed for agriculture would apply to forestry, except that the trees could be grown safely on dredged material with higher contaminant levels. No documentation has been found of tolerance levels for heavy metals that may limit growth.

7.4.6 Amenity

Amenity in this context means the improvement or provision of facilities that are designed to be enjoyed by people. It includes regeneration of derelict land, landfill, landscaping and creation of recreation areas. A particularly appropriate use is for landscaping, the deliberate creation of contoured sites. See the case study on Amager beach in Annex B.

To protect groundwater against infiltrating contaminants, special attention should be given to the bottom layers. Care should be taken to avoid the presence of vegetable matter in this layer because it can decay and leave voids and drainage paths in the substrate. A layer of sand on top of this layer will then act as a drain for any leachate from the dredged material above it.

Already there are a number of examples in the USA and also in Europe, for instance along the River Elbe in Germany and near Doel in Belgium, that give actual proof to local communities and policy makers that placement of fine-grained sediments can be turned into an ecological and socially acceptable project.

7.4.7 Habitat

This section covers various ways in which dredged material has been used to create or maintain wildlife habitats. It includes five types:
- aquatic habitats for fish and benthic organisms;
- bird habitats (upland habitats and nesting islands);
- wetlands;
- saltmarshes; and
- intertidal mudflats.

It should be noted that the creation of any new habitat means replacing an existing one. The development of this technology has been most extensive in the USA, although a few projects are underway in Europe. A full environmental benefit study is always necessary before making a decision and any scheme will inevitably mean some losses as well as gains.

Aquatic habitats
Aquatic habitat development is the establishment of biological communities on dredged material at or below mean tide level in coastal areas and in permanent water in lakes and rivers. Fisheries can be improved by appropriate placement of dredged material. For example, bottom relief created by mounds of dredged material, especially when coarse grained, may provide refuge habitat for fish.

Seagrass habitat
Seagrass can be used to stabilise dredged material, either sands or silts, through the binding of roots and rhizomes and by dissipating wave and current energy. Seagrass also enhances the habitat value.

Gravel bar habitat
Gravel and cobble-sized materials provide points of attachment and anchorage for aquatic organisms such as larvae, snails and worms. Coarse-grained particulates stabilise fine substrate and allow colonisation by long-lived invertebrates such as freshwater mussels. Particle size distribution, degree of embeddedness and presence of attached organic matter and plants determine the characteristics of invertebrate communities in flowing water systems. When gravel shoals are dredged to improve river navigation the material can be placed inside channels to create potentially valuable habitat for a number of riverine fishes and invertebrates, including ecologically and commercially valuable species.

Oyster beds
Dredged material can be used to develop oyster habitats, particularly in intertidal areas. These areas offer to oysters a competitive advantage by reducing predation pressures and enhancing growth. The most useful dredged material is sandy with some shell. The idea is to replace soft beds with hard intertidal sand beds. This can be achieved by gradually raising the level over a number of dredging cycles. The site can either be left to be colonised naturally or treated with oyster shell cultch to encourage growth.

Dredged material containment areas

Commercial fish farming in ponds has been in existence for many years. Dredged material containment areas provide some new sites for such development. Diked placement areas share many features with aquaculture ponds: level sites, good foundation soils, water-holding capability and a water control structure.

Artificial reefs

Artificial reefs include quarrystone rock, prefabricated concrete units, obsolete oil rigs and steel ship hulls. They are generally thought of as hard structures that provide three-dimensional relief. Although dredged material does occasionally consist of rock rubble, it more typically consists of sands, silts and clays. It can potentially increase habitat diversity by forming reefs that differ from surrounding substrate with respect to sediment type.

Bird habitats

This refers to islands and upland habitats for birds. Some of the most useful experience comes from the US Army Corps of Engineers' "Environmental Effects of Dredging" programme (Landin, 1986). One hundred years of dredging and open-water placement operations have resulted in the creation of over 2,000 artificial islands throughout USA coastal waters, inland waterways and the Great Lakes. A recent example in Europe is the bird island Iliot Reposoir created in the estuary at Le Havre as part of the Port 2000 project. The complex island was designed for the hosting of marine birds and built using material dredged to improve the main navigation channel. At High Water it is three islands totalling 1.5 ha. At Low Water the islands join to form a single area of 5 ha (Scherrer, 2006) (see Figure 7.7).

Figure 7.7 Iliot Reposoir is a bird sanctuary created in the estuary at Le Havre (France) as part of the Port 2000 expansion project.

The PIANC guide (1992) lists seven technical criteria:

- gradually sloped shorelines;
- suitable substrate for nests and young chicks;
- access to the water/shoreline;
- not less than 1–3 m above highest water level to prevent nest washout. At least 0.3 km from the mainland to prevent egg- and chick-eating predators from swimming to the island;
- suitable vegetation (or lack of vegetation) that meet a species' nesting requirements;
- close proximity to feeding grounds and brood cover so that chicks do not have to travel long distances to obtain adequate food items for nest-bound chicks;
- isolation, or at least restrictions, to prevent high human use during the nesting season; and
- for bare substrate, no plantings are necessary; rather the removal of excess plants is recommended.

Wetlands

The third type of habitat is the creation or restoration of wetlands. Wetland habitats are very different and various different ecosystems develop accordingly. Dredged material has been used extensively to restore and establish wetlands over 16,000 hectares in the USA (see Figure 7.8). It can be used to stabilise eroding natural wetland shorelines or nourish subsiding wetlands. Dewatered dredged material can be used to construct erosion barriers and other structures.

An important feature of using dredged material is that hydric soil conditions (i.e. containing hydrogen) are necessary and the literature suggests that dredged material may take 15 years or more to achieve this state. Thus it will probably be necessary to import hydric soil and wetland vegetation as well as creating the right hydrologic conditions.

Saltmarshes

Where saltmarshes are in decline, such as on the east coast of the UK, the loss may be detrimental to navigation, sea defence, aesthetics, conservation and recreation. Any economically viable scheme that slows or reverses the loss may be deemed to be beneficial (Figure 7.8).

Intertidal mudflats

Intertidal mudflats are an essential source of the invertebrates on which many species of wading bird, such as Dunlin and Redshank, feed during migration. They support soft-shell clams (Mya arenaria) and baitworms (sandworm, *Neris virens* and bloodworm, *Glycera dibranchiata*). They also provide feeding grounds for commercially important fish species such as winter flounder (Ray *et al.*, 1994).

The development of many estuaries, including the construction of tide excluding barrages, has reduced the extent of such mudflats. The deliberate creation of new mudflats is one means of compensation being tried (see Figure 7.9). Because it is new technology there is little available in the way of design guidance.

Figure 7.8 Dredged material being used to improve saltmarsh habitats at Westwick Marina, on the Crouch Estuary in the UK. The barriers placed in the marsh to retain the dredged material are visible in the background.

Figure 7.9 Mudflats made from dredged material in the River Orwell, UK.

7.5 UNCONFINED AND SEMI-CONFINED AQUATIC PLACEMENT

Unconfined aquatic placement is re-introduction of dredged sediments in the water system from which they originate, oceans, estuaries, rivers and lakes such that the returned sediments become again part of the natural sediment cycle. This involves placement of clean or mildly contaminated sediments on flat or gently slop-ing waterbeds in the form of mounds (see Figure 7.10). The placement sites can be dispersive or non-dispersive (retentive) depending on whether the sediment is trans-ported out of the site by currents and/or wave action or remains within the desig-nated boundaries.

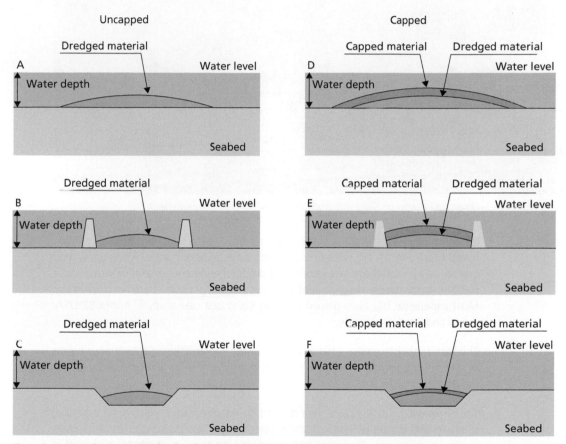

Figure 7.10 Unconfined, semi-confined and confined aquatic placement (adapted from Bray et al., 1997). A: Unconfined (unrestricted) aquatic placement; B and C: Semi-confined aquatic placement (with lateral containment); D to F: Confined (contained) aquatic placement.

Semi-confined placement is when some sort of lateral confinement is used to limit or prevent the spread of deposited sediment on the waterbed (see Figure 7.10). Placement in natural or artificial depressions such as abandoned borrow pits (see Figure 7.11) or purpose-constructed pits or behind constructed berms or dikes may provide effective means. In many cases filling of borrow pits with (contaminated)

Figure 7.11 Borrow pit filling with dredged material Kaliwaal, The Netherlands.

dredged material offers opportunities for environmental reinstatement or improvement. Many sub-sea pits are quite deep and have an anoxic zone, which limits their ecological value. In these cases storage of dredged sediments creates more attractive conditions for nature development and can have a positive effect on biodiversity. Most experience has been gained in Hong Kong (see case study in Annex B), USA, Japan and The Netherlands.

The increase in placement costs as a result of the construction of subaqueous berms or dikes may be reduced by using material from the same or another nearby dredging project.

The sediment is usually placed via direct pipeline discharge, released from hopper dredgers and barges Direct mechanical placement is also possible to areas adjacent to the dredging site. Placement and associated dredging methods are discussed in detail in Chapter 6.

Potential environmental effects of open-water relocation and disposal depend on the characteristics of the site, the properties of the placed sediments, and the behaviour of the sediment during and after placement. Site and sediment characteristics and related investigations are described in Chapter 5. The remaining parts of this section focus on the behaviour of sediments during and after placement and the resulting effects on the receiving environment. Available control measures and monitoring requirements are also discussed.

7.5.1 Behaviour of sediments and associated contaminants during and after aquatic placement

Physical behaviour

Short-term behaviour

This is defined as the behaviour during and in the first few hours after the discharge. Processes of interest are: convective descent, collapse and mound formation and passive dispersion.

Convective descent, collapse and mound formation

Pulled by gravity, the released material falls as a concentrated cloud or discharged jet (convective descent) and finally hits the bottom with the formation of a horizontal surge (dynamic collapse). As the discharge continues a mound develops. The physical characteristics of the discharged sediment, which in turn depend on the type of discharge, greatly influence these processes:

- The material from direct pipeline discharge is a liquid slurry which may contain clay balls, gravel or coarse sand material. The coarse material quickly settles, while the mixture of process water and fine particles descends to the bed to form a fluid mud mound. Some fine material may remain in suspension as a turbidity plume.
- Hopper dredger discharges are a mixture of water and solids. At the placement site, the hopper doors in the bottom of the ship's hull open and the entire hopper content is released within minutes. The material falls through the water column as a well-defined jet of high-density fluid. Upon hitting the floor most of the material comes to rest. A smaller portion may expand radially away from the impact area. Some hopper dredgers have pump-out facilities, which have an effect on the material similar to that of pipeline discharges.
- Mechanically dredged cohesive material discharged from barges is nearly at in-situ density, has its in-situ structure and often remains in clumps. The released material descends rapidly to the sea floor and only a small amount remains suspended. The mounds tend to be tighter as compared to pipeline or hopper discharges.
- With direct mechanical placement, the cohesive material falls directly to the bottom as consolidated lumps.

The behaviour of material as a function of discharge method is illustrated in Figure 7.12.

Prediction of mound geometry – footprint, height, slope – is critical to ensure that the footprint does not extend beyond specified site boundaries, and that the height does not exceed the limit (navigation and/or erosion constraints). It is also needed to identify holding capacity and, if the mound is to be capped, to design the cap and to calculate the volume of capping material.

To date, predictions of mound geometry have been based on experiences with similar material dredged and placed in a similar way. Most mounds created from unconfined placement of mechanically dredged contaminated material (LBC projects) in the USA have had largely consistent geometry: round or elliptical shape; defined, relatively flat

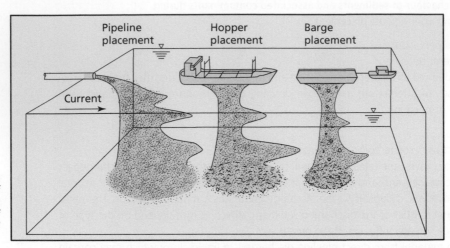

Figure 7.12 Short-term physical behaviour of discharged sediments as a function of discharge method (EPA/USACE, 2004).

crest; a main mound side slope (inner flank); sometimes an outer flank with a more gentle slope; a thin outer apron. The apron of fine-grained material is usually 1 to 15 cm thick and can extend up to several hundred metres beyond the main mound flanks. Computer models are available to predict mound characteristics and to facilitate the design of optimal placement scenarios (Lillycrop and Clausner, 1998).

Passive dispersion
Fine sediments that remain in the water column are dispersed laterally as a suspended plume by ambient currents, wave and tide action. The plumes can exist for several hours following the discharge. Fine sediments have low settling velocities and therefore they may travel well beyond the disposal site boundaries. However, experience shows that, generally speaking, only a small part, 5 to 20%, of the solids remaining in suspension is exposed to current velocities capable of transporting the material out of the placement site. Mechanically dredged cohesive material from barges results in less water column dispersion than hydraulically dredged material from hoppers or pipelines. The degree of physical dispersion (and possible release of any contaminants) is important in deciding whether the placement is acceptable in terms of water column impacts. Computer models are available to predict the development of suspended solids concentrations in time and space following the discharge.

Long-term behaviour
This is defined as the behaviour over months or years. The processes of interest are mound consolidation, resuspension and erosion of deposits, and transport and re-deposition of eroded mound material.

Mound consolidation
Mound consolidation is caused by self-weight (and with capped mounds, by the cap load). As the mound consolidates, pore water is expelled, mound elevation decreases and additional volume becomes available for placement (so long as the mound has not been capped). For fine-grained sediments, a 50% reduction of initial deposit thickness from self-weight consolidation alone is not uncommon (Rollings and

Rollings, 1998). Prediction of consolidation rate and magnitude is required to determine site capacity. This is especially important when sites are intended to be used for large amounts of material from several dredging projects over years.

For contaminated material, consolidation predictions provide data to evaluate the potential pore water movement and contaminant flux out of the deposited sediment into the overlying water or cap layer. For capped mounds, consolidation data enable one to decide whether cap replenishment is needed. When height changes are caused by consolidation rather than by erosion, no cap replenishment is necessary (unless other circumstances make it necessary). Computer models have been developed and provide reasonably accurate predictions provided there is good geotechnical information available.

Resuspension and erosion
Depending on the potential for re-suspension and erosion, two types of sites are distinguished: non-dispersive (retentive) and dispersive sites. Re-suspension and erosion are influenced by bottom current velocity, potential for wave-induced currents, sediment grain size and cohesion. Wave-induced excess pore pressures can destabilise mounds resulting in large submarine slides sometimes extending more than a kilometre. In Hong Kong such processes have been observed during the passage of typhoons waves which can destabilise a 1-in-20 mound to move to a 1-in-200 slope over a much larger area (Whiteside, 1998). Hydraulically dredged sediments have an increased erosion potential owing to the high water content.

Predictions can be made by conventional sediment transport methods. These range from simple techniques to indicate potential for site erosion, to sophisticated numerical modelling capable of producing more specific predictions. These latter models indicate whether a site is predominantly dispersive or non-dispersive and also calculate erosion thickness and mound migration for given current and wave conditions, mound geometry and sediment characteristics.

"Bio-erosion" can also take place if colonising crabs, lobsters, fish or other organisms burrow into the mound surface. Gas production (e.g. as a result of the decomposition of organic material) may also affect the stability of the mound.

As fine particles are washed away from the surface, the mound may become "armoured" because the coarse residue of sand, shells, and gravel are more resistant to erosion. Once equilibrium has been reached, the compacted, smoothed and armoured deposit may be subjected to further erosion only during severe storms.

The likelihood of erosion increases with decreasing water depth. Setting a maximum mound height may, in some cases, be a way to limit erosion.

Transport and re-deposition of eroded mound material
Depending on the hydrodynamic conditions and the type and size of the eroded particles, these may be transported far away from the placement site. Eventually the particles may come to rest in a low-energy, stable environment or become integrated

into the natural sediment transport processes of a region (e.g. coastal drift). The location and thickness of re-deposition can be predicted by the same hydrodynamic and sediment transport models that can be used for assessing the erosion potential.

Physico-chemical and bio-chemical behaviour

At low-energy open-water sites very little change occurs in the physico-chemical nature of the material during dredging and discharge. The relocated sediments commonly remain anoxic with a near neutral pH. Thus, most of the contaminants remain chemically immobile. In the long-term bioturbation may introduce oxygen from overlying oxygen-rich waters into the deeper anoxic zones of deposited material. At ocean pH the oxidising conditions generally result in the formation of low solubility metal salts, i.e. the heavy metals remain bound. For more details see Section 5.6.8.

7.5.2 Potential environmental effects of unconfined or semi-confined aquatic placement

These can be grouped in two main categories physical effects and contaminant-related effects. Both types can occur in the water column and on the benthos. Potential effects are summarised in Figure 7.13.

There will almost always be physical effects associated with open-water placement. These may arise from covering the waterbed at the placement site itself and/or at adjacent areas, from the increase of suspended solids concentration and turbidity in the water column.

A temporary local increase of suspended solid concentrations in the water column during and after discharge may affect water column organisms but generally it does not cause more harm than storms, high flows or human activities such as fishing and swimming. There are cases, however, when high-suspended solids concentrations, even if temporary, may impede mobility and interfere with respiration or feeding. The increase in turbidity may affect aesthetics and recreational values. It may reduce photosynthesis, which may be a problem if it causes reduction in food supply for plankton feeding organisms. However, in waters with eutrophication problems this may be seen as a positive effect. High turbidity may adversely affect visual predators as a result of reduced visibility.

Covering the bed may smother benthic organisms that may not be able to migrate upwards. Owing to differences in the physical and/or chemical properties of the deposited sediment organisms which re-colonise the site may be different from those present before. This can have an effect on the fish population frequenting the area. These changes may be seen in a different light by various stakeholders, depending on their particular point of view. Dispersion of sediment particles into spawning grounds, breeding areas, oyster reefs, coral beds and areas with sea grasses may result in harmful effects.

Computer models are used to predict mound geometry, dispersion and transport of sediment particles during and after placement and to assess possible physical

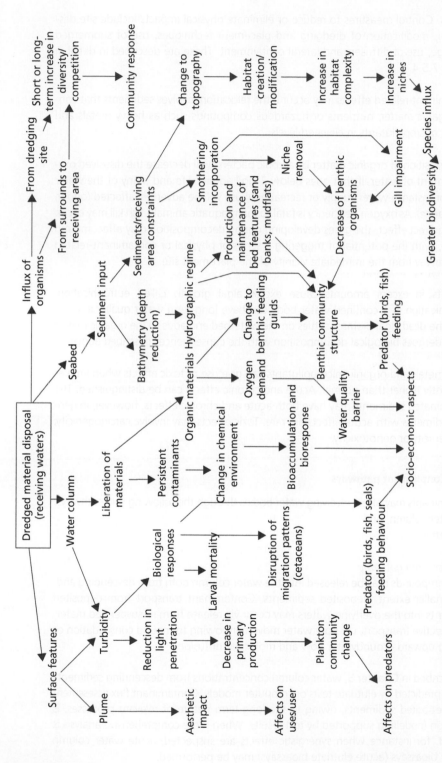

Figure 7.13 Potential impacts of unconfined and semi-confined aquatic placement of dredged sediments (PIANC EnviCom, 2006a).

impacts. Control measures to reduce or eliminate physical impacts include site designation, modification of dredging and placement techniques, use of submerged discharge, use of diffusers and lateral containment. These are described in detail in Section 7.5.4.

Contaminant-related effects may occur if the relocation involves sediments that contain organic matter, nutrients or hazardous compounds such as heavy metals and organic micropollutants in elevated levels.

Decomposition of organic matter by aerobic bacteria will decrease the dissolved oxygen content of water. If that goes below 2 mg/l aquatic life and many of the water uses, for instance water supply or recreational use will be adversely affected (odour, appearance). As oxygen deficiency is fatal to many aquatic animals, fish kill may occur. As an indirect effect, the gases developing during decomposition may affect mound stability, with the potential of triggering a series of physical or contaminant-related effects away from the immediate vicinity of the placement site.

Nutrients in excess amounts cause excess algal growth called eutrophication. Eutrophication in a confined water body can have long-term water quality implications. The dead algal biomass settles on the waterbed and adds to the organic matter that undergoes biological decomposition with the consequences described above.

Heavy metals and organic micropollutants can give rise to toxic effects when intakes are slightly higher than normal. Acute and chronic effects can be distinguished. TBT contaminated sediments may have both acute and chronic effects, however, in general, sediments with acute effects are rare. Toxic effects may involve carcinogenicity, mutagenicity or genotoxicity.

7.5.3 Contaminant pathways

Contaminants may affect receiving water bodies through the following two pathways:
- water column, and
- benthic.

Water column pathway
Toxic compounds may be released into the water column both from descending and, to a smaller extent, deposited sediments. Contaminant transport from deposited sediments into the overlying waters may occur via release from re-suspended material, advective transport, i.e. pore water movement owing to mound consolidation or existing upward groundwater flow and molecular diffusion.

As described in Chapter 5, water column concentrations from descending sediments can be predicted by elutriate tests or computer models. Contaminant flux assessment from deposited sediments, owing to the long-term nature of relevant processes, is based on modelling supported by test results. When more comprehensive analysis is needed, for instance when synergistic effects are suspected, acute water column toxicity bioassays (acute elutriate bioassays) may be performed.

Benthic pathway
The greatest potential for environmental impact from dredged material is in the benthic environment. This is because deposited sediments are not mixed and dispersed as rapidly or as greatly as the particles remaining in the water column, and bottom-dwelling organisms live and feed in and on deposited material for extended periods. Therefore, normally the evaluation efforts focus on this pathway. Benthic toxicity and benthic bioaccumulation are assessed by benthic bioassays and bioaccumulation tests (see Chapter 5).

7.5.4 Control measures

Direct physical impacts may be reduced and site capacity may be increased by a variety of control measures such as the use of enviromental windows, modification of dredging and placement operations including the use of submerged discharge, and thin layer placement. The same measures plus capping can be considered for contaminant control. A brief description of these measures follows. It is important to note that not all measures work in all cases and under all conditions.

Modification of dredging and placement operations
These measures may be effective for controlling physical effects (covering the water bed, increased suspendng solids concentrations and turbidity) and for water column and benthic contaminant pathways. The following modifications may be considered:
- using mechanical instead of hydraulic placement methods, in particular for fine-grained material;
- reducing discharge rate;
- reducing vessel velocity;
- changing vessel approach direction; and
- for hopper dredgers, coming to a stop during discharge.

Environmental windows
If the season where the environmental risk is high can be determined (e.g. when fish spawn), it can also be avoided. Restricting placement operations to particular seasons may reduce both physical and contaminant impacts (US NRC, 2001). On a shorter time scale dispersion may be reduced by avoiding placement during spring tides or high river flows. The direction of dispersion may be controlled by selectively discharging only during flood or ebb tides. In considering these measures it should be born in mind that the use of favourable environmental windows may increase costs as dredging contractors have restricted time, thus reducing overall utilisation and requiring higher production rates. Dredging operations may be prolonged to such an extent that the solution can be worse than the issue (Holliday, 1998).

Submerged discharge
Submerged discharge methods reduce both physical and contaminant-related water column impacts. By isolating the material from the water column for part of its descent they reduce dispersion. This is particularly useful if gas or air is present in the sediment. This technique may also reduce mound spread owing to a reduction in the entrainment of site water during descent. In capping material placement operations,

submerged discharge increases control and accuracy, thus decreasing the amount of material needed. Methods include submerged pipelines for conventional pipeline dredgers or pump-down pipes for hopper dredgers. Submerged diffusers provide an additional control by reducing the material exit velocity. Diffusers can be used with any hydraulic dredging operation such as hydraulic pipeline dredgers, pump-out from hoppers or reslurried pump-out from barges. Chapter 6 gives further details on submerged discharge.

Silt curtains

Another possibility to reduce water column dispersion with associated physical and contaminant related effects is provided by silt curtains. These trap the sediment particles dispersed upon discharge and descent and keep them within the placement area. They are effective mainly for medium- to coarse-size silt and sand and can only be used in calm (low-energy) waters. Ooms (1997) points out that if not carefully designed and applied, the curtain will disturb the seafloor and cause sediment resuspension (see also Jin et al., 2003). The net effect may even be detrimental to the placement operation (Shaw et al., 1998).

Thin layer placement

Placement of material in thin layers (150–600 mm) reduces the smothering effect of placement by allowing benthic organisms to burrow up. It also increases the rate of recolonisation of the placement site, especially if the placed material is similar to the bed material. The acceptable thickness is very site specific (Bray et al., 1997). In projects surrounded by shallow waters, thin-layer placement can reduce the environmental impacts to an acceptable level.

Capping

Capping is mainly used for contaminated sediments to control the benthic contaminant pathway but it may also be considered for clean sediments to reduce mound erosion. Capping provides control through the following mechanisms (Thibodeaux et al., 1994):
- physical isolation of contaminated sediments from the benthic environment and overlying water;
- chemical isolation of contaminated sediments from the benthic environment and overlying water by increasing the transport path length between the contaminated layer and the water column, and by providing additional sorptive capacity for contaminants; and
- stabilisation of (contaminated) sediments, preventing resuspension and off-site transport.

Capping as a control measure is not acceptable if:
- water column dispersion of contaminants during placement (with controls) exceeds the predefined levels;
- cap placement cannot be provided with the required precision and thoroughness; and
- cap stability cannot be maintained in the long term.

The following sections briefly discuss the main design considerations.

Capping materials
Clean sediments or soils are normally used. Clean sediments from other nearby dredging projects are often available. Since preferably low-energy sites are selected, erosion is rarely a problem and thus caps are usually constructed from one type of material. In those cases when cap erosion is of concern, armouring with coarse sand, gravel or stone may give protection.

Cap thickness
The required thickness depends on:
- the physical and chemical characteristics of the contaminated and capping material;
- the bioturbation potential;
- the consolidation potential of the contaminated material and associated pore water expulsion;
- the consolidation and erosion potential of the capping material; and
- the operational factors (e.g. ice gauging, anchoring, unevenness of material placement).

Current design practice presumes that cap thickness components are additive, i.e. no dual function of cap layer components is taken into account. This means that, depending on the intended cap functions and the above-listed factors, a cap component may have to be included for:
- bioturbation;
- chemical isolation;
- operational factors;
- cap consolidation; and
- cap erosion.

Generally, a cap thickness of 50 cm to 1 m is required. A thicker cap means more material but is easier to place. A balance in terms of the environment and economics needs to be maintained.

Long-term stability of capped mounds
Long-term stability of capped mounds depends on the consolidation and erosion processes as with uncapped mounds (see Section 7.5.1). Differential settlement of capping and placed contaminated material may result in movement, deformation or even disruption of the cap.

Compatibility of equipment and placement techniques
The evaluation of the ability of the capped deposit with a given shear strength (which may be initially lower than the in-situ shear strength) to support the cap with given material characteristics and placement techniques needs comprehensive geotechnical analysis.

Slope stability and/or bearing capacity need to be defined. So far no such investigations have been undertaken and capping project designs have been largely based on field experience with past projects.

Rollings and Rollings (1998) urges that this needs to be changed. Current practice is limited to the evaluation of compatibility of equipment and placement technique for contaminated and capping sediments with sediment characteristics. Table 7.2 may be used as an initial guideline to avoid displacement of the previously placed material or excessive mixing of capping and contaminated material. Accurate vessel location is essential.

		CAPPING MATERIAL								
		Barge			Hopper			Pipeline		
CDM	Hopper Spreading	Sandy	Clumps	Silt/ Clay	Sandy	Clumps	Silt/ Clay	Sandy	Clay Balls	Slurry
Pipeline										
Slurry	I	I	I	I	I	I	I	C	I	C
Clay Balls	C	C	C	I	C	C	C	C	C	C
Sandy	C	C	C	C	C	C	C	C	C	C
Hopper										
Slurry	I	I	I	I	I	I	I	C	I	C
Clay Balls	C	C	I	C	C	C	C	C	C	C
Sandy	C	C	C	C	C	C	C	C	C	C
Barge										
Silt/Clay	C	I	I	C	I	I	C	C	I	C
Clumps	C	C	C	C	C	C	C	C	C	C
Sandy	C	C	C	C	C	C	C	C	C	C

I: Generally incompatible   C: Generally compatible

Table 7.2 Compatibility of capping and contaminated material placement options (Palermo 1994)

Case studies and technical guidance on design and implementation of capping projects are given in Palermo et al. (1998). Detailed case studies are also presented in Whiteside et al. (1996) and Shaw et al. (1998).

7.6 PLACEMENT IN CONFINED PLACEMENT (DISPOSAL) FACILITIES – CDFs

This means the placement of dredged sediments within the confinement of a special structure – mostly a diked enclosure – that isolates the sediments both during and after placement from the surrounding waters or soils. The basic objective of this type of placement is to retain dredged material solids and allow the discharge of process water from the confined area. For sites receiving contaminated material, an additional objective is to reduce or prevent the transport of contaminants out of the site. CDFs are designed and constructed specifically for dredged sediments, taking into account their special properties and are not solid waste landfills (although there are similarities).

CDFs may be constructed in water as island CDFs, nearshore using the coast as one of the sides of the containment facility and on land. With island and near shore facilities, the dikes must be constructed above the mean high-water elevation to prevent direct interchange with the adjacent waters. On land unused borrow pits may provide good opportunities for CDF creation in particular in areas characterised with no or upward groundwater flow. CDFs in which the dredged material eventually dries out through the whole depth are often referred to as upland CDFs. Figure 7.14A and 7.14B shows the basic CDF options.

CDFs are the most commonly considered placement alternatives for contaminated sediments that cannot be used or are unsuitable for unconfined or semi-confined aquatic placement. Broadly speaking this applies to some 5–10% of sediments originating from navigation dredging and obviously to the major portion of material from clean-up dredging. In this latter case placement in CDFs may serve either as final destinations or as temporary re-handling sites. Some CDFs are used to confine clean sediments, where other options are not feasible either for economic or environmental reasons (e.g. physical impacts of open-water relocation is unacceptable).

Dredged material is usually placed hydraulically, via direct pipeline, hopper dredger pump-out or re-slurrying the material from barges (which may have been filled mechanically). Direct mechanical placement from barges or trucks is possible but rare. Placement and associated dredging methods are discussed in detail in Chapter 6. Most CDFs receive material periodically over an extended period of time.

Finding areas for confined placement is becoming increasingly difficult, especially in coastal areas, where the need is greatest. CDFs are visible to the public and compete with other land uses such as housing, recreation, nature reserves and so on. This affects project budgets and the intensity of public and political feelings. Use of the sites, once placement operations have been completed, becomes an indispensable element of many placement projects to facilitate public acceptance. Numerous examples, for instance, of habitat development and landscaping, can be found in the literature both for water-based and land-based CDFs. Storage of dredged material in abandoned pits can lead to win-win situations regarding recreation, flood management and mineral extraction.

Many CDFs are used periodically: The complete filling of a site may take years. Between placement operations, birds may be attracted to the site or other natural habitats may develop. Continued placement may cause conflicts with conservation interests, especially where CDFs are near coastal ecological sites (e.g. RAMSAR areas – areas specified by the Convention on Wetlands of International Importance). This should be considered during CDF design.

To date, CDFs have been designed and managed to achieve efficient containment. Treatment of contaminants placed inside has not been an objective. Current research is looking into the potential of natural processes to degrade contaminants and thus extend CDF functions. Case studies for island and upland confined placement areas are provided later in this section.

UPLAND CDF	
Description A upland CDF is a facility in which the dredged material is stored above the groundwater level. A dike is constructed on dry land to confine the dredged material. There are two main considerations for upland CDFs: 1) the hydraulic head of water in the facility acts as a permanent driving force with the potential to cause the water to flow down to the groundwater, and 2) the contaminated dredged material may dry and become oxidized, changing the potential for release of contaminants. It may be necessary to use watertight liners to prevent emissions into the groundwater.	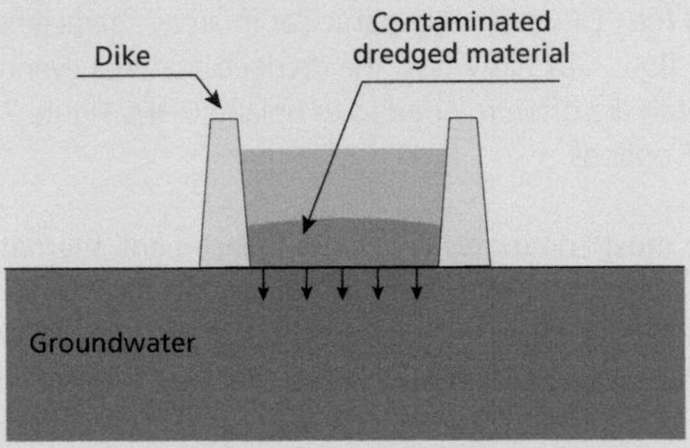 Contaminated dredged material Dike Groundwater
Advantages	**Disadvantages**
Highly visible, therefore its presence will be apparent to the local community in the long term and unintentional disturbance is unlikely.	**Isolation measures** may be needed to reduce advective transport of contaminant during the life time of the CDF.
Monitoring is relative simple, and it is easy to determine and access the locations to monitor.	De-watering of the CDM will result in **oxic conditions**, increasing the possibility of the mobilisation of heavy metals to the surface and ground waters.
	An upland CDF is relatively **expensive to fill**. Island and near-shore CDFs can be filled hydraulically, which is economical. In most case it is difficult to transport the material with barges.
	Dikes may be large and the public may raise objections based on aesthetics.

Figure 7.14A Confined (CDF) placement options (based on PIANC 2002) in upland CDF.

The potential effects of CDFs on the receiving environment depend on the characteristics of the site, the design and operation of the CDF, the properties of the placed sediments and the behaviour of the sediment during and after placement. Detailed information on CDF design and operation can be found in PIANC (2002) and USACE (1987). Site and sediment investigations are discussed in detail in Chapter 5.

The remaining parts of this section focus on the behaviour of sediments during and after placement and the resulting effects on the receiving environment. Available control measures and monitoring requirements are also discussed. The other mentioned aspects are discussed in the remaining parts of this section.

INLAND/NEARSHORE CDF	
Description An island/nearshore CDF is a diked placement facility constructed in water in which the dredged material is at least partially stored under the water level. Compared with that of the upland type, the hydraulic head of the contaminated water is much smaller. Sometimes a pit is excavated to increase the storage capacity. The main pathway for contaminants to the surrounding water requiring control will be the effluent. The quantity of the effluent is about the same as the amount of dredged material placed in the CDF.	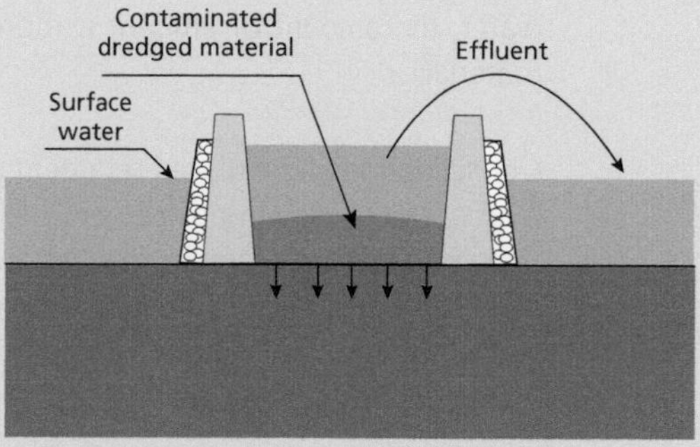 Contaminated dredged material — Effluent Surface water
Advantages	**Disadvantages**
If the CDM remains saturated and under **anoxic conditions**, the heavy metals will remain immobile.	**Ship traffic** may be effected negatively.
Highly visible, therefore its presence will be apparent to the local community in the long term and unintentional disturbance is unlikely.	**Highly visible dikes**. Often lower dikes are requested from surrounding communities.
There is **less dispersion to the surface water** because of the surrounding dikes. There is only a discharge via the effluent. The amount of effluent is approximately the amount of stored CDM.	
The waterhead, which is the driving force of advective transport of contaminants, can be controlled.	
Monitoring is relatively simple; the locations to monitor are easy to determine and also easy to access.	

Figure 7.14B CDF options in inland/nearshore.

7.6.1. Behaviour of sediments and associated contaminants during and after placement in CDFs

Physical behaviour

When dredged material is placed in a CDF hydraulically, the coarser fraction rapidly falls out near the inlet point and forms a mound. By careful management of the inlet locations – moving the inlet points – a continuous line of mounds can be constructed. The fine-grained fraction continues to flow through the containment area and settles out slowly resulting in a layer of deposited material overlaid by a clarified supernatant. The clarified water, called effluent, is discharged over a weir or through an outlet structure.

Material placed in the CDF mechanically is at or near its *in situ* water content. The settling behaviour of the material is less important. If the CDF is constructed in water, the amount of effluent is more or less equal to the water displaced by the material.

Consolidation of the placed sediment begins right after placement and continues for long periods following placement, decreasing the volume occupied. The consolidation process is governed by the characteristics of the placed material such as density, compressibility and, in particular, permeability. The consolidation rate will be affected by the permeability of any applied liner, drainage conditions, geohydrological conditions and bio-gas production. Long-term storage capacity depends on the degree of consolidation.

Unless deliberately prevented, eventually the surface begins to dry. Depending on adjacent water levels, the deposited material may dry out through the whole vertical profile. The surface may crack and it may vegetate immediately. Salts may accumulate on the surface, in particular on the edges of the cracks.

Physico-chemical and bio-chemical behaviour
Potential contaminant activity and receptors potentially at risk outside the CDF depend on the physico-chemical and bio-chemical processes in the CDF. These processes are governed by the availability of oxygen which in turn depends on the water saturation conditions within the facility. Dredged sediments, when placed, are by rule water saturated, anoxic, reduced, conditions under which most contaminants remain immobile, bound to the sediment particles. As soon as the dredged material starts drying and becomes exposed to the atmosphere, oxidation begins. The slurry water, initially dark, becomes light grey or light yellowish brown. The material itself turns light grey.

In the presence of oxygen, aerobic bateria degrade organic matter. Sulphide compounds will be oxidised to sulphates. If the sediment is rich in sulphur, organic matter, and/or pyrite – constituents that generate acidic conditions – and poor in carbonates – constituents that tend to neutralise acidity – oxidation may result in the formation of highly acidic conditions. Oxidation and decreasing pH, especially in marine and estuarine sediments, and consequent chemical transformations can significantly enhance the release of particulate associated metals into the water. Aerobic microbial degradation of organic matter may enhance the desorption of organic micropollutants from sediment particles.

Dry (unsaturated, oxic) conditions through the whole vertical profile of a CDF will exist only in on-land facilities constructed above groundwater level, after the dredged material has dried out. In other cases three distinct physico-chemical environments may exist simultaneously within the same facility:
- dry, unsaturated, oxic, in layers above the tidal elevation or groundwater-table;
- intermediate, partially or intermittently saturated, in layers influenced by fluctuating water level, and

- aquatic, i.e. totally saturated, anoxic/anaerobic, in layers below the tidal elevation or groundwater-table.

In aquatic environments most of the contaminants remain chemically immobile.

In addition to its effect on contaminant mobility, biodegradation of organic matter, both naturally present and from organic contaminants, has another significant effect with environmental consequences, namely, gas production. The gases produced are mainly methane and carbon dioxide. In the top layer of the Slufter disposal facility in Rotterdam Harbour, The Netherlands, up to 20% of the volume is occupied by gas at atmospheric pressure (Deibel et al., 1996).

7.6.2 Potential environmental effects of CDFs

These can be grouped in two main categories:
- direct physical impacts during and after the construction, filling and use (if exists), and
- contaminant-related impacts.

Direct physical effects may include removal or burial of existing habitat, alteration of topography, changes in hydrological conditions (e.g. circulation patterns in surface water, ground water flow). Further impacts may include disturbances to navigation and other legitimate water uses, interference with land use, degradation of aesthetic or cultural values of the receiving area, noise pollution during construction and filling.

Contaminants can leave the containment area through various routes and can pollute surrounding soil, groundwater, surface water and air. Exposure to these substances may cause adverse effects to ecological and human receptors. Non-toxic contaminants such as oxygen-consuming substances or nutrients may deteriorate the quality of surface and ground waters. The effect of toxic contaminants include acute, lethal responses or chronic, sub-lethal responses in flora, fauna and humans impairing survival or reproduction. Human health effects include carcinogenic, nervous system and other non-carcinogenic responses.

Contaminants can leave the confines of a CDF through the following five possible mechanisms:
- effluent discharge to surface waters during filling and subsequent settling and dewatering;
- rainfall surface runoff into surface waters following filling;
- leaching into adjacent soils, groundwaters and surface waters;
- gaseous and volatile emissions to the atmosphere; and
- direct uptake by plants and animals living on the dredged material and subsequent movement in the food web.

Pathways involving movement of large masses of water such as effluent and surface runoff have the greatest potential for transporting significant quantities of contaminants out of the CDF.

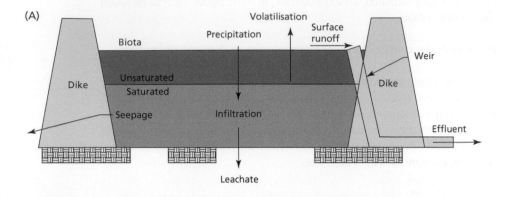

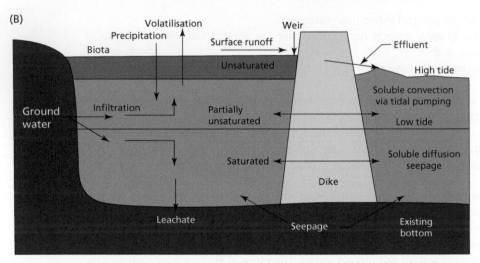

Figure 7.15 Contaminant migration pathways for (A) upland CDFs and (B) nearshore CDFs (Brannon et al., 1990).

Potential contaminant pathways for land-based CDFs constructed above groundwater and pathways for nearshore CDFs are shown in Figures 7.15.

The relative importance of contaminant pathways for a nearshore CDF differs from an upland CDF owing to the fact that in a nearshore CDF the placed sediment may remain saturated, reduced and anoxic with minimal contaminant mobility. The fluctuation of water level however, through pumping action, may result in soluble convection through the exterior dike in the partially saturated zone and soluble diffusion from the saturated zone through the dike. Pathways for island CDFs are similar to nearshore sites. The portion of a nearshore or island site above the mean high water level functions as an upland CDF.

7.6.3 Contaminant pathways

Effluent
Effluent is the supernatant water discharged from the CDF during filling and initial dewatering. It consists of carrier water, water displaced by the placed dredged material

(with nearshore or island sites), expelled pore water during consolidation and precipitation on the site during filling. There will always be effluent, regardless of the method of filling, especially if the CDF is built in water.

Supernatant waters are discharged from the CDF after a retention period allowing for settling of sediment particles. The retention time may vary from a few hours to several days, depending on the design of the ponded area and the placement of the discharge weir. Effluents are typically discharged into nearby surface waters. In view of the relatively large volumes discharged during filling operations, the short-term impacts on the receiving waters are of major concern. Negative effects may arise from the high suspended matter content and associated contaminants. Contaminants may also be present in colloidal or dissolved form. The suspended solids concentration in the effluent depends on the settling characteristics in the ponding area, resuspension near the sediment inlet point, because of upward water flow through the slurry mass during settling and wind- and wave-induced erosion on the surface.

The evaluation of the effluent may be required in terms of water quality and water column toxicity. Procedures based on the equilibrium partitioning principles are available for screening (Schroeder et al., 2006). An elutriate test is available to predict chemical releases in effluent for comparison to water quality criteria, and water column bioassays are available to determine potential toxicity of effluent (Chapter 5; USACE, 2003).

Rainfall surface runoff
Surface runoff is water discharged from the CDF following precipitation. It becomes of potential concern when the dredged material becomes exposed to precipitation after the ponded water is decanted. Unlike the effluent pathway, the runoff pathway may continue to be of concern through the life of the CDF, as long as the dredged material surface is exposed to precipitation. Surface runoff is usually discharged into surface waters. It may also be released onto the surface of surrounding soils. The quality of surface runoff from a fully or partially saturated CDF is similar to that of the effluent: Owing to rainfall-induced erosion it may contain high levels of suspended solids concentrations. As anoxic, reduced conditions prevail, most contaminants are adsorbed to the particles. Nutrient levels, as a result of the anaerobic conditions, may be high.

Runoff from dry, oxidised dredged material may differ significantly from the quality of the effluent: It may contain lower levels of suspended solids as rain erodes less material. As a result of the oxic conditions and especially if the dry layer is acidic, the surface runoff may contain elevated levels of heavy metals. Organic contaminants become tightly adsorbed onto soil and organic particulates and remain associated with suspended solids in surface runoff water. Rainwater may dissolve and remove salt that accumulates on the surface as it dries. Factors influencing runoff quality, in addition to the saturation state of the material, include the method of filling (as it influences the sensitivity of the dredged material to erosion) ponding depth, runoff rates and vegetation coverage.

Concern focuses on short-term impacts in the receiving water in terms of water quality and water column toxicity. As for the effluent, screening procedures based on equilibrium partitioning have been developed for both oxic and reduced conditions (the latter is the same as used for effluent). Rainfall/runoff simulation procedures are available to predict surface runoff quality. If necessary, water column toxicity test can be conducted (Price, 2002).

Leachate

Leachate is the water with associated dissolved and colloidal materials that seeps through dredged material in a CDF and subsequently through dikes or foundation. (Ponded water that may seep through dikes is considered effluent.) Leachate may be produced from three potential sources of water: Original pore water, rainfall infiltration and in the case of water-based sites, groundwater or surface water seepage into the site.

Concerns focus on leachate that may reach the groundwater. Leachate generally does not contain solid particles. It can however contain high levels of dissolved contaminants: Leachate from dry, acidic dredged material may be rich in heavy metals. Anaerobic leachate, on the other hand, may be rich in organic micropollutants. Estuarine dredged sediments may exude salt.

Leachate generation and transport within and out of the CDF and the migration of contaminants through the foundation soils and aquifer to potential receptors, for instance a well, are complex processes (Figure 7.16). These processes are influenced by a broad range of parameters, specific to the placement site, placed sediments, aquifer and so on.

Evaluation of the leachate pathway is complex and includes prediction of leachate contaminant concentrations and if necessary, leachate flow from the CDF to the receptor.

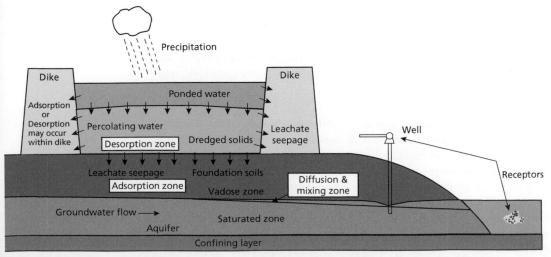

Figure 7.16 Potential leachate pathways (after Schroeder and Aziz, 2004).

Screening procedures (Schroeder, 2000) and laboratory tests (USACE, 2003) are available to predict the water quality of leachate at the boundary of the CDF. A variety of groundwater attenuation and/or mixing or dispersion models from 1-D to 3-D are available to predict leachate flow and contaminant concentration at the receptor. Contaminant mass flow (leachate flow combined with leachate concentration) is important for evaluating leachate management actions.

Gaseous and volatile emissions
Volatilisation is the movement of dissolved volatile contaminants from the water phase into the air. Volatilisation may occur from the dredged material exposed to the air, from plant-covered dredged material, from ponded water and from effluent. The rate of contaminant transfer is the highest in the first few hours after the surface of the dredged material becomes exposed. After initial drying the rate will be lower than that with ponded surfaces. As ponded conditions may remain for a longer period, it may result in a higher overall contaminant mass flux. Therefore the ponded condition is likely to be the most critical for most CDFs. This pathway is only of concern if the sediment contains elevated levels of volatile organic compounds (e.g. benzene, toluene, and such). Concerns focus on exposure of humans on or adjacent to the site.

Contaminant flux predictions can be made by using theoretical models or laboratory tests. The results of these can be converted into exposure concentrations taking into account site specific atmosperic conditions such as wind speed, mixing, temperature and so on.

Plant and animal uptake
Colonisation of CDFs by plants and animals begins during the early stages of filling and becomes established when filling stops and the site slowly takes on the characteristics of a natural area. Aquatic, wetland or terrresrial habitats may occur in different parts of a CDF at the same time or in succession through its life time. Plant uptake is bioaccumulation of contaminants in tissues of plants growing on the dredged material. Animal uptake is bioaccumulation of contaminants in tissues of animals that are exposed to dredged material either directly or indirectly via the food chain. In general, plant and animal bioaccumulation are of concern only if via contaminant transfer through the food chain they affect organisms outside the site. In terms of total mass loss, this pathway is probably insignificant compared with pathways involving movement of large amounts of water. Biological uptake, however, can mobilise contaminants in ways that water cannot.

Plant and animal uptake/bioassay tests may be conducted. Observed parameters are growth, phytotoxicity or animal toxicity (as appropriate) and bioaccumulation. Only pathways which give reason for concern need be included in the assessment procedure and only for target contaminants.

7.6.4 Control measures

If the predicted direct physical impacts or contaminant releases through the various pathways are not acceptable control measures can be used.

Table 7.3 gives an overview of applicable control measures for the various contaminant pathways.

Direct physical impacts may be reduced by decreasing the volume of the material before or after placement and hence reducing the need for larger or additional sites. Measures to achieve this include:

able 7.3 Overview of control measures for CDFs

Controlled pathway		Contaminant control measures
Effluent		
To reduce effluent water volume	OCM	Using mechanical dredging and discharge methods rather then hydraulic techniques; Increasing the dry solid content of dredged material by optimising dredging techniques, and Reuse of effluent as carrier water.
To reduce suspended solids and associated contaminants	OCM	Improvement of settling conditions in the placement facility via: • compartmentalisation; • increasing the retention time of the effluent before discharge (increased ponding water depth, weir length, or weir number), and • creation of stimulated sedimentation zones. Control of dredged material discharge into the facility via the use of: • diffusors (a reduction in suspended solids by a factor of 3–6 can be achieved); • decreased discharge rate, and • multiple discharge points.
	TM	Control of wind-induced erosion via: • increased dike height, and/or ponding water depth, and • buffer zones. Effluent treatment: plain sedimentation, clarification, filtration
To reduce dissolved contaminants	TM	Effluent treatment: coagulation/flocculation, nitrification (potentially followed by denitrification), adsorption, ion exchange.
Surface runoff		
To prevent erosion of deposited sediments	OCM ECM	Maintaining ponded conditions Planting vegetation Surface cover
To reduce suspended solids and associated contaminants	OCM	Improvement of settling conditions in the placement facility via increasing the retention time of rainwater in the facility before discharge
	TM	Runoff collection and treatment (plain sedimentation, clarification, filtration)
To reduce soluble metal losses	OCM	Maintaining anaerobic condition in to the layer of the sediment deposit Liming the surface to prevent acidification
	TM	Creation of a flocculation zone beyond the discharge weir Runoff collection and treatment (adsorption, Ion exchange)

Table 7.3 (Continued)

Controlled pathway		Contaminant control measures
Leachate		
To reduce leachate production	OCM	Dewatering before placement Enhanced dewatering after placement (surface and subsurface drainage); Planting vegetation to increase drying, and Enhancing consolidation thereby reducing permeability of the material (e.g. placing the material in thick layers). Control ponded water to: • reduce hydrostatic head
	ECM	• to maintain negative hydraulic gradient causing seepage into the facility Surface cover
To reduce contaminant mobilisation/dispersion	OCM	Selective placement configuration (reducing contact surface area with subsoils; enhancing consolidation); Maintaining anoxic conditions; Maintaining basic pH in oxic conditions, and Selective placement of dredged material. • sandwiching • self-sealing Planting vegetation to stabilise contaminants
	TM	Treatment of dredged material solids (immobilisation) Leachate collection and treatment (adsorption, ion exchange, chemical oxidation, etc.) Natural bioremediation
	ECM	Liners Surface cover Cutoff walls Reactive barriers
Direct plant and animal uptake		
To prevent mobilisation of contaminants into food web	OCM	Maintaining ponded conditions Liming or chemical treatment
To minimise contaminant uptake	OCM ECM	Planting selective vegetation Surface cover
Atmospheric contaminants		
To reduce gas emissions or dust	ECM	Surface covers
To reduce dust	ECM	Physical, chemical or vegetative stabilisation of surface soils

OCM: Operational Containment Measure; ECM: Engineered Containment Measure; TM: Treatment.

- modifications in dredging and placement operation such as reduction in over-dredging depth, increase of pipeline concentration or use of mechanical dredging and placement instead of hydraulic techniques;
- treatment of sediments before placement such as dewatering or separation of sand and silt (Section 7.7);
- improved dewatering of the sediment deposit through surface trenching, compartmentalisation, and vertical or horizontal underdrainage (gravity or vacuum-assisted), placement of sandy layers; and
- removal of material from the surface of the deposit after allowing for consolidation and, perhaps, natural treatment, and subsequent use.

Direct physical impacts can be compensated by the creation of alternative habitats, use of completed sites and landscaping. Such measures can improve the overall environmental appraisal of placement sites.

Contaminant releases can be reduced or eliminated by a variety of measures that can be grouped into two main categories:
- treatment of dredged material solids and water flows, and
- containment measures (operational and engineered).

Treatment of dredged material solids and water flows

Treatment of dredged material solids before placement or in the CDF itself may be considered if it provides a cost-efficient alternative to the treatment of other contaminant release pathways. The available methods are described in Section 7.7.

Water flow discharges, effluent, surface runoff, leachate and drainage water may contain a large variety of contaminants, the large majority being associated with the sediment particles (with the exception of leachate). Common waste water treatment technologies can be used to reduce the contaminant concentrations of these streams to acceptable levels. Many of these processes concentrate contaminants into another phase which may require further special treatment or disposal.

As suspended solids removal offers the greatest benefit in terms of quality improvement, suspended solid removal is the most common objective for effluent and surface runoff. Processes used are plain sedimentation, chemical clarification and filtration. Ammonia can be reduced by nitrification, possibly followed by denitrification. Reduction of dissolved heavy metals can be achieved by chemical precipitation and ion exchange. Organic micropollutants can be removed by carbon adsorption and ozonisation. Air and steam stripping could be used for volatile contaminants (see also case histories in Annex B for examples of storage and/or treatment of dredged material).

The more sophisticated treatment processes are very expensive and rarely applied.

Operational containment measures
Selective placement of different types of dredged material layers

The alternating layers of clean and contaminated material (also called sandwiching) provide for attenuation or containment of contaminants (sorption, ion exchange, filtration, biodegradation, and so on) hence improve leachate quality. Placement of fine grained clayey material with low permeability at the bottom of the site or on the inside face of the dikes will act as a seal, reducing seepage into the ground soils or through the dikes. Placement of suitable dredged material as the final layer in a CDF may act as a cover or may enhance vegetation of the site should that be an objective.

Selective placement configuration

This is placement of contaminated sediments within the CDF where contaminants remain relatively immobile. This is possible with respect to water levels for nearshore or in-water CDFs to maintain water-saturated, anoxic conditions which reduce the mobility of many contaminants (especially heavy metals). Selective placement below groundwater level is possible with on-land CDFs with the same effect. Existing Dutch guidelines for large-scale placement sites for contaminated sediments recommend solutions which are based on this principle. This measure mainly affects the leachate pathway. Reducing the "footprint" of the site by using greater depths and smaller surface area, thereby reducing surface runoff, plant and animal uptake and volatilisation, is also possible.

Control of ponded water

This can be used to reduce hydrostatic head or to maintain a negative hydraulic gradient causing seepage into the CDF as opposed to seepage out of the CDF.

Reuse of effluent as carrier water (van der Pluijm et al., 1995; Netzband and Rohbrecht-Buck, 1992)

This measure decreases the volume of effluent and consequently the load of suspended matter and associated contaminants. For dissolved contaminants the concentration in the effluent depends on a combination of physical, chemical and biological processes. For instance, owing to the increased retention time of water in the system, soluble nitrogen (nitrate and ammonium) may accumulate in the ponded water resulting in elevated concentrations in the effluent when it is eventually discharged.

Creation of stimulated sedimentation zone (van der Pluijm et al., 1995)

The effluent discharge point is isolated from the rest of the CDF surface by a steel or concrete structure to form a "still" zone with reduced wind effects and improved conditions for sedimentation. Controlled flocculent dosage can further increase sedimentation efficiency. Decaying flocculents are now available which do not have long-term effects on the environment.

Creation of buffer zone (van der Pluijm et al., 1995)

During windy periods causing elevated levels of suspended solids concentrations in the effluent, the effluent discharge is temporarily stopped. When conditions return to normal, the retained effluent is discharged at an increased rate. A detailed description of the above measures with case studies from all over the world is given by Palermo and Averett (2000).

Some of the operational measures may affect several pathways simultaneously in a conflicting way and may also conflict with other management objectives. For instance, maintaining ponded conditions may have a positive effect on the effluent, surface runoff, volatilisation and plant and animal uptake pathways and at the same time, it may increase the movement of leachate through the site. Also, it may conflict with site capacity requirements.

Operational approaches together with prudent site selection should be first considered for CDFs where containment is needed to meet applicable environmental standards. If these measures do not provide the required containment level, engineered containment measures may be considered.

Engineered containment measures
Engineered containment measures are used for facilities that receive highly contaminated material or are situated in environmentally sensitive areas.

Liners
These can be applied in the bottom or on the inside face of the dikes to reduce or prevent leaching of contaminants out of the site. Natural and artificial liners may be used.

Natural liners include clay, peat, and the conditioned silt fraction of dredged material. In addition to physical isolation these have a good adsorption capacity for many contaminants. Because of its high organic carbon (humus) content, peat is an especially effective adsorbent. Natural liners are generally attractive, both financially and technically. Their life span (hundreds to thousands years) is another significant advantage over synthetic liners. According to our current knowledge synthetic liners do not provide absolute impermeability longer than 30 to 50 years. With regard to clay, note that organic and inorganic acids and bases may dissolve a portion of the structure.

Artificial liners include cement concrete, bituminous concrete, bentonite, bituminous membranes and synthetic membranes. A wide variety of synthetic membranes are in use. They vary not only in physical and chemical properties but also in installation procedures, costs and chemical compatibility with waste fluids. HDPE (High Density PolyEthylene) and PVC (PolyVinilChloride) are the most commonly used (see Figure 7.17). Synthetic membranes are usually used for dry deposits only as they tend to swell when exposed to fluids with consequences such as increased permeability, and loss of tensile and mechanical strength. A comprehensive survey of synthetic membranes is given in Male and Cullinane (1989).

The choice of liners depends on the type of subsoil, groundwater level, required permeability, and mechanical and chemical resistance.

Surface covers
These are placed on top of the consolidated material (see Figure 7.18). Surface covers are effective in controlling leachate generation, surface runoff, plant and animal uptake. The same materials may be used as for lining. With artificial covers, a gas withdrawal system should be installed to avoid serious damage to the cover.

Lateral containment
These are used to control contaminant release through lateral seepage through the dikes or through the foundation soil layers beneath the CDF. Possible options include

Figure 7.17 Treated and recycled dredged material is placed over a contaminated site which is covered by a synthetic membrane.

Figure 7.18 Covering of placement site with dewatered dredged material (Gelderland, The Netherlands).

a layer or core material in the dikes to reduce permeability, cutoff walls such as slurry walls or sheet piles placed either within the dike or outside the dike perimeter in foundation layers.

Reactive barriers
These can be placed within the dikes to provide treatment for seepage by filtration, sorption or biological treatment.

7.7 TREATMENT

In this section treatment technologies for contaminated material are reviewed. The processes include those for reducing the amount of material for placement and those for reduction, removal or immobilisation of contaminants.

7.7.1 General

There is no single "cure all" technology and each sediment requires proper analysis, assessment and prescription for treatment. This may involve more than one treatment process to deal with a range of contaminants in varying proportions. These may include: heavy metals, oil and oil products, organochlorine compounds, PAHs and PCBs.

Treatment processes may be classed as follows:
- pre-treatment;
- physico-chemical;
- biological;
- thermal;
- electrokinetic; and
- immobilisation.

It is important to recognise that treatment processes have their own environmental impacts. These may include discharges of waste waters and gas emissions. Some of the techniques produce highly dangerous concentrated residues which need to be handled and/or stored with extreme care. Some processes have high energy needs, others use large areas of valuable land. Many treatment processes use water that ultimately needs to be purified; to limit this quantity water reuse should be considered.

No treatment process can be regarded as inexpensive and some are very expensive. Proper consideration of all of these aspects is necessary.

7.7.2 Pre-treatment methods

The aim is to reduce the volume of the contaminated dredged material that requires further treatment or special placement. In the main, these techniques are based on the preferential adhesion of contaminants to the fine, cohesive fraction of sediments. Thus, if the fines can be separated, the remaining coarse fraction may be used or safely placed as described in previous sections. Pre-treatment is of little or no value where the proportion of coarse sediment is small (e.g. <30%). The separation is a hydro or mechanical process and the technology is well established. Before separation large objects may often need to be removed by sieving.

Several methods are used to reduce the volume of contaminated dredged material:
- separation basins;
- hydrocyclones;
- flotation; and
- dewatering.

Separation basins
Coarse material is separated from fine material by flushing with water jets (e.g. Zwakhals *et al.*, 1995) or by longitudinal-current classification (e.g. Detzner and Knies, 2004).

Hydrocyclones
One of the most established forms of processing is the use of the hydrocyclone. Hydrocyclones have wide use in the sand, gravel and mineral processing industries. Their primary use is for separating different density or weight materials within a slurry mixture.

Operation of the hydrocyclone is based on the principal of centrifugal force. The hydrocyclone has no moving parts and requires relatively low energy to perform its primary function. The technique has been used in some European ports to increase solids concentrations in dredged slurries.

A slurry mixture is introduced to the feed chamber under pressure. The tangential entry causes the slurry to rotate at a high angular velocity, forcing coarser or heavier particles to the sidewalls where they continue downward with increasing velocity to the bottom of the cone section. This material then exits through the apex as a denser, higher percentage solids material called the underflow.

The cyclonic flow creates a centrally located low-pressure vortex, causing the lighter, finer-grained sediments and water to flow upwards and exit the top through the vortex finder. This finer-grained, reduced-percentage solids slurry is called the overflow. As an example a hydrocyclone separating at about 63 microns would have a diameter of about 0.6 m and would process about 100 litres/sec at up to 50% solids content.

Flotation
Particles are separated from the bulk liquid using density differences. The method has been used for the separation of oil and metal sulphides. A process capacity of 500 m³/hr has been reported (van Germet *et al.*, 1994).

Dewatering
Dewatering improves the texture of the material and make it smaller in volume and easier to handle, thereby reducing costs (see Figure 7.19). For fine-grained sediments, silts and clays, this can be a major task. Filter presses and thickening agents may

Figure 7.19 A separation and dewatering plant. Residue from the cleaning process is stored at De Slufter while the clean sand is sold for construction purposes.

be used. In most cases dewatering fields offer a relatively inexpensive solution (e.g. Antwerp and Rotterdam). The water itself is likely to be contaminated and will require special handling and/or treatment. In the case of dewatering fields (see Figure 7.20), care must be taken not to allow seepage into groundwater.

Figure 7.20 Dewatering lagoons in Krankeloon, on the River Scheldt, Belgium.

Magnetic separation
Large magnets are used to remove magnetic material in contaminated soil remediation and may have application in some dredged materials.

7.7.3 Treatment methods

These are:
- Physico-chemical which include extraction techniques – such as acid, complexation, solvent and supercritical fluid extraction; immobilisation techniques; wet air oxidation; base catalysed decomposition; and ion exchange;
- Biological;
- Thermal – such as thermal desorption; incineration and thermal immobilisation; and
- Electrokinetic.

Physico-chemical
These treatment processes involve the use of chemical processes to remove, change or stabilise the contaminants.

Extraction techniques
These techniques are usually based on the use of solvents. They do not destroy contaminants but concentrate them into a reduced volume and transfer them into an aqueous phase.

Four types of extraction techniques are described briefly:
- Acid extraction is particularly applicable to heavy metals but is not always effective enough with cadmium. Undesirable compounds may form if organic matter is present. Aeration would reduce organics but increases costs. Hydrochloric acid, biochemically produced sulphuric acid and citric acid have all been investigated for use in The Netherlands.
- Complexation extraction involves using organic chemicals that form complexes with metals, thereby extracting them from the sediment matrix and putting them into solution. The technique shows promise at laboratory scale.
- Solvent extraction is used in the USA. Known as BEST (Basic Extractive Sludge Treatment), it involves stripping and removing organic contaminants to form a solid or liquid matrix using selected solvents.
- Supercritical fluid extraction has not been demonstrated at full scale for dredged material. It is, however, proven technology for certain industrial processes.

Immobilisation techniques
The material is treated so that the contaminants become fixed to the solid matter and become immobile. The fact that the contaminants are not destroyed implies a possible long-term risk of release.

Muddy dredged material containing contaminants may be stabilised and solidified by adding an appropriate proportion of cement or lime-kiln dust (about 5–10% by weight). This has been demonstrated to be effective in dealing with:
- halogenated semi-volatiles;
- non-halogenated semi-volatiles and non-volatiles;

- volatile metals;
- non-volatile metals;
- asbestos;
- radioactive materials (low level);
- inorganic corrosives; and
- inorganic cyanides.

It is expected (but not demonstrated) to effectively isolate:
- PCBs;
- pesticides;
- dioxins/furans;
- organic cyanides; and
- organic corrosives.

Wet air oxidation

This has been shown to be effective at destroying PCBs and more effective for PAHs. It is established technology for municipal waste-water treatment and sewage sludges. Being a wet process it is not hampered by the large quantities of water in dredged material but it does require elevated temperatures and pressures which will make it costly at bulk scale.

Base-Catalysed Decomposition (BCD)

This has only been tested at bench scale. It has been tested on New York/New Jersey sediments and is reported to be effective for destroying PCBs, chlorinated hydrocarbons, chlorinated pesticides and dioxins.

Ion exchange

Technologies using selective cation exchange may be applicable to dredged material and would be used mainly for metal reduction. To be effective the applications need to be selective for particular metals. The need to adjust pH and the need for long contact times (up to one week) make continuous processing difficult.

Biological

Biological degradation aims to enhance the natural breakdown of organic contaminants into harmless compounds by micro-organisms. It can be undertaken *ex-situ* by land farming or using a bio-reactor. The effectiveness depends on temperature, humidity and nutrients.

Toxic metals and other non-biodegradable substances will not be affected, as they are recalcitrant. If present in high concentrations they cause problems by destroying the microbial population. The application for sediments requires oxygen. It is a low-speed process, requiring controlled environmental conditions for several weeks to a year for highly contaminated sediments.

Many firms offer proprietary biological mixtures that are claimed to clean up contaminated sediments, but there is no substitute for pilot-scale trials.

Thermal

Thermal treatment can be used to remove, destroy or immobilise certain contaminants. It usually entails dewatering and drying of the sediment first. For large-scale operations a cheap source of abundant heat will be required. Any process involving heating the material is liable to give rise to emissions of chemicals. These must be considered and properly controlled.

Thermal desorption

Heat is applied to volatilise and remove organic contaminants. The process condenses the volatilised contaminants and collects them as an oily residue. The contaminants are not destroyed by this method, only extracted from the sediment. This is proven technology for hazardous wastes, sludges and soils. Some pilot studies have been carried out to reduce PAHs but the results were not conclusive.

Incineration

Rotary kiln incineration is tested technology. Operational conditions are set depending on the target contaminant. Incineration at high temperature (1200°C) and pressure effectively removes PCBs and dioxins. The technology is available but at very high cost. Incineration does not destroy heavy metals. They are concentrated in the ash, which may require controlled disposal. In addition, the gases given off may require recombustion and washing to avoid air pollution.

Thermal immobilisation

At about 1400°C the sediment smelts. A variety of usable products can be obtained by controlling the rate of cooling. Flue gas treatment will be required as with incineration. Obstacles to large-scale use of this method include high cost and consumer resistance to contaminated material (however safe it is declared to be).

Electrokinetic

A constant DC current applied to a saturated soil containing contaminants will lead to metal ions and other cations collecting at the cathode while anions head towards the positive anode.

There are four mechanisms of ion transport:
- electro-osmotic advection;
- advection under hydraulic potential differences;
- diffusion owing to concentration gradients; and
- ion migration as a consequence of electrical gradient.

The method has the attractive potential to be used *in situ* to remove heavy metal contaminants, thus avoiding the need for special handling and containment.

Several firms offer this technique for land remediation, but it is relatively untried for river or estuarine sediments. The main difference is the amount of moisture present in the sediment. Pilot-scale tests are in progress in the UK in an attempt to remove very high concentrations of mercury from canal sediments. The choice of material for the electrodes is important as the process removes a number of other elements, in addition to the target metal, depending on their relative position in the periodic table.

Treatment of TBT contaminated dredged material
Tributyltin (TBT) is given special attention because of particular recent concerns about the effects of anti-fouling paint. In fact many of the above pre-treatment and treatment technologies have been used as reported in Van Passen *et al.* (2005), HTG (2006).

Characterisation of the sediment contaminated with TBT is necessary to select a suitable technique. Three types of TBT contamination can be identified: Large paint flakes, small paint flakes derived from high pressure hosing of the ship hulls and finally the TBT that is slowly released and adsorbed to the sediment.

CHAPTER 8

Monitoring, Measuring and Management

8.1 INTRODUCTION

Dredging projects worldwide are facing increasingly stricter environmental constraints and requirements from authorities, project owners and banks. Gone are the days where environmental considerations were second to economic interests and all other interests for that matter. Today, environmental impact assessments are mandatory in most cases, and with environmental criteria and objectives established, monitoring of environmental parameters is likewise becoming mandatory.

Monitoring of the environment before, during and after dredging is important for a number of reasons:

- to document the state of the environment and the seasonal, as well as geographical, variations before operations start;
- to document the short- and long-term impacts caused by dredging/disposal;
- to ensure that the operations are carried out according to environmental requirements established by authorities or other parties, e.g. the financial institution(s) or the project owner;
- to document the environmental impact predicted in the Environmental Impact Assessment (EIA);
- to document the recovery of the environment from the temporary impacts caused by operations; and
- to learn and gather experience at all levels for future projects.

Environmental monitoring programmes must be specifically targeted towards variables that the EIA has identified as important for the local ecological system, and at the same time at variables that are socially and economically important for society. The scale of a monitoring programme should reflect the size of the dredging project and the potential impact.

One important aspect relating to any environmental monitoring programme is the scope for combining broad monitoring objectives for separate parameters into a single survey. For example, although monitoring of sediment plumes and changes in seabed morphology are directly linked to assessing potential impacts on physical processes, the results are also relevant to impacts on all the other parameters, including marine ecology, fish and shellfish resources, and commercial fisheries, as shown in Figure 8.1.

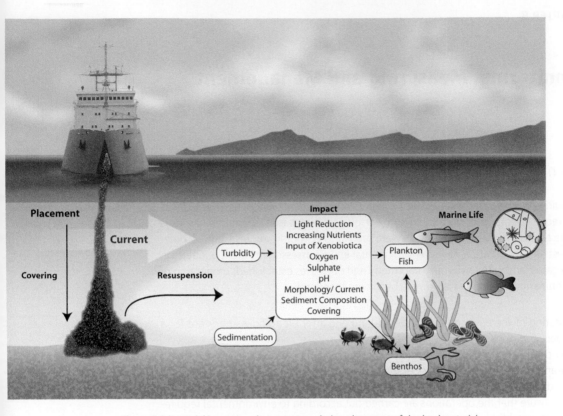

Figure 8.1 Potential impact on the ecosystem during placement of dredged material.

Monitoring of large projects will generate a vast amount of data on several environmental parameters. To ensure correct interpretation and to secure easy access to the data by all stakeholders and authorities, an environmental information system should be established. An environmental information system includes a database containing all the monitoring data, as well as information on ongoing and past construction activities and other relevant data on meteorology, hydrographics and other subjects. Such systems can be based on Geographical Information Systems (GIS) to allow the users to view the geographical distribution of data.

Proper selection of the most appropriate dredging and/or disposal techniques for a given project is, of course, of major concern when attempting to perform environmentally acceptable work. However, with each possible solution dredging and disposal activities must be controlled and monitored, so as to evaluate the results and effects of the equipment and execution methods selected.

Monitoring the progress of dredging and landfill, geotechnical sampling and other monitoring necessary for contractual purposes are often also used as support parameters for the environmental monitoring programme. For example, bathymetric surveys done as in- and out-survey can also be used in the environmental monitoring programme to map the exact extent of dredged areas. In this book these monitoring activities are described with respect to their role in the environmental monitoring programme. More detailed technical explanations and information on appropriate tolerances and such may be found in textbooks, e.g. Bray *et al.* (1997).

8.2 OBJECTIVES OF THE MONITORING ACTIVITIES

A programme to monitor or control environmental effects of a dredging process should be based on an Environmental Management Plan (EMP) which in some cases is an integrated part of the EIA and in others is an independent paper upon which the environmental aspects of a dredging project is regulated. The EMP will define the site-specific monitoring programme required and also define the role of the players involved. In some countries the EMP/EIA will be a part of the permit to carry out the specific works in other circumstances the EMP will be approved by the financing agency, e.g. the World Bank, before credits can be opened (see Annex A, Appendix 5).

Monitoring programmes can be categorised into three types depending on their objectives: Surveillance monitoring or BACI monitoring; feedback monitoring or adaptive monitoring; and compliance monitoring.

Surveillance monitoring or **BACI monitoring** (Before-After-Control-Impact) is monitoring which assesses temporal and spatial changes to selected parameters between the prior condition and the current condition. This type of monitoring is the most used and simplest to design. The preceding EIA study will have identified and predicted the impact on the relevant parameters for the BACI monitoring programme. The objective of a surveillance monitoring programme is verification of the hypotheses made during the project preparation. These hypotheses can be of a very different nature and can relate to:

- oceanographic conditions at the dredging or relocation site (e.g. current, wind, waves, depth);
- environmental background conditions at the site (e.g. suspended sediments, salinity, existing contamination level); or
- operational parameters related to the dredging equipment (e.g. suspended sediments from overflow with the TSHD, accuracy of the selected dredging equipment, turbidity generation at the dredging site).

Making these verifications is critical, especially during the first phases of the project's execution, in order to check the validity of the assumptions used as a basis for the project planning and the environmental impact assessment. This will allow one to adapt the execution method immediately when the project conditions change or when unexpected impacts are detected during the monitoring campaign.

Feedback monitoring (Europe) **or adaptive monitoring** (USA) is a special form of surveillance monitoring where a few fast-reacting and predictable environmental variables are forecast by modelling and then monitored continuously during the dredging and landfill operations. The purpose of this monitoring is to ensure that possible exceedance of environmental criteria can be forecast in such good time that dredging plans can be altered accordingly and costly down-time avoided.

The feedback monitoring programme is by far the most comprehensive and costly to carry out, while the compliance and BACI monitoring programmes normally are smaller and less costly. The feedback monitoring programme should only be adopted for projects with very strict environmental criteria or where legal binding limits for impact have to be observed.

A monitoring programme is by no means intended only as *post-factum* control of a (dredging) project, to be used as a basis for applying penalties when parameters and criteria are not met during actual execution. To optimise environmental effectiveness, a monitoring programme should also provide as much direct information as possible, not only to the project management team, but also to the individual operators and workers. This will enable them to adjust, wherever possible, the working procedures in order to achieve even better environmental effects.

Such feedback of information from the monitoring programme is not only useful for the project management team, but also for the crew on board the dredger. They are the key people in the overall success of an environmental protection plan. To give a few examples of factors that can influence environmental effects, suppose that the crew decides on the swing speed and the cutting depth or the trailing speed. Providing the crew with a direct reading from a turbidity meter installed close to the cutting/trailing head can encourage the operators to adapt the dredging process so as to mitigate environmental effects.

Compliance monitoring means ensuring compliance with contractual restrictions. A major objective in planning a control and/or monitoring programme is to ensure that the dredging process is executed in accordance with the various restrictions, which are legally or contractually imposed. Restrictions can vary markedly from one project to another depending on the prevailing human and ecological conditions at the site. They can be either physical (e.g. dredging depth, location or transport mode, limitation on turbidity or sedimentation rate at a vulnerable site nearby), seasonally related (e.g. special restrictions during breeding season) or quality oriented.

Taking into account the cost of an extended monitoring campaign as well as the cost of some of the mitigation measures, it is very important that the environmental restrictions imposed on a (dredging) project are formulated in an accurate way based on the local conditions and a careful analysis of the expected impacts as defined during the preparatory study phase. Keep in mind that most of the environmental parameters vary stochastically in nature. Therefore it is probably best to formulate the restrictions in the same way: e.g. define thresholds that should not be exceeded during a predefined fraction of the execution period. When necessary, different thresholds can be defined with a different allowable percentage by which these thresholds may be exceeded. Also the monitoring requirements have to be formulated in a clear way (locations, frequency and duration) to avoid later discussions on these parameters to "manipulate" the monitoring results by different planning of this activity.

One additional objective of any monitoring programme is to increase knowledge about the environmental conditions and effects of a given dredging process. This knowledge serves as a basis for a better assessment of the environmental effects during future dredging projects.

8.3 MONITORING STRATEGY AND RESPONSIBILITY

When considering the environmental regulations relating to a specific project, a clear definition and agreement should be determined amongst the parties involved about who is responsible for ensuring compliance. The structure around a large dredging project consists normally of three levels:

- the Authorities, local and/or central (strategic level), who are often responsible for surveillance monitoring;
- the Project Owner (tactical level), who is often responsible for feedback monitoring; and
- the Contractor (operational level), who is often responsible for compliance monitoring.

8.3.1 Strategic level

Setting the goals in relation to a specific project is normally the responsibility of the national and/or local environmental authorities. They will also be responsible for ensuring compliance at the strategic level with general environmental goals. The environmental authorities will formulate the goals and other requirements, which must be fulfilled by the project proponent. The authorities will receive reports from the project proponent describing the observed impacts on the environment and descriptions of any remedial actions taken.

8.3.2 Tactical level

The project owner is responsible for ensuring compliance at the tactical level. The project owner will receive requirements from the strategic level and method statements and work plans from the contractor. Based upon this and other relevant information the project owner, or the owner's agent, will be responsible for setting up the specific requirements imposed on the contractor and to approve methods and work plans.

8.3.3 Operational level

The issues of interest at the operational level are closely related to the activities of the contractor. Monitoring variables are only related to physical parameters such as spill rates, noise, and so on, where the contractor will be in a position to react immediately to the results. The contractor's monitoring will normally be included in the contractor's contract and in the event of non-compliance with the agreed limitations, the contractor will be obliged to adapt corrective measures to the previously agreed-upon construction methods.

The project owner (tactical level) will normally be responsible for ensuring that the monitoring programme is carried out and will have to report the results to the strategic level. The contractor (operational level) can be made responsible for monitoring some environmental parameters, e.g. spill rates, but third parties are sometimes employed by the project owner to carry out this work.

8.4 PLANNING AND SELECTING A MONITORING STRATEGY

Owing to the very large number of parameters involved, a monitoring programme can be time-consuming and costly. In order to limit the amount of energy and money spent on such control, a monitoring programme, based on the clearly defined environmental objectives for the project, should be carefully planned.

Studies conducted for the EIA (as well as supplemental studies if needed) must provide an effective baseline against which future monitoring results can be assessed. However, recognising that factors other than those associated with the dredging activity can influence the parameters being measured is crucial. For that reason, when devising the study area, all potential activities must be considered and adequate reference sites must be established so that natural occurrences and unrelated activities can be discounted from the monitoring results.

First of all, in order to predict possible changes in the seabed morphology and the benthic environment that result from natural occurrences such as seabed mobility and increases in suspended sediment because of storm activity, the natural variability within a system needs to be determined, as far as that is possible. These factors will also need to be accounted for at the control (reference) stations to ensure that the results are not biased by activities occurring outside the sphere of the dredging project.

Also, before selecting a monitoring strategy for a dredging project the regional, national and local environmental regulations must be carefully examined. In many countries the national and local authorities are implementing their environmental regulations in accordance with the framework of international conventions (e.g. the London Convention) and regional conventions (e.g. the Helcom Convention). Annex A gives a good overview of international conventions within the field of the aquatic environment and dredging.

If the environmental criteria are strict, and the size of the planned dredging operation is significant with respect to potential impact on the local environment, a feedback monitoring programme is recommended. At an early stage in the project, the potential environmental effects of each operation, possibly sub-divided according to each phase of the project, should be defined. During the planning stage and initial phase of the monitoring and control programme, the most critical operational phases, those which generate the highest environmental effect, should be identified. The principles of a feedback monitoring programme are outlined in Figure 8.2.

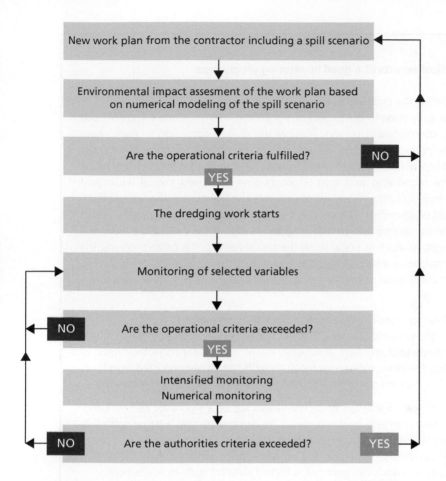

Figure 8.2 Principles of feedback monitoring.

Application of numerical models to test for maximum parameter levels, e.g. sediment spill from the various operations, is a very powerful planning tool. Modelling impact in hindcast and forecast enables the monitoring team to reduce the extent of the monitoring and control programme during the bulk of the work and to focus, as much as possible, on the critical items of the project. (See Text Box on page 256.)

So that reaction can be quick and effective, a hierarchy of responses to adverse monitoring results should be established beforehand. The level of response can be targeted to the receptor and its sensitivity. Responses could include the following steps, ranked by implication and cost, commencing with the least costly:
● Continue with dredging under the existing regime.
● Modify the dredging regime to reduce the actual effect on a sensitive parameter.
● Modify the dredging activity to increase efficiency while still having acceptable effects on key parameters.
● Cease dredging for a period within an area until further information is gathered.
● Cease dredging for a period within an area until recovery is seen.
● Cease dredging and implement active recovery measures.

Practical aspects of a good monitoring programme

● Define the critical activities

At an early stage in the project, the potential environmental effects of each piece of equipment on site, possibly sub-divided according to each phase of the project, should be defined. (Up-to-date computer dispersion models are a useful tool in this respect.) During the planning stage and initial phase of the monitoring and control programme, the most critical equipment and project phases, those, which generate the highest environmental effect, should be identified. This enables the monitoring team to reduce the extent of the monitoring and control programme during the bulk of the work and to focus, as much as possible, on the critical items of the project. Clearly, the assumptions made during the preparatory phase should be controlled by point check measurements during the initial stage of the actual monitoring campaign.

● Take an overall view instead of point measurements

Generally speaking, the exact location of the greatest environmental effects (e.g. suspended sediment generation) during a particular project is difficult to predict. Therefore, monitoring measurements should be conducted, in as far as possible, over the complete area where a significant effect can be expected.

Most of the parameters to be monitored are of a stochastic nature. Long-term measurements or simultaneous measurements over a larger surface or perimeter are necessary in order to assess the stochastic variation of a monitored parameter. Therefore, it is advisable to consider carefully the type of equipment, which can generate sufficient data for such an assessment.

● Measure the critical parameters only

Environmental effects can be found in a wide range of fields. One standard procedure to control every potential parameter and location rarely works and is certainly too expensive to implement. Therefore, during the preparatory studies and planning phases, an environmental impact assessment should be carried out in order to define the vulnerable elements at and nearby the dredging and relocation sites, as well as to define the parameters that can be influenced by the dredging process. From these parameters, the most critical should be selected in order to define an ecological and cost-efficient monitoring programme.

Based on the monitoring data, the monitoring programme could be adjusted to:
● reduce the level of monitoring because no effect was observed;
● continue with the existing monitoring programme to gain further clarification of the response; or
● expand the monitoring programme by increasing the frequency or including additional surveys or sites.

8.5 MONITORING VARIABLES

Great care should be exercised in selecting the right parameters for environmental monitoring of dredging projects. Environmental parameters selected must as a minimum be:

- Measurable.
- Relevant with respect to the actual environment and predicted impact.
- Able to give significant results.
- Predictable in response to impact by using numerical model tools.
- Measurable response within reasonable time (short response time) and cost.

Obviously, selecting monitoring variables, which are either extremely difficult, irrelevant, uncertain and/or extremely costly to monitor is not useful.

A clear and predictable connection should exist between observed effects and the impact caused by the dredging activities. Environmental parameters are not constant but vary considerably in time and space, depending on many other natural as well as human-induced factors other than dredging activities. The effect on a specific parameter caused by dredging must be included in numerical model tools so that impact can be forecast. The selected parameters should have response times that are in the same order of time scale as the time frame of the dredging project. Imposing limitations on a parameter for which the contractor is responsible, where the measurable impact can be seen only years after the dredging contractor has handed over the project to the owner is not good practice. Still, authorities can have legitimate and sensible requirements for monitoring parameters that will have long response times. In such cases, responsibility for such parameters should then be placed at the strategic level or at least at the tactical level.

Generally speaking, the exact location of the greatest environmental effect during a particular project is difficult to predict unless numerical modelling is used in the planning phase. The initial monitoring can be used to verify the model predictions and the following programme can be specifically targeted to areas where impact is predicted. If numerical modelling is not done, monitoring measurements should be conducted, as far as possible, over the complete area where a significant effect can be expected.

Most of the parameters to be monitored are of a stochastic nature. Long-term measurements or simultaneous measurements over larger surfaces or perimeters are necessary in order to assess the stochastic variation of a monitored parameter.

In this section attention is focused on the variables, which have a significant effect on the environment during the dredging process, as well as on the measurement strategy. More detailed discussions are given in Chapter 5. The following variables have been identified:

- Depth.
- Suspended sediment concentration.
- Spilled sediment accumulation.
- Hydrographic parameters.
- Leakage from CDFs.
- Chemical and biological parameters.

8.5.1 Depth

The first variable to be measured is the depth of the sea- or riverbed. As this is the prime parameter to be achieved or changed by the dredging works, accurate measurement of this depth is a prerequisite. In the case of a sandy bed, this measurement is not much of a problem. However, in the case of a weak muddy bed, serious problems can arise, as there is no clear-cut boundary between soil and water. Instead, gradual changes take place going from clear water to suspended sediment clouds, to fluid mud and then to consolidated material.

An exact definition of how a certain depth is to be achieved is a prerequisite for proper control of dredging works. The concept of the "nautical bottom", in which the low-density mud is left in place without endangering navigation, leads to considerable reductions in dredging effort, effects and costs, as well as lower disposal volumes. However, specific measurements are needed to put this concept into practice.

8.5.2 Suspended sediment concentration

Suspended sediment content can be measured in various ways. One way involves traditional water sampling combined with filtering and occasional measurement of the grain size distribution. This is the oldest technique, which gives only point results and is slow and labour intensive. To get a clear picture of the spread of suspended sediments, a large number of samples have to be taken. Simultaneous measurement of such a large number of samples is physically impossible or at least economically prohibitive.

Other systems for measuring suspended sediments, which give a direct reading of suspended sediment content after calibration of the measuring probe, have been developed. With this equipment the measuring process can be made much faster, permitting more frequent measurement of a vertical profile (with a multiple of measuring results) compared with the sampling method.

However, a new development, using Acoustic Doppler Current Profilers (ADCP) can measure suspended sediment content together with water flow velocity over the complete vertical. By slowly sailing along a certain route, a relatively rapid view of a vertical plane throughout the water can be achieved. Until recently the processing of the backscattered sound measurements, required for suspended sediment assessment, was a complex task and the results were only considered as an indication. State-of-the-art systems combined with sophisticated computer programmes have resulted in accuracies that are in the same range as with more traditional equipment. Because of the large quantities of data made available, this equipment produces a better picture of the spread of suspended sediments at a given site.

Satellite images or aerial photography can also be used to obtain an overview of the suspended sediment dispersion in a specific area. The main advantage of this is the simultaneous view of the suspended sediment content over a complete surface; the disadvantage is that the measurements currently possible are relatively difficult to calibrate with field measurements. Undoubtedly the technique for measuring and computer-based interpretation will improve within the next decade. The concept of a spill monitoring vessel and its many facilities are shown in Figures 8.3 and 8.4.

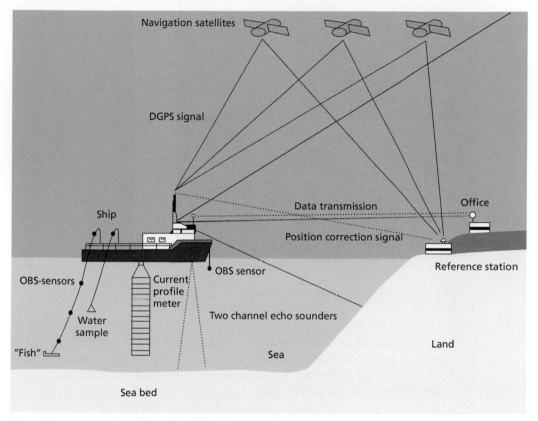

Figure 8.3 Schematic drawing of a spill monitoring vessel and its survey equipment.

8.5.3 Spilled sediment accumulation

Measurement of spilled material accumulating on the seabed is normally not a prime concern for most dredging projects provided sufficient water depth is sufficient, irrespective of the spill layer thickness. However, with environmentally sensitive projects, spill has to be prevented in order to avoid burial or spreading of contaminated sediments.

Bathymetric surveys using different frequencies as well as side scan sonar give an initial indication of the remaining spill layer. The bathymetric survey if repeated throughout the dredging operation can be used to measure changes (accumulation) on the seabed outside the dredged area to an accuracy of 100 mm. Side-scan sonar gives an area

Figure 8.4 A spill
monitoring vessel at
work.

covering picture of the seabed where changes in seabed morphology outside the dredged area could indicate accumulation of spilled sediment.

For remedial dredging projects, where great accuracy is required, this is not sufficient. In those cases only geotechnical sampling (using, for example, Beeker samples) is feasible, albeit very slow and labour intensive. Such measurement is therefore only undertaken when spill effects are very significant (high contaminant content). Currently geo-electrical probes specially developed for this type of measurement are at the testing stage. In addition, optical methods, involving the use of half-buried transparent tubes and cameras operating on a time-lapse basis are being used. (Germano et al., 1994)

8.5.4 Hydrographic parameters

Changes to hydrographic parameters caused by dredging are seldom measurable with any statistical significance as the natural variation is far greater than the dredging imposed impact can be. Such parameters are measured to allow assessment of other environmental parameters and measurements, to provide data for numerical models and forecast systems.

Measurement of physical variables is well known in the field of oceanography. Point measurements of hydrographic parameters using moored buoys or bottom-mounted equipment like Acoustic Doppler Current Profilers (ADCP) should preferably be supplemented with area or line-covering ship-based measurements of hydrographic parameters. Radar-based area-covering measurement of surface currents and waves is now a well-proven technology, but it can only work in areas where the wavelength is sufficiently long to reflect the radar waves (minimum 6 m). Satellite-based radar measurements are also a possibility, but this technique is also restricted to the open ocean with relatively large waves.

These new technologies broaden the scope for better monitoring and control of the physical conditions at a dredging or relocation site.

8.5.5 Leakage from CDFs

Placement and post-placement monitoring
Monitoring objectives for CDFs can be grouped into four categories and a brief list of the objectives for each CDF is given here:
- Monitoring of facilities. Factors of interest include the engineering/physical integrity and functioning of containment structures, structural elements (e.g. drains, pipes, weirs), isolation structures (e.g. liners, covers), sampling facilities (e.g. gauges), treatment facilities and infrastructure (e.g. roads, drainage system).
- Monitoring of placement operations. Factors of interest include the rate of discharge, degree of filling, mound development and ponding depth.
- Monitoring of processes. Processes to be monitored include consolidation/compaction, decomposition of organic matter, gas formation and colonisation by flora and fauna.
- Monitoring of emissions. Efforts should include monitoring of water flows (i.e. effluents during filling, surface runoff and leachate), receiving water bodies (surface and ground water), noise, dust and smell. The chemical quality of biota is also of interest.

The use of special, purpose-built sampling devices to ensure reliable measurement results is not uncommon. The sludge sampler developed by Rotterdam Public Works Department to take undisturbed samples at depths of 30 m is one example. Detailed descriptions of monitoring efforts for a nearshore CDF can be found in Nijssen *et al.* (1997) and for an upland CDF in Harms (1995).

8.5.6 Chemical and biological parameters

Chemical and biological variables are generally measured by taking water, biota and/or sediment samples at the location and water depth that have been chosen for the monitoring process. Samples are analysed in the laboratory or directly on site. The equipment required for the analyses varies widely, depending on the actual variable that has to be measured. This can range from a rather simple measurement of the content of a heavy metal in the water (or silt) sample to a complicated leaching test and/or mortality for a combined group of species.

Biological variables can also be surveyed over larger areas using video or side scan sonar equipment to detect/estimate coverage and biomass. Fixed, mounted sonar can also be used to monitor fish migration and radar has been use to monitor bird behaviour.

8.6 CONCLUSIONS

A well-planned and well-executed monitoring and control programme is necessary to ensure that predicted effects are really being met and that mitigating measures actually fulfil the requirements of the project.

Any monitoring effort must be in reasonable proportion to the potential impact caused by the dredging project. Monitoring parameters must be relevant with respect to environmental impact and measurable within reasonable significance and cost.

Sampling, analysing, quality control and reporting of the results must be carried out in a professional way under an organisational structure, which is embedded in the contractor's organisation on equal terms with the other parts of the construction organisation. Good quality subcontractors (e.g. analytical laboratories) capable of delivering good quality results to tight deadlines should be preferred, rather than the subcontractor with the lowest price.

For every dredging and reclamation project, an environmental impact of some nature will always occur, but this impact should be viewed in the light of the benefits gained by executing the project. As emphasised in previous chapters, a balance must be struck between acceptable and non-acceptable impacts and this decision is based on a broad number of advantages *versus* disadvantages.

CHAPTER 9

Frameworks, Philosophies and the Future

9.1 INTRODUCTION

This book has evolved from the IADC/CEDA series comprising seven guides on the Environmental Aspects of Dredging published from 1996 through 2001. When the guides were originally conceived, they were perceived to have a number of uses:
- to provide a balanced view of the whole subject of environment and dredging;
- to lead the reader through the state-of-the-art environmental evaluation process;
- to provide a source of information on dredging and environmental matters; and
- to provide more detailed references and reading.

As the series evolved, and the environmental evaluation process was set out, it became clear that a state-of-the-art of environment and dredging cannot be totally defined. The whole process is continually evolving in line with technology, public thinking and consensus. Some subject areas still require research. Others, of a more philosophical nature, are still being debated. With this in mind, the Editorial Board decided that the final chapter, like the final guide, should highlight the more philosophical areas of interest and comment on those areas where consensus has not yet been reached.

This chapter is presented in three parts:
- a general comment on the environmental aspects of dredging projects;
- a discourse on some of the philosophical points raised in the other chapters and other points which have not been mentioned previously or have only been mentioned in passing; and
- thoughts for the future.

9.2 ENVIRONMENTAL ASPECTS

The Foreword to this book begins by stating: "by its very nature, the act of dredging and relocating dredged material is an environmental impact". Accepting that statement means acknowledging that whenever we contemplate a dredging activity, we are indeed considering making an impact on the environment of a positive, negative or mixed type. We are considering change, altering nature, altering the conditions in which we humans and all other organisms live. We may even be fundamentally altering local areas in an irreversible manner.

Impacts and change may be good or bad, positive or negative, temporary or permanent. They may even be neutral. Whether you perceive them to be good for you, or not, depends on your viewpoint. That viewpoint may in turn depend on your lifestyle, how you make your living, your experience and even how you make your own moral and value judgements. What is certain is that for every proposal to make change, i.e. to dredge, there will be people who are in favour, those who are against and those who are indifferent. All these disparate views of these various people must be recognised. Any evaluation must attempt to allow those assessing the project to review all these different viewpoints in an unbiased way, with the object of arriving at a balanced assessment of their merits and their weaknesses. Thus, as a fundamental rule to achieve useful results, a proper environmental evaluation should be as objective as possible.

9.2.1 Dredging and the environment – a global issue

Since the original guides were conceived, WODA (The World Organization of Dredging Associations, comprising the Western, Central and Eastern Associations (WEDA, CEDA and EADA) has issued its "Environmental Policy on Dredging". WODA has recognised the importance of setting dredging activities in their proper environmental context.

The full Environmental Policy is as follows:

The World Organization of Dredging Associations (WODA) recognises that carefully designed and well-executed dredging conducted in an environmentally sound manner contributes to a stronger economy. WODA believes that dredging projects can be conceived, permitted and implemented in a cost-effective and timely manner while meeting environmental goals and specific regulatory requirements. WODA is committed to the development and implementation of appropriate environmental safeguards and performance guidelines for construction, maintenance, mining and remedial dredging. Beneficial use of dredged materials is encouraged. Open lines of communication amongst stakeholders, such as port interests, dredging contractors, regulatory agencies, other business interests, environmental interest groups, and the public, should be standard elements of any project. WODA encourages investment in and expeditious transfer of new technologies, and the development of new, more efficient techniques for improving the evaluation and safe handling of dredged material.

It is anticipated that this book will assist in implementing the above-stated policy.

Yet great care should be taken when translating this guidance into practice, particularly when setting environmental controls (see Text Box on page 265). There have been a number of recent cases reported where totally inappropriate environmental controls have been set by engineers in misguided attempts to protect the environment and their client's interests. The unfortunate results in these instances have been loss of client's funds for no added environmental benefit.

Establishment of environmental controls for dredging works

This IADC/CEDA guidance contains much information on the environmental aspects of dredging culled from case histories, research results and practitioners in the field. It promotes particular approaches to environmental evaluation and procedures for carrying them out. It provides the reader with a vast amount of relevant background reading and many sources for further research. It is unbiased and favours neither the promoter of dredging works nor those who are opposed to dredging being carried out.

It does not, and never set out to, prescribe specific controls for dredging works on a generalised basis. It is not intended to be used in this manner and, in fact, should not be used in this manner. It is intended to be an aid to planners, engineers and environmental scientists when they are making their own individual assessments.

The setting of suitable environmental controls for a specific dredging project should be carried out with great care. Like matrimony, it should not be entered into unadvisedly, lightly or wantonly. In fact, it is often the single most important aspect of the planning of the project as a whole.

The golden rules for establishing environmental control are:
- The promoter/designer/regulatory authority should set the limits or levels on the various environmental parameters of concern.
- The set parametric limits or levels should be based on local conditions and local environmental sensitivities – not transplanted from another climate or site, whose characteristics will invariably be different.
- The contractor/operator should be allowed to operate his equipment in any reasonable manner that allows the environmental goals to be met.

If you are a planner/designer/regulatory authority do NOT try to tell contractors/operators how to operate their equipment. Doing so will almost always cause unnecessary aggravation and will often result in the works being executed in an overly cautious and irresponsibly expensive manner.

9.2.2 The IADC/CEDA and PIANC responses

Both IADC/CEDA and PIANC have published reports containing guidance on the environmental aspects of dredging and dredged material management. The way in which the latter is covered in various complementary publications is illustrated in Figure 9.1. It should be noted that these publications are not intended to give guidance on the quantification of environmental control, as occurs in contractual specifications, because this is too specific a matter. They are intended to assist engineers and planners to achieve their objectives.

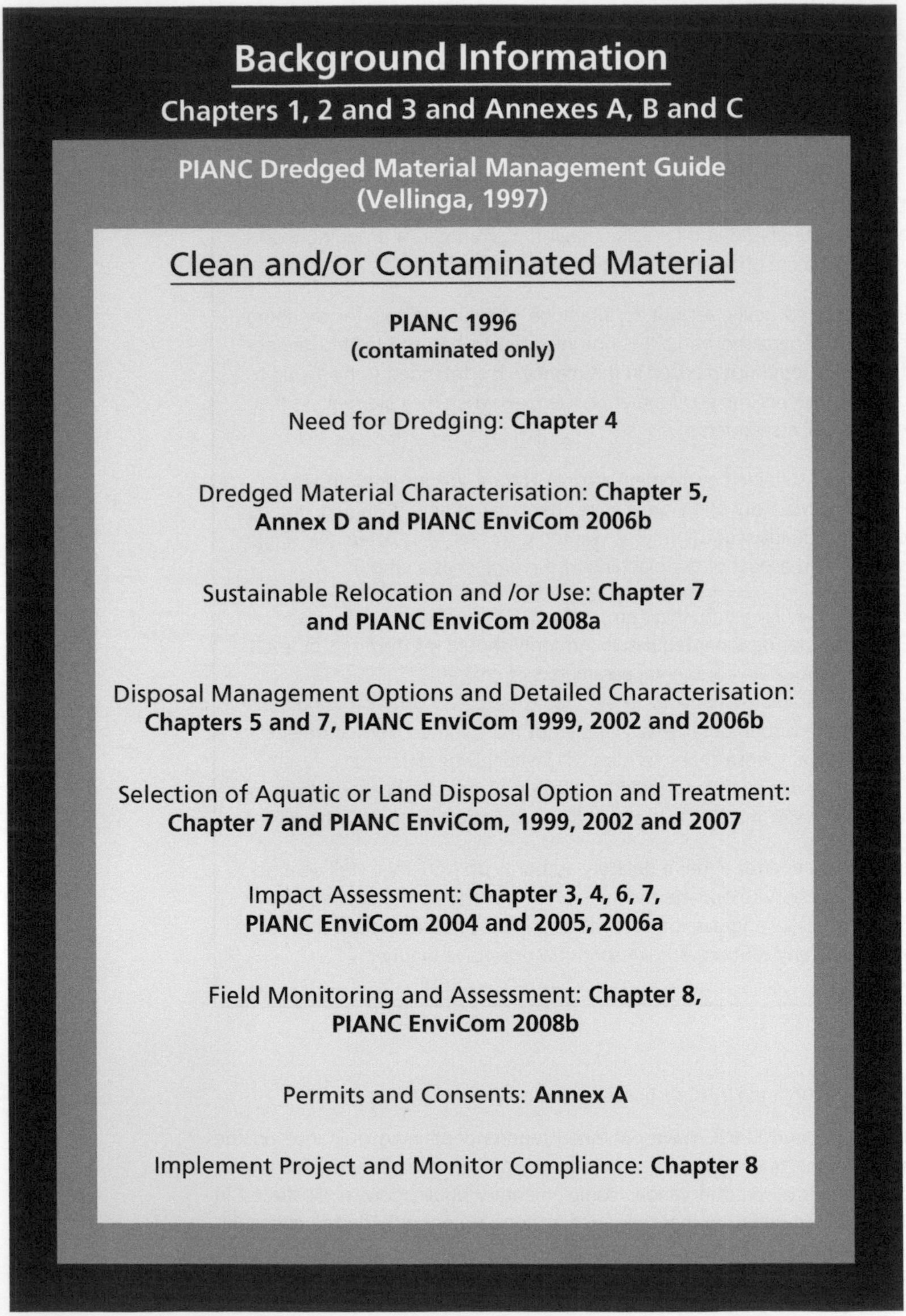

Background Information
Chapters 1, 2 and 3 and Annexes A, B and C

PIANC Dredged Material Management Guide
(Vellinga, 1997)

Clean and/or Contaminated Material

PIANC 1996
(contaminated only)

Need for Dredging: **Chapter 4**

Dredged Material Characterisation: **Chapter 5,**
Annex D and PIANC EnviCom 2006b

Sustainable Relocation and /or Use: **Chapter 7**
and PIANC EnviCom 2008a

Disposal Management Options and Detailed Characterisation:
Chapters 5 and 7, PIANC EnviCom 1999, 2002 and 2006b

Selection of Aquatic or Land Disposal Option and Treatment:
Chapter 7 and PIANC EnviCom, 1999, 2002 and 2007

Impact Assessment: **Chapter 3, 4, 6, 7,**
PIANC EnviCom 2004 and 2005, 2006a

Field Monitoring and Assessment: **Chapter 8,**
PIANC EnviCom 2008b

Permits and Consents: **Annex A**

Implement Project and Monitor Compliance: **Chapter 8**

Figure 9.1 IADC/CEDA and PIANC publications in response to environmental concerns.

9.2.3 International legislation and country response

A particular difficulty experienced by many promoters of dredging projects is what to do with the dredged material. In the past, considerable volumes of dredged material have been placed in the sea at disposal sites, but environmental awareness and concern for the protection of the world's oceans and seas have made placement of any matter at sea increasingly unacceptable.

The sea has always been regarded as clearly international, and freedom to move through it, fish in it and enjoy its wealth has always been treated as a basic human right. It was not surprising, therefore, that the international community took steps in 1972 to control material being placed in the sea by establishing the London Dumping Convention (later, London Convention 1972). Concurrently, various regional conventions came into being, also concerned with the placement of materials in the sea (e.g. the Oslo Convention 1972 and the Paris Convention 1974). These conventions have produced frameworks within which the contracting countries are obliged to operate with respect to their handling of materials destined for placement in the sea. Interestingly, almost the only materials nowadays accepted for placement at sea are certain categories of dredged material. Nearly all other materials must now be placed on land.

Placement at sea
With respect to sea disposal, it is noteworthy that for financial reasons every country has interpreted and applied the international conventions in different ways (see Annex A). This is partly a result of the differing coastlines, geology and the hydrodynamics of each nation's coastal waters, but there are also fundamental differences in their approaches. Some countries adopt physical and chemical contaminant levels as a direct method of classifying their dredged materials and then have specific regulations relating to the placement methods to be used for each class. Other countries determine the levels of contamination and then decide on the disposal method on a case-by-case approach. A third method is to determine the eco-toxicological effects of the contaminants and then to decide the best approach.

In addition to the variations in approach, described above, countries also define the areas of jurisdiction of the conventions in different ways. Some countries maintain their legislative independence over coastal waters (which others would have classified as seaward of the baseline) by deeming them to be "inland waters".

Significantly, all the above approaches to regulating dredged material placement depend on a series of sampling and testing procedures. Still, inconsistencies in how these sampling and testing procedures are conducted continue to exist. These procedures are not inexpensive, but since industrialised countries, which tend to be the richer nations, have developed them they assume a certain level of available funding, which in a developing country might be impossible to obtain. For this reason, the implementation of the conventions in developing countries is more difficult to progress and tends to be at a less refined stage than in the developed world.

Placement on land

Placement of dredged material on land is not governed by international conventions and is controlled by each individual country. Some countries have developed their own systems of control but these systems are only applicable nationally. At present no accepted international standard exists for determining how land disposal should be evaluated.

While recognising that countries vary significantly in their climates, geology, soil and sediment characteristics, and that public perceptions of acceptability also vary, certainly an important step forward would be the standardisation of the methods used to measure contamination, analyse its effect and determine the environmental accept-ability of different placement options. This is not to say that contaminant level acceptability standards should or could be applied worldwide (because natural geo-chemical variations would make this inapplicable), but that international definitions for the application of these standards would be beneficial.

9.3 PHILOSOPHIES

In this section, a number of philosophical points are discussed relating to dredging projects and environmental effects.

9.3.1 Scope of environmental evaluation

Before an environmental evaluation is carried out, the scope of the evaluation should be determined. In conventional terms, this is achieved by defining the proposed proj-ect and then identifying those areas of the environment which will be affected by the project, both after construction and during project implementation. A commonly adopted procedure is to carry out a two-stage scoping exercise; the first stage is a coarse grid approach (to determine the areas of concern), and the second is a more detailed appraisal of those areas identified in the first stage as being of major impor-tance (see Chapter 3).

This two-stage scoping procedure is too often carried out in isolation. That is, other activities of people or nature, both in the past and continuing, are not reviewed at the same time. A few anonymous real-life examples are given below:

- A proposed aggregate dredging activity will result in modification of limited areas of the seabed. The same area is heavily fished, using bottom-trawling techniques, which also disrupt the seabed. The effects of the trawling are completely ignored when the environmental evaluation of the proposed aggregate dredging is carried out.
- A nearshore dredging activity, which will give rise to modest levels of suspended sediment in the vicinity of the works, is proposed. Very close to this site a licence has been issued which permits a port authority to dispose of millions of tonnes of silt annually from maintenance dredging operations. The disposal site is dispersive and will not accumulate the silt. The environmental evaluation of the first men-tioned nearshore dredging activity takes no account of the disposal operations.
- Maintenance dredging operations in a busy waterway are to be controlled to such an extent that no discharge of silty water or overflow is to be permitted.

In the same waterway, re-suspension of sediment by propeller wash and low keel clearance occurs frequently during vessel manoeuvring operations. The environmental evaluation relating to the former ignores the latter.

- A dredging activity which will give rise to suspended sediment plumes is proposed in an area where tidal and weather processes regularly put very large sediment loads into suspension. The effects of weather conditions are not taken into account when assessing the dredging activity.

Although no doubt many similar examples of this type of approach to environmental evaluation could be identified (see also the situation described in the Text Box), this is not meant to be an excuse for the promoters or executers of dredging projects to do the same. The specific examples are not important. What is important is that an agreed, standardised method of evaluation be adopted which puts all human (and therefore controllable) activities on the same level playing field and views them together. This is a part of the so-called "cumulative effects" analysis.

Cumulative effects analysis

Another type of environmental misconception is the idea of a "significant effect". If one project is adjudged to have an insignificant effect on an area (probably because the area appears to sustain a high population of species and the impact is small), it does not mean that a dozen similar projects will also have an insignificant effect. Natural systems are often prone to gradually encroaching human-made degradation, or "death by a thousand cuts", merely because no one realises the significant cumulative effects of many projects with "insignificant effects".

The concept of cumulative effects analysis, together with its application to a specific activity, is complex and frequently not clearly defined. To some, it appears to mean the summation of the effects on the local coastal cell regime of a number of similar activities, such as a number of offshore aggregate dredging operations. To others, it appears to be the definition of the overall background situation caused by humans and nature, and the assessment of the effects of the proposed activity on the status quo. Clearly, both the definition of cumulative effects and the way in which these effects are determined need to be established. Guidance on this is given in EPA (1997) and CEAA (1992).

Natural perturbations in a discrete water body are a fact of life. Human-induced effects may have been licensed and/or accepted for years. To modify the existing situation, or background, may require legislation and control on a national or regional level. Perhaps more appropriately regional bodies should be founded to determine the status quo by carrying out regional studies. Promoters of projects within the region could then provide the regional body with details of their proposed projects and any environmental effects predicted to arise from them. In this way a regional body could start to evaluate the various demands and consequences of each activity against the environmental background of the whole region.

Carrying out a "cumulative effects" environmental impact evaluation is not easy. Before attempting to start a review of different activities on a level playing field, the extent of the field must be defined. This may be relatively simple in a river or an estuary, but offshore, large areas of sea may be involved. The data collection required is considerable and costly, and the number of activities taking place may be numerous (see Text Box, Suspended solids) so the impetus to do this may be lacking. If, however, a rational approach to evaluating the importance of each activity and the severity of its impact were required, the "cumulative effects" evaluation would seem to be the only answer. (Note: Some agencies call this the "in-combination" evaluation.)

Suspended solids

When studying the cumulative effects of suspended solids (see beginning of Chapter 3), it is important to realise that when two plumes of elevated suspension levels meet, the combined effect cannot exceed the highest level in one or other of the plumes. In general, the TSS of the mixture will be the average of the two plumes. Thus, if the absolute level of TSS is an important environmental parameter, two plumes may well be no worse than one. If, however, the time of exposure is important, then two plumes may affect a receptor for a longer period than one.

To progress the "cumulative effects" analysis further, some thought should be put into the following areas:
- How can the data required for this approach be collected in the most cost-effective manner?
- Can existing human activities be given precedence when considering the effects of all activities? i.e. is there an advantage in being there first, or should all activities be evaluated on the same basis of importance?
- How is the technique to be applied when necessary data are not yet available and cannot reasonably be made available?
- How is this type of analysis to be implemented and funded? In many countries there are few, if any, regulations under which an obligation exists for cumulative effects to be studied. (For instance, in the United Kingdom the only requirement for "in-combination" analysis is under the Habitats Regulations, as a result of the EU Habitats Directive, and possibly for strategic development plans which will also be subject to strategic environmental assessment.)

9.3.2 Data collection

Data collection is an essential prerequisite for dredged material characterisation, evaluation of alternative disposal sites, evaluation of the environmental effects of dredged material use and numerous other environmental impact evaluations. In fact virtually no environmental impact assessment can be carried out without some collection of data.

However, in many cases the collection of data is a time-consuming and expensive operation (see Figure 9.2). Even for environmental assessments involving clean materials, examining the hydrodynamics and sedimentology of large areas of the sea or of complex estuarine processes may be necessary. When contaminated materials are to be dredged, the whole process of investigation is likely to be far more expensive owing to the added costs of sampling and testing for the contaminating substances. In addition, to assess the eco-toxicological properties of these substances on the local ecology even more tests may be needed.

Figure 9.2 Towing a site investigation jack up out to sea in the UK: Such data collection methods can be costly.

A further difficulty in collecting representative data from marine sites is posed by the occurrence of extreme events. Many sites rest in a fragile dynamic equilibrium for the majority of the time, but occasionally major events such as storms, tidal surges and the like, cause fundamental changes to occur. These events are part and parcel of the environmental background. The local benthos has to cope with these events. Because of the difficulties of collecting data during extreme events, they are often ignored during the data collection process, yet the impact of a dredging project may be more akin to that of an extreme event than to the normal day-to-day background conditions. If one wished to compare the sediments re-suspended by a five-day storm with those of a particular dredging activity, it is now likely (with recent developments in predictive techniques) that the effects of the dredging activity will be better known than the effects of the storm.

Even in highly developed countries, the cost of site investigation and data collection is an onerous part of the full cost of carrying out dredging works. In a developing country, the costs are likely to be prohibitive. However, if a contracting country to the London Convention wishes to carry out dredging works according to the convention, it is bound by the convention to conduct its operations in the spirit of the Dredged Material Assessment Framework (DMAF). This involves characterising the material to be dredged, reviewing potential use of the material and the environmental effects of each use, and evaluating alternative disposal sites on an environmental basis. Clearly, in this case, the cost of data collection is a pronounced disincentive.

Bearing in mind that the collection of data is expensive, but that some data is better than no data, trying to prioritise data collection is a necessity. The promoter of a dredging project with a limited budget is then able to use the available funds most effectively. At the same time, the result is that the standard adopted by this particular promoter is less than that specified by the convention. What is missing at the present time is guidance on this aspect of applying the DMAF. For, in general, it is better for the DMAF to be applied in limited form, than not to be applied at all. In the June 2007 Meeting of the Scientific Group of the London Convention, a CEDA proposal for addressing this issue was brought forward. This proposal was welcomed by the groups concerned and steps are being taken to raise the necessary funding and have the necessary guidance developed into a formal document.

Involvement of all the stakeholders in a project is discussed in Chapter 3 of this book. However, it is relevant to point out that, through discussion and planning with interested parties, necessary environmental data acquisition and assessment can be more easily targeted.

A suggestion is made that the following questions should be answered:
- Given limited funds, which areas of investigation are the most important – noting that this may vary according to the type of project being carried out and the site characteristics?
- Which types of investigation are most important in the priority areas selected?
- How will the DMAF be applied with the limited information available?

In the UK, CIRIA (Construction Industry Research and Information Association) has made a start at improving the accessibility and management of coastal data (CIRIA 2000).

9.3.3 Use of dredged material, impact and evaluation

"Beneficial use" is the term used by the legislators to denote a use for dredged material other than placing it at a so-called "disposal site". The term implies that use of the material, rather than disposal, will be of benefit to the community. A number of examples may be found in the PIANC publication "Dredged Material as a Resource" (PIANC EnviCom, 2008a). However, most people do not define what exactly is meant by beneficial use, nor do they define who the beneficiary is. In Chapter 7, use is defined as "any use which regards the material as a resource". Since there are now many examples of projects where dredged material *has* been used, even when it is classified as a waste, in this book the term "beneficial" has been dropped.

A more pro-active approach is to assume that the dredged material is a resource and seek a use for it. Examples of this have been described by Pellissier (1999) in Rhode Island, USA and Jeschke and Jeschke (1999) in Bremenhaven, Germany. When a use for the material is actively sought, the chance of finding a project which needs the material is far greater. When this happens a "win/win" situation arises which benefits both projects (see Figure 9.3). This pro-active attitude to utilising dredged material is to be recommended because it often enables two projects to get off the ground, and frequently leads to a very positive overall environmental and/or economic benefit.

- In practice, all dredging projects involve the movement of material from one location to another location. Whether the material is being viewed by the promoter of the project, or anyone else, as an unwanted material or as a resource is really of no consequence in relation to the environmental evaluation. What really matters is how this movement of material from one place to another affects the community as a whole. In an ideal world, the process for deciding how to manage dredged material would be as follows:

  - select a number of different options for the ultimate destination of the dredged material;
  - evaluate all the environmental effects of each option, including the economic, social and political effects (see Chapter 4);
  - determine the sum total of all the benefits and dis-benefits of the alternative schemes; and
  - select the optimum scheme and compare its effects with the benefits to be accrued by carrying out the dredging works.

Figure 9.3 A win-win situation: This constructed wetland at Wakodahatchee in Florida benefits the community by combining water treatment, wildlife habitat and public recreation.

In the final analysis (see Chapter 3), if society deems that the benefits accruing from the dredging works are worth the environmental impacts which are predicted to occur, then the project is worth implementing. Such terms as "best environmental option" and "best available technology at reasonable cost" are all attempts to take the overall view of acceptability and would be covered in the above procedure.

It should be noted that the "best environmental option" is not necessarily either socially or environmentally acceptable. For instance, a scheme may have three alternative options, all of which are environmentally less attractive than the "do nothing" scenario, and are thus technically unacceptable, especially if safety is at stake (e.g. to cater for sea level rise). But the "do-nothing" scenario may be socially unacceptable, because of the need to provide work for the local community and serve the needs of the country. If overriding public interest requires that one of the other options is selected, then the accepted scheme is the best environmental option even though it is environmentally unsatisfactory.

At the present time, an integrated environmental evaluation as described above is difficult to perform in its entirety. Although many countries prescribe specific types of sampling and testing and general procedures for evaluation, there are no standard procedures for comparing disparate aspects of the same project. For instance, how are the livelihoods of a few subsistence farmers or fishermen to be valued against the economic benefits to be accrued from a major channel deepening project? How valuable is an endangered species? What value is put on the aesthetic and cultural aspects of a locality?

Obviously, carrying out a completely integrated evaluation of a project is a lengthy process. Clearly, since many values are subjective, values in one country may not be the same as values in another. What is important is that the framework for the whole evaluation process should be set out so that each country can adopt it in its most suitable form.

Before leaving this subject a few words of caution: Every action has a reaction. A dredged material use project also requires an environmental assessment.

The story goes that in a particular project dredged material was used to make a new island for wading birds. The island was so successful in attracting birds that the local fish population fell because of the additional beaks to feed. The local fishermen, who suffered reduced catches as a result of the falling fish population, demanded action to be taken. The island was removed.

9.3.4 Preservation, conservation and sustainability

Many attempts have been made to develop guidelines for environmental evaluation of proposed works. In so doing, a number of expressions and principles have come into common parlance, such as "sustainability", "conservation" and the "precautionary principle". With respect to the environment, what most people are generally concerned about is the "natural" world and whether and in what ways the dredging works will affect it. In this context, the natural world also seems to include heritage (or evidence of and changes made by ancient peoples).

Many other expressions exist and the possibility of becoming confused by what all these mean, particularly in the context of dredging works, is real. Take for example, preservation versus conservation. Preservation, which at first sight might seem appropriate, only becomes important when change is irreversible, for instance, if a dredging project was going to destroy an archaeological resource. In many cases, preservation is not desirable in the natural world because natural systems are constantly changing and preventing those changes from taking place is not acceptable. Conservation allows natural changes to take place. If a proposed project is predicted to alter the rate of change, then this is likely be treated as a negative environmental impact and steps might be taken to mitigate any damaging effects.

Sustainability is less concerned with environmental evaluation of immediate effects and has more to do with the future effects of the project being proposed. Sustainability is an attempt to ensure that a development undertaken now does not prejudice the options available to future generations with respect to resource use and development. It is about ensuring a better quality of life for everyone, now and for generations to come (DETR, 1998).

However, the principle of sustainable development assumes that predicting the consequences of a proposed project going ahead is possible, which may or may not be the case (Brooke, 1998a). Where any uncertainty in respect of such predictions exists, this uncertainty must be acknowledged. If the "unknown" or "uncertain" consequences are potentially significant, a "precautionary approach" should be adopted and nothing should be done which might prejudice or cause damage to the resource or to resources elsewhere unless clear and unequivocal justification exists for so doing (Brooke, 1998a). Sustainable development and the precautionary principle are, thus, often intrinsically linked.

The precautionary principle as described above has increasingly been incorporated into legislation to protect the environment from anthropogenic impacts. Recent suggestions for redefining the principle would entail changing from a truly precautionary approach to a risk-based approach to environmental protection (Santillo *et al.*, 1998).

9.3.5 Mitigation and compensation

During the environmental evaluation of any proposed dredging or reclamation project frequently, in the first instance, a few potentially adverse environmental impacts are identified. These might be:
- alteration to the site hydrodynamics and consequential changes in erosion and accretion patterns;
- noise of the dredging equipment and its effect on the local community;
- temporary effects of dredged material disposal at the reclamation site; and
- loss of valuable habitat at the dredging and/or reclamation site.

Many other impacts might be identified. These are described in detail in Chapter 4.

Irrespective of the positive impacts of any scheme, these adverse impacts should always be mitigated where possible. Mitigation may be defined as "avoiding, reducing, remedying or compensating for an adverse impact" and a hierarchy of mitigation options in order of preference should be produced (see Figure 9.4).

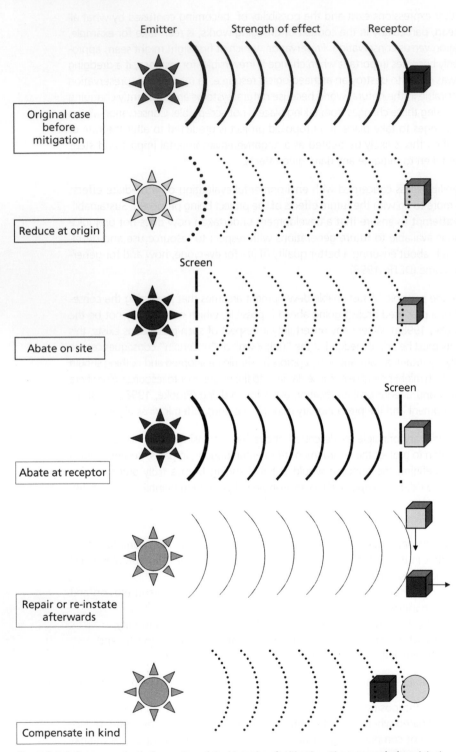

Figure 9.4 A diagrammatic representation of the hierarchy of mitigation. The top row is the original project before any kind of mitigation has been applied.

The word "source" is often used to denote the point from which the undesirable environmental effect emanates. When dealing with contaminated materials, however, the source is often a remote industrial plant. It may even be in another country. For this reason the word "origin" is used here to denote the origin of the undesirable effect at the site.

Examples of mitigation options are given in Figure 9.4 and are explained below. The argument could be put forward that "compensation in kind" is not a true mitigation measure, because compensation is something that is done when mitigation is not possible (Brooke, 1998b).

Avoid at origin
Avoidance at origin may apply to the actual project itself, such as the formation of a dredged channel or the construction of a transport link, or it may apply to the construction process. In the first instance, the objective of total avoidance of impact may be built into the design by a so-called "zero effect" clause in the contract documents. Examples of this may be found in both the Storebaelt (The Great Belt Fixed Link) and the Øresund road and rail crossings in Scandinavia, where the design specification stipulated that the project was not to have any effect on the flow of water through the straits across which the crossings were to be built (see Annex B).

In the second instance, the design may stipulate that certain environmental parameters are not to be exceeded (which may or may not exclude certain types of equipment) or that the dredging or reclamation activities may only be carried out in a particular season.

Reduction at origin
Here the adverse effect may be reduced to an acceptable level, for instance, by the installation of acoustic reduction equipment on noisy vessels or land-based plant, or the imposition of restrictive working practices during the dredging or reclamation operations. Prevention of overflow during the dredging of silts by trailing suction hopper dredger is an example of imposing restrictions.

Abatement on site
In the case of a dredging process which results in a significant increase in suspended sediment levels, the installation of a silt curtain (see Figure 9.5) might prevent the suspended materials from leaving the vicinity of the dredging works (although it should be noted that silt curtains are really only feasible in very low energy locations, such as some sheltered or inland waters). Alternatively, the dredging might be carried out within a temporary bund which prevents egress of turbid waters from the site, the temporary bund being removed when the suspended material has settled.

Abatement at the receptor
In this case mitigating the impact at origin would be impractical or excessively expensive, but the receptor or sensitive receiver may be effectively protected from the impact. For instance, the dredging of an ornamental lake may be quite acceptable

Figure 9.5 Abatement on site: A floating silt curtain being deployed during the reclamation activities for the construction of Sydney, Australia's new runway (C.P. Herbert and T.P. Biggart, 1996).

for the majority of the flora and fauna but may produce an unacceptable impact on a rare resident fish population. The cost of temporarily removing the fish, storing them in appropriate conditions and returning the fish to the lake after the dredging has been completed may be significantly cheaper than trying to mitigate the impact at the dredging site.

Reparation or re-instatement after completion of the work
In this instance, the dredging or reclamation may cause significant modification or loss of valuable habitat. However, this modification or loss may be temporary – because the habitat will recover – or the habitat can be recreated upon completion of the works, thereby ensuring that after a recovery period the modification or loss will have been repaired or re-instated.

Compensation in kind
Sometimes a dredging project will cause a habitat or an amenity to deteriorate or will even modify it to a completely different habitat, or convert it to a non-environmental use (e.g. construction fill). In such circumstances, if the project is to proceed, compensating for the loss of the resource by re-creating it in a different location may be an acceptable solution. For instance, if a dredging project in an estuary involves removal of some valuable marsh or wetland, then the dredged material could possibly be used to re-create this type of habitat in another location in the same estuary.

In such cases, however, conservation agencies are increasingly requiring that the replacement habitat be in place, and functioning, before any loss of existing habitat occurs. This can create logistical problems, and often increased costs, for the developer if the dredging programme has to be altered or if material has to be brought in from elsewhere. Some agencies operate on the basis of "no net loss", such that no project will obtain consent unless it can be demonstrated that the basic stock of the resource (habitat) is not being depleted.

Another example of compensation in kind is the creation of underwater berms for spawning areas when nearby spawning areas are to be affected.

Compensation by other means
An example of this is when a fishing area is affected by dredging or reclamation works and a significant drop in the value of future catches is predicted. The fishermen who use the affected area may be given a monetary compensation related to their predicted loss of future income.

Other examples may be less obvious. A project may modify or destroy a locally important wildlife site used for recreation and the local population may be given a leisure centre to compensate for the loss. Although both the above examples may be deemed acceptable solutions for the local community, it is doubtful whether either of them should properly be classified as mitigation measures.

Enhancement
Enhancement is achieved by providing some local resource or facility, which did not exist before the project took place, i.e. a resource, asset or facility which is provided over and above measures taken to restore the pre-development or "status quo" situation. The compensatory enhancement may not be connected to environmental mitigation at all in the sense used above, but may be provided to please a local community who have permitted a development to take place for the benefit of the region, but who are not going to benefit much from it themselves. As such it might be considered as project mitigation but not environmental mitigation.

In terms of natural habitats, enhancement may refer to the creation or restoration of a resource over a larger area than that requiring mitigation – the excess area may, thus, be classified as an enhancement.

9.3.6 Impact of climate change

The impact of climate change is coming into the forefront of public awareness at an increasing pace. Up to now, the effects of climate change in coastal areas, for instance in Denmark (Erichsen, 2001), have focused on the impacts of a rising sea level on coastal morphology and the marine infrastructure. Now, however, preliminary evaluations of the possible effects of climate change on the productivity of the coastal waters stress the importance of including ecological considerations in future assessments of potential impacts.

In addition, when designing ports and coastal protection measures, it has now also become evident that long-term climatic variability, whether caused by humans or nature, must be taken into account. Sea-level rise as much as 3 mm per annum has been observed in recent years and expectations are that a rise of 28–43 cm will take place over the next hundred years (IPCC, 2007). This will increase the need for construction materials, e.g. to build new dikes in areas that previously had not been exposed to flooding and to increase the height and strength of existing dikes and breakwaters. Clearly, climate change will increase the need for dredging. For example, looking at beach nourishment, additional amounts of sediment will be required to main-tain the equilibrium beach profile that relates to a higher water level (Mangor, 2004).

Climate changes may also affect coastal ecosystems, because of changes in salinity lev-els and interface, increase in water temperature, altered dynamics in phytoplankton production and increased water level (CONWOY, 2006). So in design and in environ-mental impact assessment, climate change must be considered in order to minimise the risk of structural failure and to prevent potential environmental impacts in an environment that might undergo measurable natural changes, i.e. change in base-line conditions, within the lifetime of the structure.

9.4 THE FUTURE

9.4.1 Taking account of climate and biodiversity

One of the reasons for publishing this book is that countries which have not yet pro-gressed very far in dealing with the environmental aspects of dredging may learn from those that are further down the line. Yet a note of caution is due at this point, because the world is complex and its characteristics vary enormously from place to place. As an example, climate has a profound effect on the subject under scrutiny.

As a general rule, climate in higher latitudes is cold, dynamic and generally less hospitable than the climate in the lower latitudes. Weather and sea conditions are generally more severe and variable. For this reason, the biodiversity of the higher latitudes is lower, and plants and animals which inhabit these regions are capable of surviving in extreme con-ditions. They are, consequently, less susceptible to sudden changes to the ecosystem.

In warmer and more benign latitudes, conditions are more conducive to sustaining life and biodiversity is greater. Plant and animal life are still subject to natural shocks, such as cyclones, monsoons and heavy rainfall, but either they are able to recover quickly or their life cycles have been developed to avoid the effects of these events. Either way, many delicate forms of life are able to survive. This may make these regions more vulnerable to environmental shock.

Significantly, many of the sites where dredging and reclamation occur are situated in rivers and estuaries, where the major ports have evolved. These sites often become heavily silt-laden in times of heavy upland rainfall and are, thus, likely to contain a benthic fauna which can accept the silt loads caused by dredging activities.

In fact, much of the development of the procedures for assessing the environmental effects of dredging activities has occurred in the higher latitudes. Therefore, the possibility exists that some of the practices acceptable in these regions may not be acceptable in the more benign areas of the world. While the procedures for dealing with the environmental effects of dredging are to all intents and purposes global, they must be applied in an appropriate manner for the climate of the country under scrutiny.

9.4.2 Partnering in future works with active public participation

Partnering, a process which is founded on the pursuit of mutual objectives (Hopman and Tate, 1995), has been used on a number of major dredging-related projects. In its comprehensive form, where public participation is encouraged, partnering is a process which brings together all the important aspects of environmental awareness under one umbrella and ensures that any project which proceeds has the acceptance of the vast majority of the people it affects.

Whether it be a means of enabling an existing practice to continue or determining whether a new project may proceed, partnering involves all of the environmental checks and balances described in this book. In its simplest form, partnering is a means of involving the whole community in the planning, evaluation, permitting and monitoring of a proposed development. As such it involves publicity, consultation, transparency of assessment and decision-making, and considerable public participation.

The essentials for partnering are:
- identifying all interested parties;
- publicising the project and inviting involvement from the outset;
- drawing all interested parties into the project to become stakeholders;
- carrying out comprehensive environmental assessment of conceptual ideas, prior to detailed engineering studies;
- ensuring consultation takes place at all stages of development, particularly at the beginning but also after environmental assessment has been carried out;
- ensuring that dissemination of information about the project and the environmental effects is presented clearly;
- involving end users in the planning, design and construction phase;
- maintaining open communications, and
- maintaining a balance between environmental and commercial goals.

A particularly good account and example of partnering at work is given in the paper by Vallis and Maynard (1999) on the Northumberland Strait Crossing Project in Canada. This major engineering project, which involved the planning, design and construction of 12.9 kilometres of bridge, 40 metres above the Straits with 65 spans and included dredging to form the foundations for bridge piers, was driven by concerns for both the social and physical environment. Numerous features of the development came as solutions to these concerns. The bridge design followed rather than led the environmental assessments. Quoting from the conclusions of the Vallis and Maynard paper:
- good engineering design solutions flow from sound environmental assessment;
- environmental considerations must be a major design criteria from the outset, and
- good environmental assessment depends on sound science and open public involvement.

Without a doubt all major and many minor dredging and reclamation projects in the future will benefit from the partnering approach. At this time, however, the methods by which all parties are involved in the early, pre-competitive stages of a project are still being developed. Attempts to obtain a wider ownership of proposed schemes may be seen in public-private partnerships and enterprise schemes.

While the partnering philosophy is essential for involving the community in a project, another aspect of partnering also needs to be developed. This is the means by which client and contractor together take both technical and environmental responsibility for the works. It is not uncommon to see a client/contractor relationship in which both parties share the benefits gained from, say, a cost saving resulting from value engineering. Less common is to see both client and contractor taking the lead in bearing the environmental responsibility for any work carried out.

Partnering and Technical Advisory Groups (TAGS) are useful tools for long-term assessment and management of dredging (Dobson, 2000 and Challino, 2007). A TAG for a port authority is made up of stakeholder representatives, including those from the local community and environmental groups. A TAG meets regularly to address any existing or future dredging issues that come up from time to time. In this way agreement can be reached on such matters as the placement of dredged material and any environmental effects can be better understood and managed.

9.4.3 Future development of environmental economics

Much of the discourse in this book will lead the reader to the conclusion that a major problem in environmental evaluation is obtaining a satisfactory balance between economy and ecology. One of the reasons for this difficulty is that people find it easy to show the economical costs and benefits of a scheme but very difficult to show the ecological losses and gains in the same format. In other words, valueing the tangible ecological attributes of an area is difficult and valuing intangible effects is even more difficult.

The fact that many environmental resources, such as habitats, species, landscapes and so on, are not reflected in market prices tends to send the wrong message to the consumer (Spurgeon, 1999). If market prices are used in economic valuations, a risk exists that these resources will suffer excessive degradation and unsustainable use. Hence an economic value for these environmental resources and the impacts upon them needs to be developed. This would then allow for the development of economic instruments to adjust the way the market works to help protect environmental resources. Measures such as altering prices by green taxes/subsidies, and creating markets, e.g. by tradeable emission permits, may then be put in place.

The total economic value of an environmental asset may be split into three parts, as follows:

1. its direct uses, such as products or for recreation;

2. its indirect uses, such as biological support, e.g. a food source, or physical protection, e.g. a saltmarsh providing coast protection by dissipating wave energy; and

3. non-use values, such as option and existence values.

The last values are unfamiliar concepts and need to be defined. The option value is the value obtained from ensuring that an environmental asset is maintained for future use. The existence value is the value obtained from simply knowing that an environmental asset continues to exist. Both these values are generally measured in terms of the public's willingness to pay for existence and options for future use.

To determine the total economic value of an environmental asset is difficult, particularly when trying to put a figure on the non-use values. A number of techniques are available. Most are imperfect, but they nonetheless provide an order of magnitude of value, generally giving minimum values only. Full valuation studies are expensive. Tables 9.1 and 9.2 (Spurgeon, 1999) show the range of techniques available in terms of the asset being valued and the type of value being assessed. Some of the techniques need to be explained further.

Valuation technique	Habitats/ species	Water quality	Noise	Air quality	Recreation	Heritage
Change in productivity	•	•	•	•	•	
Preventative expenditure	•	•	•	•		
Replacement cost	•					•
Hedonic pricing	•	•	•		•	•
Travel method cost	•	•			•	•
Contingent valuation	•	•	•	•	•	•

Table 9.1 Range of valuation techniques by environmental parameter.

Valuation technique	Direct uses	Indirect uses	Non-use values
Change in productivity	•	•	
Preventative expenditure	•	•	
Replacement cost	•	•	
Hedonic pricing	•		
Travel method cost	•		
Contingent valuation	•		•

Table 9.2 Range of valuation techniques by usage.

Change in productivity

This is based on the cause and effect on market prices of an environmental good. For example, the loss of a salt marsh may lead to losses in fishery benefits, or air pollution may lead to an increase in health costs. This technique is readily applicable, but estimating the relationship between cause and effect is often very difficult.

Replacement cost

This technique assumes that the replacement cost of an asset is a proxy for its value. For example, a salt marsh may act as a form of coast protection. Putting a value on the coast protection function is possible. Putting a value on the salt marsh itself (or the cost of re-creating it) is also possible. This technique is also readily applicable, but replacement costs may vary enormously from place to place.

Hedonic pricing

Hedonic pricing is based on the principle that the value of an asset is comprised from the sum of its attributes and that these can be disaggregated. For example, the price of a property will be higher in an area of outstanding natural beauty than the aggregate value of the land and construction cost of the building. In a noisy area, the value may be lower than these costs. Applying this principle is difficult and may be complex and expensive to use, although some basic variations may be useful.

Contingent valuation

This technique relies on determining the value of an environmental asset by asking a representative sample of people how much they are willing to pay for it. Examples of this technique are the value to society of nourishing a sandy beach, or their willingness to pay for increased road tolls relating to the cost of environmental compensation during road construction. This technique is also readily applicable and highly flexible, but may be contentious and expensive.

The environmental economics techniques described above can be used:
● to provide an accountable means of assessing who benefits and who loses, and by how much, from any use of resources;
● to improve economic and financial returns from resource use, and
● to help achieve sustainable development and optimum use of environmental resources.

9.4.4 Future research

The following list of research topics has been developed from this book and from other sources. These are subjects which IADC and CEDA have identified as requiring further research. The list does not imply any particular priority, nor is it exhaustive.

a) Standard specifications for sampling and testing
   Methods of testing for contaminants vary greatly, with little consistency in the procedures carried out. Analytical methods need to be standardised. Some analytical methods for the measurement of hazardous substances still need to be developed.

b) Placement of capping material on weak sub-aqueous sites
   Design and construction methods suitable for the placement of capping material on weak, sub-aqueous sites need to be researched and publicised. Designs for capping have been carried out and some placement techniques have been developed. They are not well known outside the industry and do not cover all types of site. Designers should be able to design capping schemes which are suitable for their environment and are constructable at reasonable cost. Currently, more data from existing sites where capping has been carried out is needed.

c) Environmentally sensitive dredging techniques
   Chapter 6 describes a number of dredgers specially designed for working in environmentally sensitive sites. New dredgers are continually being developed, as well as

new techniques for environmental mitigation. A research project is required to identify and describe mitigation measures. These measures might be specific on-site examples of mitigation techniques or more generic "greenhouse effect" measures, which are already being applied in some countries, such as reducing fuel consumption by fuel-efficiency methods, the pros and cons of electric power, use of low sulphur fuels and so on. These measures need to be promoted on a worldwide basis.

d) The development of Cumulative Effects Analysis
    An agreed, standardised method of evaluation should be adopted which puts all human (and therefore controllable) activities on the same playing field. This is a part of the so-called "cumulative effects" analysis. Research should focus attention on cumulative impact analysis; on who, ultimately, should evaluate the environmental status quo of the region; on who should pay for the data collection required and how the various competing economic and environmental pressures should be balanced. Note that analysis should be on a coastal cell, watershed (catchment) wide basis.

e) The effect of extreme events on environmental analysis
    The degree of environmental impact from a dredging operation is often less than that of an extreme natural event, such as a storm or large flood. These extreme events, however, are often poorly recorded and many environmental parameters are not recorded during the extreme event because of the cost and/or difficulty. A project is required to devise means for better recording environmental parameters during extreme events and for incorporating these data into the environmental evaluation.

f) Development of inexpensive contaminant screening tools
    At the current time, the screening for contaminants can be a costly business, leading to the problems outlined in Section 9.2.3. To alleviate this problem, further research is required to develop inexpensive methods of screening for contaminants.

g) Assessment of chronic/sub-lethal effects of contaminated dredged material
    Research is required to identify the levels of contamination in dredged material which cause chronic or sub-lethal effects on various types of organisms.

h) Assessment of the real impact of physical changes to the environment
    As mentioned in Section 9.4.1, considerable differences exist in the biodiversity and sensitivity of ecologies in differing lattitudes and types of climate. The real impact of changes in environmental parameters needs to be investigated for the different categories. This investigation could go a long way to ensuring that regulations are appropriately applied to the appropriate situation, rather than indiscriminately on a worldwide basis.

9.5 EPILOGUE

This IADC/CEDA book on the Environmental Aspects of Dredging is aimed at promoting a responsible attitude toward dredging and dredging-related activities in accordance with the WODA Mission Statement, quoted at the beginning of this chapter.

The Editorial Board recognises that a number of challenges arise in attempting to accomplish this task:

- since the whole field of play is constantly evolving, the book is aiming at a moving target;
- the book needs to be read carefully to be understood, and the messages contained in it need to be considered thoroughly when implementing future projects. This requires that a concerted effort be put into disseminating this knowledge in trainings and seminars, and
- the book include the facts, views and opinions of numerous authors, reviewers and commentators. Feedback is required to improve them or to get the relevant messages across more effectively.

CEDA and IADC will continue the WODA mission. Our intention is to update this book periodically or to produce additional material to complement it, in an effort to keep the aim on target. Our target audiences in fact are all those involved in dredging projects – stakeholders of all varieties, the public, scientists and financiers. For some of these readers, the book may raise more questions and we stand ready to try to provide more information.

These and all other readers are invited, in fact, encouraged, to comment on this publication. They can constructively add to the debate by sending us their views on any of the subjects raised and by providing us with details of interesting projects carried out. People who have any reflections on the more philosophical issues presented in this chapter are encouraged as well to send us their thoughts and concerns.

IADC may be contacted at: info@iadc-dredging.com and CEDA may be contacted at: ceda@dredging.org.

Seminars and training
To communicate the various messages presented in this book, we also need to promote the idea of training in this field. Both IADC and CEDA run seminars and conferences on dredging and reclamation and have jointly produced a specific two-day seminar/workshop on the Environmental Aspects of Dredging, which has proved to be popular to a wide audience. Often they have organised this seminar at the request of third parties. Other members of the WODA community also have seminars on environmental matters. In addition, the International Maritime Organisation (IMO) runs a Technical Co-operation and Assistance Program for developing countries.

This book on the Environmental Aspects of Dredging is specifically intended to be used for these types of training courses and seminars. If you are running a suitable training operation, contact IADC and CEDA and everything possible will be done to make this book available at a reasonable cost.

References/Bibliography

AS4360 (2003). *Australian Standard for Risk Management.*

ASTM (1994). *Standard Guide for Collection, Storage, Characterization, and Manipulation of Sediments for Toxicological Testing*. American Society for Testing and Material (ASTM) 1994 Annual Book of Standards. Vol. 11.04, E1391–90. Philadelphia, PA, USA.

Bailey, S.E. *et al.* (2006). *Evaluation of chemical clarification polymers and methods for removal of dissolved metals from CDF effluent*, DOER-R10, U.S. Army Engineer Research and Development Center, Vicksburg, MS, USA.

Baretta-Bekker *et al.* (1998). *Encyclopedia of Marine Sciences*, 2nd Edition, Springer-Verlag Berlin Heidelberg New York.

Bates, A.D. (1981). Profit or Loss Pivot on Pre-Dredging Surveys. *World Dredging and Port Construction*, pp 21–23.

Bearman, G. (Ed.) (1989). *Waves, Tides and Shallow-water Processes*. The Open University Pergamon Press, Oxford, UK.

Blake, R.R. and Mouton, J.S. (1970). *The Managerial Grid*. Gulf Publishing Company, Houston, TX, USA.

Bolam, S.G. and Whomersley, P. (2003). Invertebrate recolonisation of fine-grained beneficial use schemes: an example from the south-east coast of England. *Journal of Coastal Conservation* 9:159–169.

Borst, W.G. *et al.* (1994). Monitoring of water injection dredging, dredging polluted sediments. *Dredging 94 Conference Proceedings, Volume 2.*

Brannon, J.M. *et al.* (1990). *Comprehensive analysis of migration pathways (CAMP): Contaminant migration pathways at confined dredged material disposal facilities*. Miscellaneous paper D-90-5, U.S. Army Engineer Waterways Experiment Station, Vicksburg, MS, USA.

Bray, R.N. (Ed.) (2004). *Dredging for Development*. 5th Edition. IADC/IAPH. The Hague, The Netherlands.

Bray, R.N. *et al.* (1997). *Dredging; a Handbook for Engineers*. Arnold Publishing, London, UK.

Brooke, J. (1998a). Strategic coastal defence planning: the role of the planning system. Paper presented to CIWEM meeting *"Sustainable development: what does it really mean and cost for rivers and coasts?"*

Brooke, J. (1998b). Environmental impact assessment: training notes. Personal communication.

BS 6349 (1991). *Code of Practice for Maritime Structures: Part 5: Dredging and Land Reclamation*. British Standards Institution. London, UK.

CEAA (1992). *Cumulative Effects Assessment Practitioners Guide*. Canadian Environmental Assessment Agency, Cumulative Effects Assessment Working Group and AXYS Environmental Consulting Ltd.

CEEC-CEDA (1994). Environmental Dredging and Treatment of Dredged Material Course at Leuven, Belgium. February 1994.

Challinor S. (2007). Public Aspects of Dredging. *Proceedings Third International Conference on Dredging*. Aviles, Spain.

CIRIA (2000). *Maximising the use and exchange of coastal data*. Construction Industry Research and Information Association. London, UK.

Collins, R.J. (1979). Dredged silt as a raw material for the construction industry. *Resource Recovery and Conservation. No4 1980*. pp. 337–362. Elsevier Scientific Publishing Co. Amsterdam, The Netherlands.

CONWOY (2006). *Consequences of weather and climate changes for marine and freshwater ecosystems – Conceptual and operational forecasting of the aquatic environment*. Danish Research Project with among others DHI Water, Environment and Health. (Book in Danish) http://www.dhigroup.com/News/NewsArchive/2006/ConsequencesOfWeatherAndClimateChanges.aspx

Csiti, A. and Donze, M. (1995). Sediment Sampling Strategy for Cleanup Projects. *Proceedings of World Dredging Congress XIV*, Amsterdam. Central Dredging Association, Delft, The Netherlands. pp. 743–754.

Dearnaley, M.P. *et al.* (1996). *Resuspension of bed material by dredging*. HR Wallingford Report SR 461.

Dearnaley, M.P. *et al.* (1995). Beneficial uses of muddy dredged material. *Proceedings of World Dredging Congress XIV*, Amsterdam. CEDA. Delft, The Netherlands.

Deibel, I.K. *et al.* (1996). Nine years' experience in filling a large disposal site with dredged material. *Proceedings CATS III Congress*. Oostende, Belgium. pp. 219–231.

DETR (1997). *Mitigation Measures in Environmental Statements*. Department of the Environment, Transport and the Regions, UK.

DETR (1998). *Sustainable local communities for the 21st century: why and how to prepare an effective Local Agenda 21 strategy*. Department of the Environment, Transport and the Regions, UK.

Detzner, H.-D. and Knies, R. (2004). Treatment and beneficial use of dredged sediments from Port of Hamburg. *Proceedings of the 17th World Dredging Congress*. Hamburg, Germany. 27 September–1 October 2004.

Detzner, H.D. *et al.* (1997). New technology of mechanical treatment of dredged material from Hamburg Harbour. *International Conference on Contaminated Sediments*. September 1997. Rotterdam, The Netherlands. Pre-prints. pp. 267–274.

Detzner, H.D. (1995). The Hamburg project METHA: large-scale separation, dewatering and reuse of polluted sediments. *European Water Pollution Control. Vol. 5., No. 5.*, September 1995. pp. 38–42.

DGE (2002). DGE (2003). *Dredged Material and Legislation*. Dutch-German Exchange (DGE) on Dredged Material – Part 1.

DGE (2002a). *Treatment and Confined Disposal of Dredged Material*. Dutch-German Exchange on Dredged Material – Part 2.

Dixon J, (1999). *Environmental Economics and Indicators*. The World Bank Group, www.worldbank.org, September. Washington, DC, USA.

Dobson, J. (2000). Personal communication.

Dobson, J. (1998). Personal communication. Eastern Dredging Association (EADA).

Donze, M. (Ed.) (1990). *Shaping the environment: Pollution and dredging in the European Community*. DELWEL. The Hague, The Netherlands.

Dortch, M.S. *et al.* (1990). Methods of determining the long-term fate of dredged material for aquatic disposal sites. *Technical Report D-90-1*. US Army Engineer Waterways Experiment Station, Vicksburg, MS, USA.

Dtv-Atlas zur Ökologie (1990). Deutsher Tashenbuch Verlag GmbH & Co. KG, München, Germany (In German).

Dutch Government (1994). Besluit Milieu-effectrapportage 1994 (EIA Decree 1994). Stb. nr. 540, 26 July 1994.

EFTEC (1996). *An economic appraisal of the disposal and beneficial use of dredged material*. Report prepared for Ministry of Agriculture, Fisheries and Food. Economics for the Environment Consultancy Ltd, UK.

EPA (1997). *Considering Cumulative Effects under the National Environmental Policy Act*. US Environmental Protection Agency. Washington, DC, USA.

EPA (1995). *QA/QC Guidance for Sampling and Analysis of Sediments, Water, and Tissues for Dredged Material Evaluations: Chemical Evaluations*. U.S. Environmental Protection Agency EPA 823-B-95-001. U.S. EPA Office of Water, Washington, DC, USA.

EPA/USACE (2004). *Evaluating Environmental Effects of dredged material Management Alternatives – A Technical framework*. U.S. Environmental Protection Agency/ U.S. Army Corps of Engineers EPA-842-B-92-008. Washington, DC, USA.

EPA/USACE (1998). *Evaluation of Dredged Material Proposed for Discharge in Waters of the U.S. – Testing Manual: Inland Testing Manual*. U.S. Environmental Protection Agency/U.S. Army Corps of Engineers EPA-823-B-98-004. U.S. EPA Office of Water (4305), Washington, DC, USA.

EPA/USACE (1992). *Evaluating Environmental Effects of Dredged Material Management Alternatives – A Technical Framework*. U.S. Environmental Protection Agency/ U.S. Army Corps of Engineers EPA-842-B-92-000. U.S. EPA Office of Water, Washington, DC, USA.

EPA/USACE (1991). *Evaluation of Dredged Material Proposed for Ocean Disposal: Testing Manual*. EPA-503/8-91/001. U.S. Environmental Protection Agency/U.S. Army Corps of Engineers. U.S. EPA Office of Water (WH-556F), Washington, DC, USA.

European Council (1985/1997). Council Directive 97/11/EC of 3 March 1997 amending Directive 85/337/EEC on the assessment of the effects of certain public and private projects on the environment. Brussels, Belgium.

Edelvang, K. *et al.* (2005). Numerical modelling of phytoplankton biomass in coastal waters. *DHI Water & Environment, Journal of Marine Systems*, 2005: 57:(1–2).

Flach, E.N. *et al.* (1997). Environmental Impact Assessment for a disposal site in the lake Ketelmeer area, The Netherlands. *International Conference on Conta-minated Sediments*. September, 1997. Rotterdam, The Netherlands. Pre-prints. pp. 881–888.

Fowler, J. *et al.* (1996). Dredged material-filled geotextile containers. *Port Engineering Management*. Vol. 14: Issue 2, pp. 27ff.

Fredette T.J. *et al.*(1990). *Selected tools and techniques for physical and biological monitoring of aquatic dredged material disposal sites*. Technical Report D-90-11. U.S. Army Engineer Waterways Experiment Station, Vicksburg, MS, USA.

Freiwald, A. *et al.* (2004). *Cold-water Coral Reefs*. UNEP-WCMC, Cambridge, UK.

Germano, J.D. and Rhoads D.C. (1984). REMOTS® sediment profiling at the Field Verification Program (FVP) disposal site. pp. 536–544. R.L. Montgomery and J.W. Leach (Eds.). *Dredging and Dredged Material Disposal. Volume 1*. American Society of Civil Engineers. New York, NY, USA.

Germano, J.D. *et al.* (1994). An integrated, tiered approach to monitoring and management of dredged material disposal sites in the New England region. *Disposal Area Monitoring System DAMOS*, USACE, New England division, Contribution No. 87.

Goossens, H. and Zwolsman, J.G. (1996). An evaluation of the behaviour of pollutants during dredging activities. *Terra et Aqua*, No. 62. March 1996. pp. 20–28.

Great Lakes Commission (2004). *Testing and evaluating dredged material for upland uses*. USA.

Grosser, R.D. (1991). *Historic Property Protection and Preservation at U.S. Army Corps of Engineers Projects*. Technical Report EL-91-11, U.S. Army Engineer Waterways Experiment Station, Vicksburg, MS, USA.

Gustavson, K. and Wangberg, S.A. (1995). Tolerance induction and succession in microalgea communities exposed to copper and atrazine. *Aquatic Toxicology*. 32. pp. 283–302.

Harms, C. (1995). Monitoring the Hamburg silt mound. *Proceedings of World Dredging Congress XIV*, Amsterdam. CEDA, Delft, The Netherlands. pp. 697–708.

Harris, L.E. (1994). Dredged material used in sand-filled containers for scour and erosion control. *Proceedings 2nd International Conference on Dredging and Dredged Material Placement*, Florida, USA. American Society of Civil Engineers. New York, NY, USA.

Herbert, C.P. and Biggart, T.P. (1993). Kingsford Smith Airport Sydney: Planning and Tendering the New Parallel Runway. *Proceedings Institution of Civil Engineers*. Paper No. 10140, November.

Holliday, B.W. (1998). Personal communication. USACE, Dredging and Navigation Branch, Operations Division.

Hopman, R.J. and Tate, K.D. (1995). Partnering is a practical dredging benefit. *Proceedings of World Dredging Congress XIV*, Amsterdam. CEDA, Delft, The Netherlands.

HR Wallingford (1996). *Guidelines for the beneficial use of dredged material*. Report SR 488. UK.

HTG (2006). *Hafentechnische Gesellschaft Fachausschuss Baggergut: Abbau von Tributylzinn/TBT in Sedimenten und Baggergut*. Literaturübersicht, Versuche, praktische Erfahrungen.

Hufschmidt *et al*. (1983). *Environment, Natural Systems, and Development – An Economic Valuation Guide*, The John Hopkins University Press, Baltimore, MD, USA.

IADC/CEDA (1999). *Environmental Aspects of Dredging. Guide 5: Reuse, Recycle or Relocate*, International Association of Dredging Companies/Central Dredging Association, The Netherlands.

IADC/CEDA (1998). *Environmental Aspects of Dredging. Guide 4: Machines, Methods and Mitigation*. International Association of Dredging Companies/Central Dredging Association, The Netherlands.

IADC/CEDA (1997). *Environmental Aspects of Dredging. Guide 2a: Conventions, Codes and Conditions: Marine Disposal*. International Association of Dredging Companies/Central Dredging Association, The Netherlands.

IADC/CEDA (1997a). *Environmental Aspects of Dredging. Guide 3: Investigation, Interpretation and Impact*. International Association of Dredging Companies/ Central Dredging Association, The Netherlands.

IMO (2005). *Guidelines for the Sampling and Analysis of Dredged Material for Disposal at Sea*. International Maritime Organization, London, UK. ISBN 9280141929.

IPCC (2007). Intergovernmental Panel on Climate Change. Climate 2007, WMO, UNEP, IPCC Secretariat, c/o WMO Geneva, Switzerland.

Janssen, R. (1991). *Multiobjective decision support for environmental problems*. PhD thesis. Free University, Amsterdam, The Netherlands.

Jansen P.Ph. *et al*. (1979). *Principles of river engineering – the non-tidal alluvial river*. Rainbow-Bridge Book Co., Teipei, Taiwan.

Jensen, K. *et al*. (1998). *Environmental Impact Assessment using the Rapid Impact Assessment Matrix (RIAM)*, Olesen & Olesen, Fredensborg, Denmark.

Jeschke, G. and Jeschke, B. (1999). Bremen's Überseehafen is reborn as reclaimed land. *Dredging and Port Construction*, March, Vol. XXVI, No. 111.

Jin, J-Y. *et al.* (2003). Behavior of Currents and Suspended Sediments around a Silt Screen. *Ocean and Polar Research* Vol. 25 (3S), pp. 399–408.

John, S.A. *et al.* (2000). *Scoping the assessment of sediment plumes arising from dredging*. Construction Industry Research and Information Association (CIRIA). RP600.

Kirby, R. (1996). Beneficial uses of fine-grained dredged material – new frontiers. *Port Engineering Management*. Vol. 14, No. 4. pp. 32–35.

Korevaar, A. (1994). An update on dredge instrumentation and automation. *Dredging 94 Conference Proceedings*. Volume 1.

Laboyrie, H. and Flach, B. (1998). The handling of contaminated dredged material in the Netherlands. *Proceedings of World Dredging Congress XV*, Las Vegas, Nevada. WEDA, Vancouver, Washington, USA. pp. 513–526.

Laboyrie, H.P. *et al.* (1995). The construction of large-scale disposal sites for contaminated dredged material *Proceedings of World Dredging Congress XIV*, Amsterdam. CEDA. Delft, The Netherlands. pp. 709–722.

Land, J.M. and Bray, R.N. (1998). Acoustic Measurement of Suspended Solids for Monitoring of Dredging and Dredged Material Disposal. *Proceedings of World Dredging Congress XV*, Las Vegas, Nevada, USA.WEDA, Vancouver, Washington, USA. pp. 105–120.

Landin, M.C. (1986). *Building, developing and managing dredged material islands for bird habitat*. Publication EEDP-07-1. US Army Engineer Waterways Experiment Station, Vicksburg, MD, USA.

Landin, M.C. *et al.* (1994). New applications and practices for beneficial uses of dredged material. *Proceedings of the 2nd International Conference on Dredging and Dredged Material Placement*. American Society of Civil Engineers. New York, NY, USA.

LaSalle, M.W. *et al.* (1991). *A Framework for Assessing the Need for Seasonal Restrictions on Dredging and Disposal Operations*. Technical Report D-91-1. U.S. Army Engineer Waterways Experiment Station, Vicksburg, MS, USA.

LC-72 (1996). *Dredged Material Assessment Framework*. LC.2/Circ.368. 28 February 1996. (Available at http://www.dredging.org.)

Lee, C.R. *et al.* (1991). *General Decision making Framework for Management of Dredged Material: Example Application to Commencement Bay, Washington*. Miscellaneous Paper D-91-1. U.S. Army Engineer Waterways Experiment Station, Vicksburg, MS, USA.

Lillycrop, L.S. and Clausner J.E. (1998). Numerical design of the 1997 capping project at the Mud Dump Site. *Proceedings of World Dredging Congress XV*, Las Vegas, Nevada. WEDA, Vancouver, Washington, USA. pp. 937–951.

Male, J.W. and Cullinane, Jr., M.J. (1989). Procedure for Managing Contaminated Dredged Material. *USACE Course, Dredged Material Management: Engineering and Environmental Advances. 13–17 February*. Vicksburg, MS, USA.

Mangor, K. (1998). Coastal restoration considerations. *Presented at ICCE '98 26th Coastal Engineering Conference, 22–26 June*. Copenhagen, Denmark.

Mangor, K. (2004). Shoreline Management Guidelines, *DHI Water & Environment*, December 2004, ISBN 87-981950-5-0.

Mathieson *et al.* (1991). Intertidal and Littoral Ecosystems, *Ecosystems of the World*, Vol. 24, Elsevier Science Publishers B.V., The Netherlands.

Mehta, A.J. and Jiang, F. (1993). *Some observations on water wave attenuation over nearshore underwater mudbanks and mud berms*. Paper UFL/COEL/MP-93/01. Coastal and Oceanographic Engineering Department, University of Florida, Gainesville, Florida, USA.

McFarland, V.A. *et al.* (1989). Factors Influencing Bioaccumulation of Sediment-Associated Contaminants by Aquatic Organisms. *Environmental Effects of Dredging Technical Notes* (Numbers EEDP-01-17, 18, 19, and 20). U.S. Army Engineer Waterways Experiment Station, Vicksburg, MS, USA.

Mintzberg, H. *et al.* (1976). The Structure of "Unstructured" Decision Processes. *Administrative Science Quarterly*, Vol. 21, pp. 246–275.

Murdoch, A. and MacKnight, S.D. (1991). *Handbook of Techniques for Aquatic Sediments Sampling*. CRC Press, Inc., Boca Raton, Florida, USA.

Murk *et al.* (1996). Chemical-activated luciferase gene expression (CALUX): a novel in vitro bioassay for Ah receptor active compounds in sediments and pore water. *Fund. & Applied Tox.* 33: 149–160.

Murray, P.M. *et al.* (1998). Monitoring results from the first Boston harbor navigation improvement project confined aquatic disposal cell. *Proceedings of World Dredging Congress XV*. Las Vegas, Nevada. WEDA, Vancouver, Washington, USA. pp. 415–430.

Myers, T.E. *et al.* (1993). *Management plan for the disposal of contaminated material in the Craney Island Dredged Material Management Area*. Technical Report EL-93-20. US Army Engineer Waterways Experiment Station, Vicksburg, MS, USA.

Nelson, D.A. and Shafer, D.J.(1996). *Effectiveness of a Sea Turtle-deflecting Hopper Dredge Draghead in Port Canaveral Entrance Channel, Florida*. (MP D-96-3). US Army Corps of Engineers Waterways Experiment Station, Vicksburg, MS, USA.

Netzband , A. *et al.* (1999). Water Injection Dredging in Hamburg – Application and Research. *Proceedings CEDA Dredging Days*, Amsterdam, The Netherlands.

Netzband, A. (1994). Dredged material management in the port of Hamburg. *European Water Pollution Control*. Vol. 4. No. 6. pp. 47–54.

Netzband, A. and Rohbrecht-Buck, K. (1992). Treatment of effluent from dredged material disposal sites/suspended solids removal and nitrification. *Water Science and Technology*. Vol. 25, pp. 265–275.

Nijjssen, J.P.J. *et al.* (1997). Monitoring environmental effects in the Slufter, a disposal site for contaminated sludge. *International Conference on Contaminated Sediments*. September 1997. Rotterdam, The Netherlands. Pre-prints. pp. 623–630.

Nilson, S. *et al.* (1998). The Boston Harbor navigation improvement project: dredging today for a deeper tomorrow. *Proceedings of World Dredging Congress XV*, Las Vegas, Nevada, USA. WEDA, Vancouver, Washington, USA. pp. 783–797.

Ooms, K. (1997). Disposal and capping of contaminated sediments – the Hong Kong solution. *Proceedings of the CEDA Dredging Days*, Amsterdam. CEDA, Delft, The Netherlands.

OSPAR (1998). *OSPAR Guidelines for the management of dredged material*. OSPAR 98/14/1-E. Annex 43 (Available at http://www.dredging.org).

Pace, S.D. *et al.* (1998). An automated surveillance system for dredged material transport and disposal operations. *Proceedings of the 15th World Dredging Congress*, Las Vegas, Nevada. WEDA, Vancouver, Washington, USA. pp. 293–307.

Palermo, M.R. (1994). Placement techniques for capping contaminated sediments. *Dredging 94. Proceedings of the 2nd International Conference on Dredging and Dredged Material Placement*. American Society of Civil Engineers. New York, NY, USA. pp. 1111–1121.

Palermo, M.R. *et al.* (1998). *Guidance for Subaqueous Dredged Material Capping. Technical Report DOER-1*. US Army Engineer Waterways Experiment Station, Vicksburg, MS, USA. (The report is available in PDF format at www.wes.army.mil/el/dots/doer/reports.html).

Palermo, M.R. and Averett, D.E. (2000). Summary of constructed CDF containment features for contaminated sediments, *Proceedings of the Western Dredging Association XX and Texas A&M University 32nd Annual Dredging Seminar*, CDS Report No. 372, Texas A&M University, College Station, TX, USA. pp. 447–456.

Panageotou, W. and Halka, J. (1994). Application of Studies on the Overboard Placement of Dredged Sediments to the Management of Disposal Sites. *Dredging 94. Proceedings on the 2nd International Conference on Dredging and Dredged Material Placement*. American Society of Civil Engineers. New York, NY, USA. pp. 349–358.

Peddicord, R.K. (1993). Managing Ecological Risks of Contaminated Sediment. *In*: Landis, W. *et al.*, *Environmental Toxicology and Risk Assessment*, ASTM STP 1179. pp. 353–361. American Society for Testing and Materials, Philadelphia, PA, USA.

Peddicord, R.K. and McFarland, V.A. (1978). *Effects of Suspended Dredged Material on Aquatic Animals*). Technical Report D-78-29. U.S. Army Engineer Waterways Experiment Station, Vicksburg, MS, USA.

Pelletier, J.P. *et al.* (1994). Evolution of the Cable Arm Clamshell bucket. *Dredging 94 Conference Proceedings*. Volume 2.

Pellissier, P. (1999). Landfill closure project enables beneficial use of dredged material. *Dredging and Port Construction*, March, Vol. XXVI, No. 111.

Pennekamp, J.G.S. *et al.* (1996). Turbidity Caused By Dredging; Viewed In Perspective. *Terra et Aqua*, No. 64, pp. 10–17.

Pequegnat, W.E. *et al.* (1990). *Revised Procedural Guide for Designation Surveys of Ocean Dredged Material Disposal Sites*. Technical Report D-90-8. U.S. Army Engineer Waterways Experiment Station, Vicksburg, MS, USA.

Pethick, J. (1997). The beneficial use of dredged sediment in estuaries: a trickle charging experiment in the Medway Estuary, Kent, UK. *Proceedings of CEDA Dredging Days*, Amsterdam. CEDA. Delft, The Netherlands. pp. 143–158.

Pethick, J. (1984). *An Introduction to Coastal Geomorphology*, Edward Arnold Publishers, Great Britain.

PIANC EnviCom (2008a). *Dredged material as a resource*. Report of Working Group 14. International Navigation Association. Brussels, Belgium (In preparation).

PIANC EnviCom (2008b). *Best Management Practices applied to dredging and dredged material disposal projects for protection of the environment*, Report from Working Group 13 of the Environment Commission, International Navigation Association. Brussels, Belgium (In preparation).

PIANC EnviCom (2007). *Management, Reclamation of Dredged Material and End Use of Existing Confined Disposal Facilities*. Report from Working Group 11 of the Environment Commission, International Navigation Association. Brussels, Belgium (Draft).

PIANC EnviCom (2006a). *Environmental risk assessment of dredging and disposal operations*. Report from Working Group 10 of the Environment Commission. International Navigation Association. Brussels, Belgium.

PIANC EnviCom (2006b). *Biological Assessment Guidance for Dredged Material*. Report from Working Group 8 of the Environment Commission. International Navigation Association. Brussels, Belgium.

PIANC EnviCom (2004). *Environmental Risk Assessment of Dredging and Disposal Operations*, Report from Working Group 10 of the Environment Commission, International Navigation Association. Brussels, Belgium (Draft).

PIANC EnviCom (2002). *Environmental Guidelines for Aquaitic, Nearshore and Upland Confined Disposal facilities for Contaminated Dredged Material*. Report of Working Group 5. International Navigation Association. Brussels, Belgium.

PIANC EnviCom (1999). *Environmental Management Framework for Ports and Related Industries*. Report of Working Group 4. International Navigation Association, Brussels, Belgium.

PIANC EnviCom (1998). *Management of the aquatic disposal of dredged material*. Report of Working Group 1 Permanent Environmental Commission. International Navigation Association. Brussels, Belgium.

PIANC (1996). *Handling and Treatment of Contaminated Dredged Material from Ports and Inland Waterways "CDM"*, Volume 1. PTC I, Working Group 17. Permanent International Association of Navigation Congresses (PIANC). Brussels, Belgium.

PIANC (1992). *Beneficial Uses of Dredged Material. A Practical Guide*. PTC II, Working Group 19. Permanent International Association of Navigation Congresses (PIANC). Brussels, Belgium.

PIANC (1984). *Classification of Soils to be Dredged*. Report of a Working Group of PTC II, International Navigation Association. Brussels, Belgium.

Poindexter-Rollings, M.E. (1990) *Methodology for analysis of subaqueous sediment mounds*. Technical Report D-90-2. US Army Engineer Waterways Experiment Station, Vicksburg, MS, USA.

PoMC (2007). *Supplementary Environmental Effects Statement, Channel Deepening Project, Port of Melbourne Corporation. Appendix 6.*

PoMC (2004). *Environmental Effects Statement, Channel Deepening Project, Port of Melbourne Corporation*, download from website: http://www.channelproject.com/global/docs/ChannelDeepeningEES_SummaryBrochure.pdf

Postma *et al.* (1988). *Continental Shelves, Ecosystems of the World*, Vol. 27, Elsevier Science Publishers B.V., The Netherlands.

POSW. (undated) *Fact Sheet No. 3.* Rijkswaterstaat Institute for Inland Water Management and Waste Water Treatment (RIZA), Lelystad, The Netherlands.

Price, R.A. (2002). *Evaluation of surface runoff water in a freshwater confined disposal facility – Effects of vegetation*, DOER-C28, U.S. Army Engineer Research and Development Center, Vicksburg, MS, USA.

Rankilor P.R. (1994). *Technical manual for the design of UTF geosynthetics into civil and marine engineering projects.* UCO Technical Fabrics NV. Lokeren, Belgium.

Ray, G.L. *et al.* (1994). Construction of intertidal mudflats as a beneficial use of dredged material. *Proceedings of the 2nd International Conference on Dredging and Dredged Material Placement*, Florida, USA. American Society of Civil Engineers. New York, NY, USA. pp. 946–955.

Reinking, M.W. (1993). The development of a special remedial dredging technique – The environment-friendly auger dredger. *CEDA Dredging Days Proceedings.* Amsterdam, The Netherlands.

Rhoads, D.C. *et al.* (1978). Disturbance and Production on the Estuarine Sea Floor. *American Scientist*, Vol. 66, pp. 577–586.

Rijkswaterstaat (2004). *Building with dredged material, daily practice*! The Hague, The Netherlands.

Rijkswaterstaat (1997). *Vierde Nota waterhuishouding. Regerings-voornemen. (Fourth National Policy Document on Water Management. Government proposals. English Summary.)* September 1997. The Hague, The Netherlands.

Rijkswaterstaat (1996–1997). *Klei uit baggerspecie. (Clay from dredged sediments. Series of reports.)* W-DWWW-95-345/96-043/96; -118/96-123/97; -032/96-121.

Rollings, M.E. and Rollings, R. (1998). Consolidation and related geotechnical issues at the 1997 New York Mud Dump Site. *Proceedings of the 15th World Dredging Congress.* Las Vegas, Nevada. WEDA, Vancouver, WA, USA. pp. 1–16.

Roukema, D.C. *et al.* (1998). Realisation of the Ketelmeer Storage Depot. *Terra et Aqua*, No. 71, June 1998. pp. 15–27.

Safei El-Deen Hamed *et al.* (1996). *Challenges in managing the EA Process*, EA Source Book UPDATE Nr. 16, The World Bank, Washington, DC, USA.

Santillo, D. *et al.* (1998). The Precautionary Principle: Protecting Against Failures of Scientific Method and Risk Assessment. *Marine Pollution Bulletin*, Vol. 36, No.12.

Scherrer P. (2006). Le Havre – Port 2000. Initiating the environmental renovation of the estuary of the Seine. *Proceedings of CEDA Dredging Days* 1–3 November 2006, Tangier, Morocco.

Schiereck, G.J. and van der Weide, J. (Unknown). *Coastal Zone Management: A Round Game*. Workshop Coastal Zone Management, Malaysia.

Schroeder, P.R. (2000a). *Confined Disposal Facility (CDF) Containment Features: A Summary of Field Experience*, DOER-C18, U.S. Army Engineer Research and Development Center, Vicksburg, MS, USA. http://el.erdc.usace.army.mil/dots/doer/pubs.cfm?Topic=TechNote&Code=doer

Schroeder, P.R. (2000b). *Leachate Screening Considerations* DOER-C16, U.S. Army Engineer Research and Development Center, Vicksburg, MS, USA. http://el.erdc.usace.army.mil/dots/doer/pubs.cfm?Topic=TechNote&Code=doer

Schroeder, P.R. *et al.* (2006). *Screening evaluations for upland confined disposal facility effluent quality.*, DOER-R11, U.S. Army Engineer Research and Development Center, Vicksburg, MS, USA. http://el.erdc.usace.army.mil/dots/doer/pubs.cfm?Topic=TechNote&Code=doer

Schroeder, P.R. and Aziz, N.M. (2004). *Dispersion of Leachate in Aquifers*, DOER-C34, U.S. Army Engineer Research and Development Center, Vicksburg, MS, USA. http://el.erdc.usace.army.mil/dots/doer/pubs.cfm?Topic=TechNote&Code=doer

SedNet (2007). Bortone G. and Palumbo L. (Editors). *Sediment and Dredged Material Treatment, 2*. Elsevier Publishing. The Netherlands.

Shaw, J.K. *et al.* (1998). Contaminated mud in Hong Kong: A case study of contained seabed disposal. *Proceedings of the 15th World Dredging Congress. Las Vegas*, Nevada. WEDA, Vancouver, Washington, USA. pp. 799–820.

Smits, J. and Sas, M. (1997). Maintenance Dredging – the Environmental Approach. *Proceedings of the 2nd Asian and Australasian Ports and Harbours Conference*. Ho Chi Minh City, Vietnam. pp. 455–464. CEDA. Delft, The Netherlands. ISBN 90-75254-07-5.

Smits, J. *et al.* (1997). Automated supervision and follow-up of dredging works. EADA *Proceedings of the 2nd Asian and Australasian Ports and Harbours Conference*. Ho Chi Minh City, Vietnam. CEDA Delft, The Netherlands.

Sorensen *et al.* (1990). Coasts. *Coastal Publication No. 1*, Renewable Resources Information Series, Research Planning Institute, Inc.

Spurgeon, J.P.G. (1993). The Economic Valuation of Coral Reefs. *Marine Pollution Bulletin* (24:11), pp. 529–536. Pergamon Press, UK.

Spurgeon, J.P.G. (1998). The Socio-economic Costs and Benefits of Coastal Habitat Rehabilitation and Creation. *Marine Pollution Bulletin* (37:8–12), pp. 373–382. Pergamon Press, UK.

Spurgeon, J.P.G. (1999a). Economic Valuation of Damages to Coral Reefs. *Coral Reefs: Marine Wealth Threatened*. Conference organised by National University of Mexico. Cancun, Mexico.

Spurgeon, J.P.G. (1999b). Environmental Economics for Ports, Harbours and Marinas. *Seaworks '99 Conference*, July 1999. Southampton, UK.

Standaert, P. *et al.* (1993). The scoop dredger, a new concept for silt removal. *Proceedings CEDA Dredging Days*. Amsterdam, The Netherlands.

Stark, T.D. and Fowler, J. (1994). Strip drains in dredged material placement areas. *Dredging 94. Proceedings 2nd International Conference on Dredging and Dredged Material Placement*, Florida, USA. American Society of Civil Engineers, New York, NY, USA. pp. 420–429.

Stern, E.M. and Stickle, W.B. (1978). *Effects of turbidity and suspended sediment in aquatic environments: Literature review*. Technical Report 0-78-21. U.S. Army Engineer Waterways Experiment Station, Vicksburg, MS, USA.

Stronkhorst, J. (1998). *Impact hypothesis and ecological monitoring of the relocation of the coastal dump site for dredged material from the Port of Rotterdam*. LC/SG 21/3/2. Paper submitted by The Netherlands for the 21st Scientific Group Meeting of LC-72.

Sullivan, N. and Murray, L. (1999) The Use of Agitation Dredging, Water Injection Dredging and Sidecasting in the UK. Results of a Survey of Ports in England and Wales. *Proceedings CEDA Dredging Days*. Amsterdam, The Netherlands.

Sumeri, A. (1995). Dredged material is not a spoil. A status on the use of dredged material in Puget Sound to isolate contaminated sediments. *Proceedings of World Dredging Congress XIV*, Amsterdam. CEDA, Delft, The Netherlands. pp. 345–377.

Thackston, E.L. and Palermo, M.R. (1998) Improved Methods for Correlating Turbidity and Suspended Solids for Dredging and Disposal Monitoring. *Proceedings of the 15th World Dredging Congress (WODCON)*. Las Vegas, Nevada, USA. WEDA, Vancouver, Washington, USA.

Thibodeaux, L.J. *et al.* (1994). Capping contaminated sediments – The theoretical basis and laboratory experimental evidence for chemical containment. *Dredging 94. Proceedings of the 2nd International Conference on Dredging and Dredged Material Placement*, Florida, USA. ASCE, New York, NY, USA. pp. 1001–1007.

Tresselt, K. *et al.* (1997). Harbour sludge as barrier material in landfill cover systems. *International Conference on Contaminated Sediments*. September 1997. Rotterdam, The Netherlands. Pre-prints. pp. 973–980.

UNECE (1991). *Convention on Environmental Impact Assessment in a Transboundary Context*. United Nations Economic Commission for Europe (February 25, 1991). Espo, Finland.

USACE (2003). *Evaluation of Dredged Material Proposed or Disposal at Island, Nearshore, or Upland Confined Disposal Facilities – Testing Manual*. ERDC/EL TR-03-1.

USACE (1996). *Natural Processes for Contaminant Treatment and Control at Dredged Material Confined Disposal Facilities. Environmental Effects of Dredging*. Technical Notes EEDP-02-19. US Army Engineer Waterways Experiment Station, Vicksburg, MS, USA.

USACE (1986). *Beneficial uses of dredged material.* Engineer Manual No. 1110-2-5026. Department of the Army Corps of Engineers. Washington, DC, USA.

USACE (1987). *Confined disposal of dredged material.* Engineer Manual 1110-2-5027. Office, Chief of Engineers, Washington, DC, USA.

US NRC (2001). *A Process for Setting, Managing and Monitoring Environmental Windows for Dredging Projects.* U.S. Transportation Research Board, Special Report 262. Washington, DC, USA.

US NRC (1997). *Contaminated Sediments in Ports and Waterways Cleanup Strategies and Technologies.* National Research Council's Committee on Contaminated Marine Sediments. National Academy Press. Washington, DC, USA.

Van Berk, A.H. and Oostinga, H. (1992). North Lake of Tunis and its shores: Restoration and Development, *Terra et Aqua*, No. 49.

Vallis, R. and Maynard, D. (1999). Engineering Solutions to Environmental Concerns and the Northumberland Strait Crossing Project. *Proceedings Øresund Link Dredging & Reclamation Conference 1999.* Copenhagen, Denmark.

Van der Pluijm, J.L.P.M. *et al.* (1995). Reduction of the emission of contaminants from CDM disposal sites by reuse of surplus water as transport water and in situ measures. *Proceedings of the 14th World Dredging Congress*, Amsterdam. CEDA. Delft, The Netherlands. pp. 731–742.

Van Diepen, H. *et al.* (1993). Dredging and the environment: new developments from the Netherlands. *PIANC Bulletin No. 80.* International Navigation Association. Brussels, Belgium.

Van Drimmelen, N.J. and Schut, T. (1994). New and adapted small dredgers for remedial dredging operations. *World Dredging, Mining & Construction.* December.

Vandycke, S. *et al.* (1996). New developments in environmental dredging: from scoop to sweep dredger. *11th Harbour Congress Proceedings.* June. Antwerp, Belgium.

Van Germert, W.J.Th. *et al.* (1994). Options for treatment and disposal of contaminated dredged sediment. *Environmental management of solid waste; dredged material and mine tailings.* W. Salomons and U. Forstner (Editors). Springer Verlag. Germany.

Van Passen, K. *et al.* (2005). Development of an integrated approach for the removal of tributyltin (TBT) from waterways and harbors: Prevention, treatment and reuse of TBT contaminated sediments. TBT Clean LIFE02 ENV/B/000341. Antwerp Port Authority. Antwerp, Belgium.

Van Raalte, G.H. and Bray, R.N. (1999). Hydrodynamic Dredging: Principles, effects and methods. *Proceedings CEDA Dredging Days*, Amsterdam, The Netherlands.

Van Someren, G.J. and Oosterbaan, N. (1988). Cleaning up the North Lake in Tunisia, *Terra et Aqua*, No. 36.

Van Wijck, J. *et al.* (1991). Underwater disposal of dredged material – a viable solution for the maintenance dredging works in the river Scheldt. *PIANC Bulletin No. 73.* International Navigation Association. Brussels, Belgium.

Vellinga, T. (1997). Dredged Material Management Guide. Special Report of the Permanent Environmental Commission. *Supplement to PIANC Bulletin No. 96.* International Navigation Association. Brussels, Belgium.

Verweij, J.F. and Winterwerp, J.C. (1999). Environmental Impact of Water Injection Dredging. 175. *Proceedings CEDA Dredging Days*, Amsterdam, The Netherlands.

Viles, H. and Spencer, T. (1995). *Coastal Problems – Geomorphology, Ecology and Society at the Coast*, Edward Arnold – Hodder Headline PLC, Great Britain.

Vogt, C. *et al.* (2007). Dredged material management in a watershed context: seeking integrated solutions. *Proceedings of 26th WEDA Conference*, San Diego, CA, USA.

Waterman, R.E. (2007). Land in Water, Water in Land: Achieving Integrated Coastal Zone Development by Building with Nature. *Terra et Aqua*, No. 107, June.

Wichman, B. (1998). Rekenmethode voor consolidatie van gashoudend slib in depots (Modelling the consolidation process of dredged material containing gas.) *AKWA Niews Brief No. 1.* May.

Whiteside, P.G.D. (1998). Personal communication. Civil Engineering Department, Government of the Hong Kong Special Administrative Region.

Whiteside, P.G.D. *et al.* (1996). Management of contaminated mud in Hong Kong. *Terra et Aqua*, No. 65, pp. 10–17.

Winsemius, P.S. and Tjeenk, W. (1986). *Considerations on Environmental Management.* Alphen a/d Rijn, The Netherlands.

World Bank (1990). Environmental Considerations for Port and Harbour Developments. *World Bank Technical paper Number 126.* The World Bank. Washington, DC, USA.

Zwakhals, J.W. *et al.* (1995). Separation of sand from dredge material. Field experience at the Slufter disposal site. *Proceedings of the 14th World Dredging Congress*, Amsterdam. CEDA, Delft, The Netherlands. pp. 125–136.

ANNEX A

International Legislation

A1 OVERVIEW

To protect the environment during dredging activities, various countries or cross-country areas have inaugurated laws or developed guidance on what may or may not be done with dredged material. The process is continuing and, as such, any information in this document must be checked with the relevant competent authorities before embarking on potentially costly decision-making processes. The information presented is as accurate and up-to-date as possible, but it is unavoidable that some material has already been superseded.

In recent years a number of countries (e.g. Australia, Hong Kong and Singapore) have also started producing complex regulations for dredging activities and this is likely to become common practice around the world over the next decade or so. This Annex does *not* cover these emerging regulations. However, an example of an Environmental Monitoring and Management Plan (EMMP) that has been developed to satisfy dredging legislation in Singapore is given in Appendix 5.

Dredged material may be used or relocated in many ways, both in the sea and on land. These are discussed in Chapter 7 of this book. This Annex seeks to provide information on the way different countries view the various practices and illustrate, where possible, typical standards, regulations and the decision-making processes relating to the fate of dredged material.

Placement at sea

Very often Placement of the material at sea will be the cheapest option. This is generally regulated by international or regional conventions with contracting parties being obliged to introduce national legislation that conforms to the conventions. This usually takes the form of a licensing procedure that includes an assessment of potential impacts.

In the 1970s, protocols for the control of Placement of dredged material were set up, two of which are the London (Dumping) Convention and the Oslo and Paris Convention. They were set up primarily to regulate the Placement of noxious substances into the oceans, but they also included (it might be said "accidentally") the regulation of dredged sediment. This inclusion was to be expected, given that the annual volume of dredged material disposed at sea greatly exceeds any other material.

In the northeast Atlantic/North Sea regions approximately 150 million wet tonnes of dredged material were disposed of in 1990 compared with 10 million tonnes of sewage sludge and less than 2 million tonnes of chemical waste. This drew the attention of those who wish to regulate dumping to dredged material. Recognising that dredged material consists mainly of natural sediment and only a small proportion of the total volume dredged annually is contaminated, a number of organisations, including CEDA, PIANC and IAPH, launched a campaign to change the image so that gradually dredged "spoil" became known as dredged "material". This term is now embedded in the conventions and has been a contributory factor in getting dredged material treated as a special case.

Placement on land and in inland waters

Many countries have navigable waterways. Some of these are wide and stretch hundreds of km and some are narrow and relatively short. Virtually all of them require dredging from time to time and in most cases Placement of the material at sea is not an option for practical, economic or environmental reasons. In most, if not all countries that have environmental legislation, placement on land (or in inland lakes and waterways) is covered by different legislation than that of placement at sea. Placement at sea is covered by various international conventions, whereas no such conventions cover placement on land. In Europe a number of pertinent European Union Directives are binding on member states but the interpretation into national legislation can be quite different amongst Members. Similarly in Australia and in the USA, national environmental legislation provides guiding principles, but individual states can have quite different interpretations in practice. This makes it unwise to generalise and for this reason detailed information is not included.

Standards

There is a distinct lack of harmonisation of standards between countries and even within countries in which local or regional authorities form part of the regulatory framework. Standards are often applied without any reference to the particle size distribution of the sediment. Most contaminants have an affinity to the fine (muddy) fraction of sediment so a sediment containing a large proportion of sand is unlikely to be assessed as contaminated when the fine (and therefore more mobile) fraction might be highly contaminated. Conversely sediment that consists only of fine material may be adversely assessed against standards based on a more average particle size distribution. Some countries have adopted a normalisation process that takes account of this; others specify the fraction to be analysed.

Environmental impact assessments

Many countries that do not have specific legislation concerning the placement of dredged material either in the sea or on land nevertheless do have environmental legislation. This usually takes the form of requiring an environmental impact assessment to be carried out on major projects.

Structure of the Annex

Annex A is structured by geographic region. Admittedly, despite best efforts, it has not been possible to obtain information from every country or region.

A2 INTERNATIONAL CONVENTIONS

At the heart of the London and OSPAR Conventions, and many other conventions as well, are two basic principles (the precise wording varies and is here abbreviated):

a) The precautionary principle, by which preventative measures are to be taken when there are reasonable grounds for concern that substances or energy introduced into the marine environment may bring about hazard, harm, damage or interference, even when there is no conclusive evidence of a causal relationship between inputs and the effects.

b) The polluter pays principle, by which the costs of pollution prevention, control and reduction measures are to be borne by the polluter.

A result of the first principle has been the development of the "reverse list" where by only substances which have been proven NOT to cause harm are permitted to be disposed at sea. The phrase "reasonable grounds for concern" allows some flexibility.

In 1993 the London (Dumping) Convention was renamed the Convention on the Preservation of Marine Pollution by Dumping of Wastes and Other Matter (London Convention 1972), hereafter LC72. There are currently 81 Parties to the Convention (i.e. nation states that have signed, ratified, and otherwise acceded to it). The 1996 Protocol is a separate agreement that modernised and updated LC72 following a detailed review that began in 1993. The 1996 Protocol will eventually replace the LC72 and so far, 28 states have acceded to the 1996 Protocol. States can be a Party to either the LC72, or the 1996 Protocol, or both.

A list of the signatories to the LC72 and the 1996 Protocol is given in Appendix 1.

The Oslo and Paris Conventions have been merged into the OSPAR Convention. This covers the northeast Atlantic and North Sea and is open to countries which border these sea areas or which have rivers passing through that discharge into them (e.g. Switzerland and Luxembourg). CEDA has observer status both at the London Convention and OSPARCON and has contributed significantly to drawing up the Dredged Material Assessment Framework (DMAF). Each signatory country is obliged to put in place legislation and infrastructure to implement the LC72 and 1996 Protocol.

The London Convention 1972 (LC72) and 1996 Protocol

LC72 has 10 articles that address the obligations of the contracting parties to ensure that the properties of the material to be disposed of at sea are in accordance with the convention requirements, that the parties encourage co-operation amongst themselves and seek the formation of regional agreements, and that measures are taken to punish any conduct in contravention of the Convention. Other articles are concerned mainly with the details of procedures for setting up and operating the Convention.

In 1993, Parties started a detailed review of the LC72, leading to the adoption of a number of crucial amendments prohibiting the dumping of all radioactive wastes or other radioactive matter and industrial wastes, as well as the prohibition of incineration at sea of industrial wastes and sewage sludge. The review was completed in 1996 with the adoption of the 1996 Protocol to the LC72. This Protocol required ratification by 26 countries, 15 of who must be Contracting Parties to the original 1972 treaty. Mexico was the 26th country to ratify the 1996 Protocol and the Protocol entered into force on 24th March 2006.

The 1996 Protocol reflects a more modern and comprehensive agreement on protecting the marine environment from Placement activities than the original LC72 and reflects the broader aims to protect the environment in general. It embodies a more precautionary approach to dumping at sea. The Articles of the 1996 Protocol are summarised below.

LC72 permits Placement of wastes to sea, with the exception of those materials on a banned or "black list". The 1996 Protocol is more restrictive and, in essence, prohibits dumping, except for those materials on a "reverse list" in Annex 1 which defines categories of wastes that may be considered for dumping, which includes dredged material. Annex 2 of the Protocol provides the framework to be used in any such consideration. Formally titled the "Assessment of Wastes or other Matter that may be considered for Dumping", it is generally referred to as the Waste Assessment Framework (WAF). Annexes 1 and 2 of the 1996 Protocol can be found in Appendix 2.

The Waste Assessment Framework applies to all substances considered for sea Placement and is available to guide national authorities in evaluating applications for dumping of wastes in a manner consistent with the provisions of the LC72 or the 1996 Protocol. Waste-specific guidance documents have been prepared to assist Parties to assess several classes of material. For dredged material the application of the LC72 and 1996 Protocol is covered by special guidelines entitled the "Dredged Material Assessment Framework" (DMAF).

The 1996 Protocol begins by setting the basis and objectives (here abbreviated and paraphrased):

The contracting parties:
- Stressing the need to protect the marine environment and to promote the sustainable use and conservation of marine resources,
- Noting in this regard the achievements within the framework of the LC72 and especially the evolution towards approaches based on precaution and prevention,
- Noting further the contribution in this regard by complementary regional and national instruments,
- Reaffirming the value of a global approach to these matters and the importance of continuing co-operation and collaboration between Contracting Parties,
- Recognising that it may be desirable to adopt, on a national or regional level, more stringent measures with respect to prevention and elimination of pollution of the marine environment from dumping at sea than are provided for in international conventions or other types of agreements with a global scope,
- Taking into account relevant international agreements and actions,
- Recognising also the interests and capacities of developing States and in particular small island developing States, and

- Being convinced that further international action to prevent, reduce and where practicable eliminate pollution of the sea caused by dumping can and must be taken without delay to protect and preserve the marine environment and to manage human activities in such a manner that the marine ecosystem will continue to sustain the legitimate uses of the sea and will continue to meet the needs of present and future generations.

Have agreed as follows:

Article 1: Definitions

Sets out definitions, including dumping.

Article 2: Objectives

Contracting Parties shall individually and collectively protect and preserve the marine environment from all sources of pollution and take effective measures, according to their scientific, technical and economic capabilities, to prevent, reduce and where practicable eliminate pollution caused by dumping or incineration at sea of wastes or other matter. Where appropriate, they shall harmonize their policies in this regard.

Article 3: General obligations

1. In implementing this Protocol, Contracting Parties shall apply a precautionary approach to environmental protection from dumping of wastes or other matter whereby appropriate preventative measures are taken when there is reason to believe that wastes or other matter introduced into the marine environment are likely to cause harm even when there is no conclusive evidence to prove a causal relation between inputs and their effects.

2. Taking into account the approach that the polluter should, in principle, bear the cost of pollution, each Contracting Party shall endeavour to promote practices whereby those it has authorized to engage in dumping or incineration at sea bear the cost of meeting the pollution prevention and control requirements for the authorized activities, having due regard to the public interest.

3. In implementing the provisions of this Protocol, Contracting Parties shall act so as not to transfer, directly or indirectly, damage or likelihood of damage from one part of the environment to another or transform one type of pollution into another.

4. No provision of this Protocol shall be interpreted as preventing Contracting Parties from taking, individually or jointly, more stringent measures in accordance with international law with respect to the prevention, reduction and where practicable elimination of pollution.

Article 4: Dumping of wastes or other matter

1. Contracting Parties shall prohibit the dumping of any wastes or other matter with the exception of those listed in Annex 1.

2. The dumping of wastes or other matter listed in Annex 1 shall require a permit. Contracting Parties shall adopt administrative or legislative measures to ensure that issuance of permits and permit conditions comply with provisions of Annex 2. Particular attention shall be paid to opportunities to avoid dumping in favour of environmentally preferable alternatives.

3. This paragraph says that anything else, not listed in Annex 1, can be prohibited by a Contracting Party if it so wishes.

Article 5: Incineration at sea

Deals with the prohibition of incineration at sea.

Article 6: Export of wastes or other material

States that Contracting Parties shall not allow the export of wastes or other matters to other countries for dumping or incineration at sea.

Article 7: Internal waters

1. Notwithstanding any other provision of this Protocol, this Protocol shall relate to internal waters only to the extent provided for in paragraphs 2 and 3.

2. Each Contracting Party shall at its discretion either apply the provisions of this Protocol or adopt other effective permitting and regulatory measures to control the deliberate Placement of wastes or other matter in marine internal waters where such Placement would be "dumping" or "incineration at sea" within the meaning of article 1, if conducted at sea.

3. Each Contracting Party should provide the Organization with information on legislation and institutional mechanisms regarding implementation, compliance and enforcement in marine internal waters. Contracting Parties should also use their best efforts to provide, on a voluntary basis, summary reports on the type and nature of the materials dumped in marine internal waters.

Article 8: Exceptions

Deals with "force majeure". This means that if there is a threat to human life or other overriding considerations the provisions of Articles 4.1 and 5 will not apply.

Dredged Material Assessment Framework

The Protocol is supported by generic WAF Guidelines for implementation of Annex 2 and by guidelines for each of the specific categories of wastes that may be considered for dumping. For dredged material the specific guideline is entitled "Dredged Material Assessment Framework" (DMAF). This is aimed at decision makers in the field of management of dredged material and sets out the basic practical, though not necessarily detailed, considerations required for determining the conditions under which dredged material might (or might not) be deposited at sea. Under

DMAF Placement of material for which contamination is not a concern will still be subject to an audit, such that sea Placement will be permitted subject to consideration of beneficial use options and assessment of Placement site impacts. A summary of the DMAF is given in Appendix 3.

Compliance with LC72 and the 1996 Protocol

The secretariat receives reports from Contracting Parties notifying annually the permits issued in the previous year, or reports stating that no permits have been issued in that year, the so-called "nil reports". While there has been an improvement in the number of Contracting Parties reporting since the first publication of this guide in 1996, many do not submit reports or contact the secretariat. Countries are encouraged to respond routinely as decisions are only based on the evidence of the compilers.

The OSPARCON

The revision of the Oslo and Paris Conventions was completed in 1992 and the new version is called the OSPAR Convention. Both the Oslo and Paris Conventions covered the sea areas of the northeast Atlantic and the North Sea. The two original conventions were:

Oslo: Convention for the Prevention of Marine Pollution by Dumping from Ships and Aircraft, 1972; and

Paris: Convention for the Prevention of Marine Pollution from Land-based Sources, 1974.

The new convention has five annexes:

1. The prevention and elimination of pollution from land-based sources.

2. The prevention and elimination of pollution by dumping or incineration.

3. The prevention and elimination of pollution from offshore sources.

4. The assessment of the quality of the marine environment.

5. The protection and conservation of the ecosystems and biological diversity of the maritime area.

The Placement at sea Annex II is set out in the form of a reverse list of materials that may be disposed of at sea, with dredged material at the top of the list (Article 3, paragraph 2 (a)).

Dredged material guidelines

As with the LC72, there are dredged material ties. In 1989 the Oslo Commission decided to review the 1986 guidelines. France organised a seminar in Nantes in November 1989 on the environmental aspects of dredging and examined ways of reducing the impact. A number of management tools were identified for incorporation in the revised dredged material guidelines.

These revised OSPAR Guidelines for the management of dredged material (Ref: 1998-20) deal with the assessment and management of dredged material Placement and provide guidance on the design and conduct of monitoring marine and estuary Placement sites. The assessment should contain a concise "impact hypothesis" which will also form the basis of subsequent monitoring.

The main constituents of the OSPARCON revised guidelines are:
– Conditions under which permits for Placement of dredged material may be issued.
– Assessment of the characteristics and composition of dredged material.
– Guidelines on dredged material sampling and analysis.
– Characteristics of Placement site and method of Placement.
– General considerations and conditions.
– Placement management techniques.

In addition, the Oslo and Paris Conventions for the Prevention of Marine Pollution Working Group on Sea-Based Activities (SEBA) 1995 requested contracting parties to submit details of their sediment quality criteria and how they are applied, in order to allow the preparation of an overview for SEBA 1996. The dredged material activities of SEBA are now undertaken by the Working Group on the Environmental Impacts of Human Activities (EIHA). A draft report on Contracting Parties sediment quality criteria was submitted to EIHA 2003 and subsequently published (Overview of Contracting Parties' National Action Levels for Dredged Material, Biodiversity Series, OSPAR Commission 2004, available at www.ospar.org). The criteria provided by contracted parties are reported for individual countries where available. Where these criteria are preliminary they appear under future trends.

OSPAR is undertaking a thematic assessment of human activities in the marine environment (the Joint Assessment Monitoring Programme – JAMP), which includes dredging and Placement, to produce a Quality Status Report by 2010.

Similarities and differences between OSPARCON and LC72

As the original version of the OSCON guidelines was drawn up in parallel with the LC guidelines they are very similar in both structure and content. An important difference between them is that OSPAR offers more flexibility in the case where concentrations of contaminants exceed "trace" levels. If it can be shown that marine Placement is the "option of least detriment" it may be permitted, whereas under the EC guidelines such material would be prohibited. Another difference is that oil and its products, listed in Annex I of the EC, are not included.

Other conventions and regional agreements

A number of other international conventions and regional treaties have been set up with the general aim of protecting the marine environment. These will also affect dredged material Placement at sea practices for signatory countries. Such agreements are given in Appendix 4.

A3 ASIA

A3.1 HONG KONG

Hong Kong classifies sediments based on their contaminant levels with reference to the Chemical Exceedance Levels (CEL) shown below.

Sediment quality criteria for the classification of sediment in Hong Kong

Contaminants	Lower Chemical Exceedance Level (LCEL)	Upper Chemical Exceedance Level (UCEL)
Metals (mg/kg dry wt.)		
Cadmium (Cd)	1.5	4
Chromium (Cr)	80	160
Copper (Cu)	65	110
Mercury (Hg)	0.5	1
Nickel (Ni)*	40	40
Lead (Pb)	75	110
Silver (Ag)	1	2
Zinc (Zn)	200	270
Metalloid (mg/kg dry wt.)		
Arsenic (As)	12	42
Organic-PAHs (μg/kg dry wt.)		
Low Molecular Weight PAHs	550	3160
High Molecular Weight PAHs	1700	9600
Organic-non-PAHs (μg/kg dry wt.)		
Total PCBs	23	180
Organometallics (μg TBT/L in Interstitial water)		
Tributyltin*	0.15	0.15

The contaminant level is considered to have exceeded the UCEL if it is greater than the value shown.

The sediment is classified into 3 categories based on its contaminant levels:

Category L: Sediment with all contaminant levels not exceeding the Lower Chemical Exceedance Level (LCEL).

Category M: Sediment with any one or more contaminant levels exceeding the Lower Chemical Exceedance Level (LCEL) and none exceeding the Upper Chemical Exceedance Level (UCEL).

Category H: Sediment with any one or more contaminant levels exceeding the Upper Chemical Exceedance Level (UCEL).

A3.2 THE REPUBLIC OF KOREA

Action list for the dredged material disposal at sea in the Republic of Korea

Parameter	1st level (Upper Level) mg/kg dry weight	2nd Level (Lower Level) mg/kg dry weight
Chromium and its compounds	370	80
Zinc and its compounds	410	200
Copper and its compounds	270	65
Cadmium and its compounds	10	2.5
Mercury and its compounds	1.2	0.3
Arsenic and its compounds	70	20
Lead and its compounds	220	50
Nickel and its compounds	52	35
Total Polychlorinated Biphenyls	0.180	0.023
Total Polyaromatic Hydrocarbons	45	4

Notes:
1. *Total polychlorinated biphenyls is the sum of contents of PCB-28, PCB-52, PCB-101, PCB-138, PCB-153 and PCB-180 congeners in a sample.*
2. *Total polyaromatic hydrocarbons is the sum of contents of naphthalene, phenanthrene, anthracene, benzo(a)pyrene, fluoranthene, benzo(a)anthracene, benzo(b)fluoranthene in a sample.*

A4 EUROPE

A4.1 EUROPEAN LEGISLATION

EC Legislation

Several Western European countries have developed their own Placement policies or guidelines, but certain EC Directives govern the Placement and/or use of dredged material in EC countries under the definition of "waste". This chapter briefly reviews both the EC Directives and the individual countries' policies.

Classification of dredged material in the EC region

Several EU Member States have defined or proposed sediment quality levels that trigger various levels of action. While definitions vary they may be generalised as:

Class 1 – Below Action Level 1: sea Placement permitted

Class 2 – Between Action Levels 1 and 2: sea Placement permitted with restrictions (e.g. monitoring)

Class 3 – Higher than Action Level 2: sea Placement permitted only under very specific conditions.

One beneficial use option is the spreading of material on agricultural land. An EC guideline concerning the use of "wastes", predominantly sewage sludge but including dredged material, in agriculture (86/278/EEC) is binding for all EC members.

The EC itself has no specific limits for contaminant levels in dredged material but a summary of various countries' Action Levels for heavy metals is given in Table E1. The last row gives the standards for placement of sludge on agricultural land.

Limit values for heavy metals (mg/kg dry matter)
(Le Quillec *et al.*, 2004)

	Cd	Hg	As	Cr	Cu	Ni	Pb	Zn
Netherlands								
L1	0.8	0.3	29.0	100	36	35	85	140
L2	2.0	0.5	55.0	380	36	35	530	480
L3	7.5	1.6	55.0	380	90	45	530	720
L4	12.0	10.0	55.0	380	190	200	530	720
Germany								
N1	0.5	0.5	10	90	40	40	50	150
N2	5.0	5.0	40	150	150	150	150	500
N3	10.0	10.0	70	250	250	250	250	1000
N4	25.0	25.0	100	500	500	500	500	2000
Portugal								
L1	1.0	0.5	20	50	35	30	50	100
L2	3.0	1.5	50	100	150	75	150	600
L3	5.0	3.0	100	400	300	125	500	1500
L4	10.0	10.0	500	1000	500	250	1000	5000
Italy								
A	1.0	0.5	15	20	40	45	45	200
B	5.0	2.0	25	100	50	50	100	400
C	20.0	10.0	50	500	400	150	500	3000
Belgium								
Reference	2.5	0.3	20	60	20	70	70	160
Limit	7.0	1.5	100	220	100	280	350	500
France								
N1	1.2	0.4	2.5	90	45	37	100	276
N2	2.4	0.8	5.0	180	90	74	200	552
ES/Barcelona								
N1	1.0	0.6	80	200	100	100	120	500
N2	5.0	3.0	200	1000	400	400	600	3000
EEC Sludges	20–40	16–25	–	–	1000–1750	300–400	750–1200	2500–4000

Marine Strategy

This is in the process of development. The EC Thematic Strategy on the Protection and Conservation of the Marine Environment aims to achieve good environmental status of the EU's marine waters by 2021 and to protect the resource base upon which marine–related economic and social activities depend. This Marine Strategy will constitute the environmental pillar of the future maritime policy that the EC is working on, designed to achieve the full economic potential of oceans and seas in harmony with the marine environment. The Marine Strategies will contain a detailed assessment of the state of the environment, a definition of "good environmental status" at regional level and the establishment of clear environmental targets and monitoring programmes. The Marine Strategy is consistent with the Water Framework Directive (2000/60/EC) which requires the achievement of good ecological status by 2015 and the first review of the river basin management plans by 2021. The strategy is at the discussion stage so the consequences for dredging and Placement activities are still uncertain but may have implications for how the Dredged Material Assessment Framework (DMAF) is applied.

EU Environmental Liability Directive

This was adopted in March 2004 and is due to enter into force shortly. The aim is to hold operators whose activities have caused environmental damage financially liable for the necessary remedial actions and as such enforces the "polluter pays" principle.

Dredging and Placement are not specifically listed in the "occupational activities", however, the Directive applies to *all* significant damage to protected species and habitats if the operator is at fault. Given the proportion of maintenance dredging that takes place in the vicinity of designated sites and the difficulty of what constitutes "significant" impact, this could be a matter of real concern (Brooke, 2005).

A4.2 BELGIUM

Sediment quality criteria for Belgium, on metals and organics in dredged material

Parameter	Action level 1 (target value) (ppm d.m.)	Action level 2 (limit value) (ppm d.m.)
Hg	0.3	1.5
Cd	2.5	7
Pb	70	350
Zn	160	500
Ni	70	280
As	20	100
Cr	60	220
Cu	20	100
TBT	3	7
Mineral oil	$14\,mg/g_{oc}$	$36\,mg/g_{oc}$
PAKs	$70\,\mu g/g_{oc}$	$180\,\mu g/g_{oc}$
PCBs	$2\,\mu g/g_{oc}$	$2\,\mu g/g_{oc}$

A4.3 FINLAND

The action levels for dredged material in Finland were adopted by the Ministry of the Environment on 19 May 2004. These values are still, however, guidance values and not binding forms. The aim is to be able to give binding norms within a few years time. All measured contaminant contents are normalised to a "standard soil" – composition (10% organic material and 25% clay). The values in the table on page 315 refer to the normalised values.

Contaminant	Action level 1 (ppm d.w.)	Action level 2 (ppm d.w.)
Hg	0.1	1
Cd	0.5	2.5
Cr	65	270
Cu	50	90
Pb	40	200
Ni	45	60
Zn	170	500
As	15	60
PaHs		
naphthalene	0.01	0.1
anthracene	0.01	0.1
phenanthrene	0.05	0.5
fluoranthene	0.3	3
benzo(a)anthracene	0.03	0.4
chrysene	1.1	11
benzo(k)fluoranthene	0.2	2
benzo(a)pyrene	0.3	3
benzo(ghi)perylene	0.8	8
indeno(123-cd)pyrene	0.6	6
mineral oil	500	1500
DDT+DDE+DDD	0.01	0.03
	ppb d.w.	**ppb d.w.**
PCB (IUPAC-numbers)		
28	1	30
52	1	30
101	4	30
118	4	30
138	4	30
153	4	30
180	4	30
tributyltin (TBT)	3	200
	ng WHO-TEQ/kg	**ng WHO-TEQ/kg**
dioxins and furans (PCDD and PCDF)	20	500

A4.4 FRANCE

If analysis shows that concentrations are less than action level 1, a general permit is given without specific study.

If analysis shows that concentrations exceed action level 2, dumping at sea may be prohibited, especially when this dumping does not constitute the least detrimental solution for the environment (particularly with respect to other solutions, *in situ* or on land). These values do not consider the toxic character and the bioavailability of each element.

If analysis shows that concentrations are situated between action level 1 and action level 2, a more comprehensive study might be necessary. The content of these studies will be established on a case-by-case basis taking account of the local circumstances and the sensitivity of the environment.

The action levels are as shown in the following table.

Substance	Action level 1 (ppm d.w.)	Action level 2 (ppm d.w.)	Substance	Action level 1 (ppm d.w)	Action level 2 (ppm d.w.)
Metals			**PCB**		
Hg	0.4	0.8	CB 28	0.025	0.05
Cd	1.2	2.4	52	0.025	0.05
As	25	50	101	0.050	0.05
Pb	100	200	118	0.025	0.10
Cr	90	180	180	0.025	0.05
Cu	45	90	138	0.05	0.10
Zn	276	552	153	0.05	0.10
Ni	37	74	Total PCBs	0.5	1.0

A4.5 GERMANY

Sediment quality criteria for the German Federal Waters and Navigation Administration on trace metals and organic contaminants in dredged material (sediment fraction <20 um)

		Action level 1	Action level 2
Arsenic	ppm	30	150
Cadmium	ppm	2.5	12.5
Chromium	ppm	150	750
Copper	ppm	40	200
Mercury	ppm	1	5
Nickel	ppm	50	250
Lead	ppm	100	500
Zinc	ppm	350	1750
PCB28	ppb	2	6
PCB52	ppb	1	3
PCB101	ppb	2	6
PCB118	ppb	3	10
PCB138	ppb	4	12
PCB153	ppb	5	15
PCB180	ppb	2	6
Sum of 7 PCBs	ppb	20	60
α-Hexachlorocyclohexane	ppb	0.4	1
γ-Hexachlorocyclohexane	ppb	0.2	0.6
Hexachlorobenzene	ppb	2	6
Pentachlorobenzene	ppb	1	3
p,p'-DDT	ppb	1	3
p,p'-DDE	ppb	1	3
p,p'-DDD	ppb	3	10
PAH[1] (Sum of 6 PAHs)	ppm	1	3
Hydrocarbons	ppm	300	1000

[1] Total of six PAH compounds: fluoranthene, benzo(b)fluoranthene, benzo(k)fluoranthene, benzo(a)pyrene, benzo(ghi)perylene, indeno(1,2,3-cd)pyrene

Action levels for the German Federal Waters and Navigation Administration on tributyltin (TBT) in dredged material

Action Level 1	Action Level 2	Unit	Valid from
20	600	µg TBT/kg total sediment	2001
20	300	µg TBT/kg total sediment	2005
20	60	µg TBT/kg total sediment	2010

A4.6 IRELAND

Guidelines for the assessment of dredged material for Placement in Irish waters have been published (Cronin *et al.*, 2006).

Provisional Irish action levels in mg kg-1dry wt

Chemical	Category 1	Category 2	Category 3
As	<10	10–80	>80
Cd	<1	1–3	>3
Cr	<100	100–300	>300
Cu	<50	50–200	>200
Hg	<0.3	0.3–5.0	>5
Ni	<50	50–200	>200
Pb	<50	50–400	>400
Zn	<400	400–700	>700
PCB (7)	<0.01	0.01–0.1	>0.1
TBT	<0.1	0.1–0.5	>0.5
Total PCB	<0.1	0.1–1.0	>1.0

A4.7 THE NETHERLANDS

The Dutch classification system for dredged material has recently been revised:
- *Target value*: Indicates the level below which risks to the environment are considered to be negligible, at the present state of knowledge.
- *Limit value*: Concentration at which the water sediment is considered as relatively clean. The limit value is the objective for the year 2000.
- *Reference value*: Is a reference level indicating whether dredged sediment is still fit for discharge in surface water, under certain conditions, or should be treated otherwise. It indicates the maximum allowable level above which the risks for the environment are unacceptable.
- *Intervention value:* An indicative value, indicating that remediation may be urgent, owing to increased risks to public health and the environment.
- *Signal value*: Only for heavy metals. Concentration level of heavy metals above which the need for cleaning up should be investigated.

Constants in the correction of measured levels for heavy metals and arsenic based on the local sediment composition (derived from reference value)

Metal	A	b	c
Zn	50	3	1.5
Cu	15	0.5	0.6
Cr	50	2	0
Pb	50	1	1
Cd	0.4	0.0007	0.021
Ni	10	1	0
Hg	0.2	0.0034	0.0017
As	15	0.4	0.4

The water sediment standards now existing have been based upon information which estimates the effects on the aquatic ecosystem. In addition, the water sediment composition influences the standards. For the availability of heavy metals and arsenic, clay fraction (lute, particle size < 2 μm) and the quantity of organic material are of particular importance. For the availability of organic compounds, the organic substance level is a determining factor. The standards are set for sediment containing 25% of lute and 10% of organic substance. Conversion towards the standard sediment composition is done in conformity to the method followed by the WOB (Water Sediments Study Group), which is also applied to calculate the reference values for soil quality.

Classification of water sediment:

Class 0 is below target value and can be spread over the land without restrictions.

Class 1 exceeds the target value, but is below the limit value and is allowed to be disposed unless the soil quality is not significantly impaired.

Class 2 does not meet the limit value, but is below the reference value, and can be spread in surface water or on land, under certain conditions.

Class 3 does not meet the reference value, but remains below the intervention value, and should be stored under controlled conditions, specific requirements can be set, depending on the storage location.

Class 4 does not meet the intervention value, and should be contained in isolation in deep pits or on land, in order to minimise the influence on the surroundings.

Target and other values

Parameter	Unit	Target value	Limit value	Reference value	Intervention value	Signal value
Arsenic	Mg/kg ds	29	55	55	55	150
Cadmium	Mg/kg ds	0.8	2	7.5	12	30
Chromium	Mg/kg ds	100	380	380	380	1000
Copper	Mg/kg ds	35	35	90	190	400
Mercury	Mg/kg ds	0.3	0.5	1.6	10	15
Lead	Mg/kg ds	85	530	530	530	1000
Nickel	Mg/kg ds	35	35	45	210	200
Zinc	Mg/kg ds	140	480	720	720	2500
PAH Total 10 PAK*	Mg/kg ds	1	1	10	40	–
PCB-28	µg/kg ds	1	4	30	–	–
PCB-52	µg/kg ds	1	4	30	–	–
PCB-101	µg/kg ds	4	4	30	–	–
PCB-118	µg/kg ds	4	4	30	–	–
PCB-138	µg/kg ds	4	4	30	–	–
PCB-153	µg/kg ds	4	4	30	–	–
PCB-180	µg/kg ds	4	4	30	–	–
Total 6 PCB	µg/kg ds	20/0	–	–	–	–
Total 7 PCB	µg/kg ds	–	–	200	1000	–
Chlordane	µg/kg ds	10	20	–	–	–
α-HCH	µg/kg ds	2.5	–	20	–	–
β-HCH	µg/kg ds	1	–	20	–	–
γ-HCH (lindane)	µg/kg ds	0.05	1	20	–	–
HCH-compounds	µg/kg ds	–	–	–	2000	–
Heptachlor	µg/kg ds	2.5	–	–	–	–
Heptachlorepoxide	µg/kg ds	2.5	–	–	–	–
Heptachlor + epoxide	µg/kg ds	–	20	20	–	–
Aldrin	µg/kg ds	2.5	–	–	–	–
Dieldrin	µg/kg ds	0.5	20	–	–	–
Total aldrin & dieldrin	µg/kg ds	–	40	40	–	–
Endrin	µg/kg ds	1	40	40	–	–
Drins	µg/kg ds	–	–	–	4000	–
DDT (incl. DDD & DDE)	µg/kg ds	2.5	10	20	4000	–
α-endosulfan	µg/kg ds	2.5	–	–	–	–
α-endosulfan +	µg/kg ds	–	10	20	–	–
Sulphate	µg/kg ds	2.5	–	100	–	–
Hexachlorobutadiene						
Total pesticides	µg/kg ds					
Pentachlorobenzene	µg/kg ds	2.5	300	300	–	–
Hexachlorobenzene	µg/kg ds	2.5	4	20	–	–
Pentachlorophenol	µg/kg ds	2	20	5000	5000	–
Mineral oil	Mg/kg ds	50	100	3000	5000	–
EOX	Mg/kg ds	–	0	7	–	–

Napthalene, benzo(a)anthracene, benzo(ghi)perylene, benzo(a)pyrene, phenanthrene, ideno(123-ad)pyrene, anthracene, benzo(k)fluoranthene, chrythene, fluoranthene

Dredged material standards for the Netherlands

ppm d.w.	Action level 1[1]	Action level 2[2]
As	29	29
Cd	0.8	4
Cr	100	120
Cu	36	60
Hg	0.3	1.2
Pb	85	110
Ni	35	45
Zn	140	365
Mineral oil (C10–40)	50	1250
Sum 10 PAHs[3]		8
Sum 7 PCBs[4]		0.1
Alpha-HCH	0.003	–
Beta-HCH	0.009	–
Gamma-HCH (lindane)	0.00005	0.02
Sum HCHs	0.01	–
Heptachlor	0.007	–
Heptachlorepoxide	0.0000002	0.02
Aldrin	0.00006	0.03
Dieldrin	0.0005	0.03
Endrin	0.00004	0.03
Sum Aldrin + Dieldrin − Endrin	0.005	–
DDT	0.00009	–
DDD	0.00002	–
DDE	0.00001	–
Sum DDT + DDD + DDE	0.01	0.02
Hexachlorobenzene	0.00005	0.02
TBT	0.000007	0.24 (100 μg Sn/kg dw)
Sum organic compounds	0.001	

Notes:
1. General environmental quality objective (water system)
2. Numerical values for the content test distribution into salt waters (2001)
3. Naphthalene, phenanthrene, anthracene, fluoranthrene, chrysene, benzo(a)anthracene, benzo(a)pyrene, benzo(k)fluoranthene, indenopyrene, benzo(ghi)perylene
4. PCBs 28, 52, 101, 118, 138, 153 and 180

A4.8 NORWAY

The Norwegian sediment criteria for Classification of Environmental Quality and Degree of Pollution (CEQDP) in fjords and coastal waters represent the basis for managing dredging and dredged material.

Dredged material standards for Norway

Parameter	Category 1 good/fair (class I & II)	Category 2 poor/bad (class III & IV)	Category 3 very bad (class V)
Metals (ppm dry weight)			
Arsenic	<20–80	80–1000	>1000
Lead	<30–120	120–1500	>1500
Fluoride	<800–3000	3000–20000	>20000
Cadmium	<0.25–1	1–10	>10
Copper	<35–150	150–1500	>1500
Mercury	<0.15–0.6	0.6–5	>5
Chromium	<70–300	300–5000	> 5000
Nickel	<30–130	130–1500	>1500
Zinc	<150–700	700–10000	>10000
Silver	<0.3–1.3	1.3–10	>10
Organic component (ppb dry weight)			
Sum PAH (EPA 16)	<300–2000	2000–20000	>20000
Benzo(a)pyrene	<10–50	50–500	>500
Sum PCB	<5–25	25–300	>300
Hexachlorobenzene	<0.5–2.5	2.5–50	>50
EPOCl[1]	<100–500	500–15000	>15000
2, 3, 7, 8-TCDD eqv.[2]	<0.03–0.12	0.12–1.5	>1.5

Notes:

1. Extractable persistent organic chloride.   2 Total toxicity potential for polychlorinated dibenzofurans/dioxins, given as equivalents of the most toxic of these components (2, 3, 7, 8-tetrachlordibenzo-p-dioxin)

A4.9 PORTUGAL

Dredged material classification for Portugal

Substance	Class 1	Class 2	Class 3	Class 4	Class 5
As	<20	50	100	500	>500
Cd	<1	3	5	10	>10
Cr	<50	100	400	1000	>1000
Cu	<35	150	300	500	>500
Hg	<0.5	1.5	3	10	>10
Pb	<50	150	500	1000	>1000
Ni	<30	75	125	250	>250
Zn	<100	600	1500	5000	>5000
PCB sum	<5	25	100	300	>300
PAH sum	<300	2000	6000	20000	>20000
HCB	<0.5	2.5	10	50	>50
Description	Clean	Vestiges of contamination	Slightly contaminated	Contaminated	Very contaminated
Fate	Aquatic environment and beaches	Aquatic environment	Aquatic environment with monitoring	Landfill with special monitoring	Landfill (residues have special treatment

Notes:

1. Concentrations are upper bounds for each class.   2. Concentrations of metals are in mg/kg dry solids (ppm)
3. Concentrations of organics are in micrograms/kg dry solids (ppb)

A4.10 SPAIN

Sediment quality criteria applicable to Spanish harbours

ppm d.w.	Action level 1	Action level 2
Hg	0.6	3
Cd	1	5
Pb	120	600
Cu	100	400
Zn	500	3000
Cr	200	1000
As	80	200
Ni	100	400
Sum 7 PCBs	0.03	0.1

A4.11 SWEDEN

In Sweden, action levels are based on the following background concentrations. Information is not provided on the possible link between these concentrations and action levels:

Substance	Background value (ppm dry weight)
As	10
Pb	10
Fe	40000
Cd	0,3
Co	15
Cu	20
Cr	20
Hg	0,1
Ni	15
Sn	1
V	20
Zn	125

A4.12 THE UNITED KINGDOM

Most dredged material in the UK is placed at sea and is governed by Part II of the Food and Environment Protection Act 1985 (FEPA).

Sediment quality criteria for the UK on metals and organics in dredged material

Contaminant	Existing Action Level 1 mg·kg^{-1} (ppm)	Existing Action Level 2 mg·kg^{-1} (ppm)	Suggested Revised Action Level 1 mg·kg^{-1} (ppm) dry weight	Suggested Revised Action Level 2 mg·kg^{-1} (ppm) dry weight
Arsenic (As)	20	50–100	20	70
Cadmium (Cd)	0.4	2	0.4	4
Chromium (Cr)	40	400	50	370
Copper (Cu)	40	400	30	300
Mercury (Hg)	0.3	3	0.25	1.5
Nickel (Ni)	20	200	30	150
Lead (Pb)	50	500	50	400
Zinc (Zn)	130	800	130	600
Tributyltin (TBT, DBT, MBT)	0.1	1	0.1	0.5
Polychlorinated Biphenyls (PCBs)	0.02	0.2	0.02	0.18
Polyaromatic Hydrocarbons (PAHs)				
Acenaphthene			0.1	
Acenaphthylene			0.1	
Anthracene			0.1	
Fluorene			0.1	
Naphthalene			0.1	
Phenanthrene			0.1	
Benzo[a]anthracene			0.1	
Benzo[a]fluoranthene			0.1	
Benzo[k]fluoranthene			0.1	
Benzo[g]perylene			0.1	
Benzo[a]pyrene			0.1	
Benzo[g,h,i]perylene			0.1	
Dibenzo[a,h]anthracene			0.01	
Chrysene			0.1	
Fluoranthene			0.1	
Pyrene			0.1	
Indeno[1,2,3cd]pyrene			0.1	
Total hydrocarbons	100		100	
Booster Biocide and Brominated Flame Retardants[1]	–	–	–	–

[1] Provisional Action Levels for these compounds are subject to further investigation

A5 MIDDLE EASTERN LEGISLATION

A5.1 QATAR

The following information has been obtained from The Environmental Guidelines and Environmental Protection Criteria for Ras Laffan Industrial City. They may be periodically updated, as new regulations, permits or standards are issued by an appropriate regulatory authority such as the Supreme Council for the Environment & Natural Reserves (SCENR).

Maximum concentration of contaminants for toxicity characteristic

Contaminant	Regulatory level mg/l	Contaminant	Regulatory level mg/l
Arsenic	5.0	Hexachlorobenzene	0.13
Barium	100.0	Hexachlorobutadiene	0.5
Benzene	0.5	Hexachloroethane	3.0
Cadmium	1.0	Lead	5.0
Carbon tetrachloride	0.5	Lindane	0.4
Chlordane	0.03	Mercury	0.2
Chlorobenzene	100.0	Methoxychlor	10.0
Chloroform	6.0	Methyl ethyl ketone	200.0
Chromium	5.0	Nitrobenzene	2.0
o-Cresol	200.0	Pentrachlorophenol	100.0
m-Cresol	200.0	Pyridine	5.0
p-Cresol	200.0	Selenium	1.0
Cresol	200.0	Silver	5.0
Dichlorobenzene	7.5	Tetrachloroethylene	0.7
Dichloroethane	0.5	Toxaphene	0.5
Dichloroethylene	0.7	Trichloroethylene	0.5
Dinitrotoluene	0.13	Trichlorophenol	400.0
Endrin	0.02	Silvex	1.0
Heptachlor (and its epoxide)	0.008	Vinyl chloride	0.2

A6 NORTH AMERICA

A6.1 CANADA

Dredged material that exceeds the Lowest Effect Level for organic compounds and mercury is not suitable for open water and will go to land. For material with metals other than mercury exceeding the Lowest Effect Level, open water Placement may be applied under certain conditions. The quality of the material and the bed sediments at the Placement site are compared and matched (see table on page 325).

Provincial sediment quality guidelines for metals and nutrients

Substance	No effect level Below this level material is suitable for open water disposal	Lowest effect level Below this level material is suitable to use as clean fill	Severe effect level* Below this level material may be placed in a landfill or CDF. Above it, material is subject to hazardous waste disposal
Arsenic	–	6	33
Cadmium	–	0.6	10
Chromium	–	26	110
Copper	–	16	110
Iron	–	2	4
Lead	–	31	250
Manganese	–	460	1100
Mercury	–	0.2	2
Nickel	–	16	75
Zinc	–	120	820
TOC	–	1	10
TKN	–	550	4800
TP	–	600	2000
Aldrin	–	.002	8
BHC	–	.003	12
Chlordane	.0002	.007	6
DDT total	–	.007	12
Dieldrin	.0006	.002	91
Endrin	.0005	.003	130
HCB	.01	.02	25
Heptachlor	.0003	–	–
Hepoxide	–	.005	5
Mirex	–	.007	130
PCB (total)	.01	.07	530
Acenaphthene	–	–	–
Anthracene	–	.220	370
Benz[a]anthracene	–	.32	1480
Benzo[b]fluorine	–	–	–
Benzo[k]fluoranthene	–	.24	1440
Benzo[a]pyrene	–	.37	2
Benzo[g,h,i]perylene	–	.17	320
Chrysene	–	.34	460
Dibenzol[a,h]anthracene	–	.06	130
Fluoranthene	–	.75	1020
Fluorene	–	.19	160
Naphthalene	–	–	–
Phenanthrene	–	.56	950
Pyrene	–	.49	850
PAH (total of 16)	–	4	10000

Notes:

Values are in mg/kg (ppm) dry weight unless otherwise specified

* The units of this column are mg/kg of organic carbon and require conversion based on the TOC value of the substance in question.

TOC Total Organic Carbon

TKN Total Kjeldahl Nitrogen

TP   Total Phosphorus

– Insufficient data

A7 REFERENCES

Brooke, J (2005). EU Environmental Directives and their implications for maintenance dredging. *Proc 2nd Intl Conf on Maintenance Dredging*, Bristol, UK. Thomas Telford Publishers. ISBN 0 7277 3288 9.

Cronin, M., McGovern, E., McMahon, T. and Boelens, R. (2006). Guidelines for the Assessment of Dredge Material for Disposal in Irish Waters, *Marine Environment and Health Series*, No. 24, April.

EuDA (2004). Regulatory aspects of the disposal of dredged material – EuDA Analytical Report, European Dredging Association, Brussels, Belgium. May.

Le Quillec, R. and Pittavino, A. (2004). "Strategie en matiere de dragage maritime". *PIANC Bulletin 115*, January.

OSPAR (2004). Overview of Contracting Parties' National Action Levels for Dredged Material. OSPAR Commission.

Appendix 1. Signatories to the London Convention and 1996 Protocol

Contracting parties to the Convention on the prevention of marine pollution by dumping of wastes and other matter, 1972 (London Convention1972) (As at 10 September 2007)

Country	Date of ratification, accession or succession	Date of entry into force or effective date
Afghanistan	2 April 1975	30 August 1975
Antigua and Barbuda	6 January 1989	5 February 1989
Argentina	11 September 1979	11 October 1979
Australia	21 August 1985	20 September 1985
Azerbaijan	1 July 1997	31 July 1997
Barbados	4 May 1994	3 June 1994
Belarus	29 January 1976	28 February 1976
Belgium	12 June 1985	12 July 1985
Bolivia	10 June 1999	10 July 1999
Brazil	26 July 1982	25 August 1982
Canada	13 November 1975	14 December 1975
Cape Verde	26 May 1977	25 June 1977
Chile	4 August 1977	3 September 1977
China[1]	14 November 1985	14 December 1985
Costa Rica	16 June 1986	16 July 1986
Côte d'Ivoire	9 October 1987	8 November 1987
Croatia	8 October 1991	8 October 1991
Cuba	1 December 1975	1 January 1976
Cyprus	7 June 1990	7 July 1990
Democratic Rep. of the Congo	16 September 1975	16 October 1975
Denmark[2]	23 October 1974	30 August 1975
Dominican Republic	7 December 1973	30 August 1975
Egypt	30 June 1992	30 July 1992
Equatorial Guinea	21 January 2004	20 February 2004
Finland	3 May 1979	2 June 1979
France	3 February 1977	5 March 1977
Gabon	5 February 1982	7 March 1982
Germany	8 November 1977	8 December 1977
Greece	10 August 1981	9 September 1981
Guatemala	14 July 1975	30 August 1975
Haiti	28 August 1975	27 September 1975
Honduras	2 May 1980	1 June 1980
Hungary	5 February 1976	6 March 1976
Iceland	24 May 1973	30 August 1975
Iran (Islamic Republic of)	13 January 1997	12 February 1997

(Continued)

Country	Date of ratification, accession or succession	Date of entry into force or effective date
Ireland	17 February 1982	19 March 1982
Italy	30 April 1984	30 May 1984
Jamaica	22 March 1991	21 April 1991
Japan	15 October 1980	14 November 1980
Jordan	11 November 1974	30 August 1975
Kenya	7 January 1976	6 February 1976
Kiribati	12 May 1982	11 June 1982
Libyan Arab Jamahiriya	22 November 1976	22 December 1976
Luxembourg	21 February 1991	23 March 1991
Malta	28 December 1989	27 January 1990
Mexico	7 April 1975	30 August 1975
Monaco	16 May 1977	15 June 1977
Montenegro	–	3 June 2006
Morocco	18 February 1977	20 March 1977
Nauru	26 July 1982	25 August 1982
Netherlands[3]	2 December 1977	2 January 1978
New Zealand	30 April 1975	30 August 1975
Nigeria	19 March 1976	18 April 1976
Norway	4 April 1974	30 August 1975
Oman	13 March 1984	12 April 1984
Pakistan	9 March 1995	8 April 1995
Panama	31 July 1975	30 August 1975
Papua New Guinea	10 March 1980	9 April 1980
Peru	7 May 2003	6 June 2003
Philippines	10 August 1973	30 August 1975
Poland	23 January 1979	22 February 1979
Portugal	14 April 1978	14 May 1978
Republic of Korea	21 December 1993	20 January 1994
Russian Federation[4]	30 December 1975	29 January 1976
Saint Lucia	23 August 1985	22 September 1985
Saint Vincent & the Grenadines	24 October 2001	23 November 2001
Serbia[5]	–	3 June 2006
Seychelles	29 October 1984	28 November 1984
Slovenia	25 June 1991	25 June 1991
Solomon Islands	6 March 1984	5 April 1984
South Africa	7 August 1978	6 September 1978
Spain	31 July 1974	30 August 1975
Suriname	21 October 1980	20 November 1980
Sweden	21 February 1974	30 August 1975
Switzerland	31 July 1979	30 August 1979
Tonga	8 November 1995	9 December 1995
Tunisia	13 April 1976	13 May 1976
Ukraine	5 February 1976	6 March 1976
United Arab Emirates	9 August 1974	30 August 1975
United Kingdom [6]	17 November 1975	17 December 1975

(Continued)

Country	Date of ratification, accession or succession	Date of entry into force or effective date
United States	29 April 1974	30 August 1975
Vanuatu	22 September 1992	22 October 1992

[1] By notification dated 12 June 1997 from the People's Republic of China, the Convention and its 1978, 1980 and 1989 amendments will apply to the Hong Kong Special Administrative Region with effect from 1 July 1997.

[2] Ratification by Denmark was declared to be effective in respect of the Faroe Islands as from 15 November 1976.

[3] Ratification by the Netherlands was declared to be effective in respect of the Netherlands Antilles and, with effect from 1 January 1986, in respect of Aruba.

[4] As from 26 December 1991, the membership of the former USSR in the Convention has been continued by the Russian Federation.

[5] With effect from 3 June 2006, when the National Assembly of Montenegro adopted a declaration of independence, the State Union of Serbia and Montenegro was dissolved. In conformity with Article 60 of the Constitutional Charter of Serbia and Montenegro, the Republic of Serbia is continuing the membership of the State Union of Serbia and Montenegro in IMO and is henceforth known as the "Republic of Serbia", with "Serbia" being used as the short form of the name.

The Republic of Montenegro was admitted as a Member of the United Nations on 28 June 2006, and the Government has since expressed an interest in becoming a Member of IMO. To date, no further information to that effect has been received. Clarification has been sought from Montenegro as to its status vis-à-vis the treaties deposited with the Organization.

[6] The United Kingdom declared ratification to be effective also in respect of:

Effective Date

Bailiwick of Guernsey	)
Isle of Man	)
Belize*	)
Bermuda	)
British Indian Ocean Territory	)
British Virgin Islands	)
Cayman Islands	)
Falkland Islands and Dependencies**	)
Gilbert Islands***	)
Hong Kong****	) 17 November 1975
Monserrat	)
Pitcairn	)
Henderson	)
Ducie and Oeno Islands	)
Saint Helena and Dependencies	)
Seychelles*****	)
Solomon Islands******	)
Turks and Caicos Islands	)
Tuvalu*******	)
United Kingdom Sovereign Base Areas of Akrotiri and Dhekelia in the Island of Cyprus	)
Bailiwick of Jersey	) 4 April 1976

* Has since become the independent State of Belize

** A dispute exists between the Governments of Argentina and the United Kingdom of Great Britain and Northern Ireland concerning sovereignty over the Falkland Islands (Malvinas)

*** Has since become the independent State of Kiribati and a Contracting State to the Convention

**** Has since become (1 July 1997) a Special Administrative Region of the People's Republic of China

***** Has since become the independent State of Seychelles and a Contracting State to the Convention

****** Has since become the independent State of Solomon Islands and a Contracting State to the Convention

******* Has since become the independent State of Tuvalu

1996 Protocol to the London Convention 1972 list of contracting States (as at 5)

	Date of signature or deposit of instrument
Angola (accession)	4 October 2001
Australia (ratification)[1]	4 December 2000
Barbados (accession)[1]	24 July 2006
Belgium (ratification)[1]	13 February 2006
Bulgaria (accession)	25 January 2006
Canada (accession)[1]	15 May 2000
China (ratification)[1]	29 September 2006
Denmark (signature)[1,2]	17 April 1997
Egypt (accession)[1]	26 May 2004
France (accession)[1]	7 February 2004
Georgia (accession)	18 April 2000
Germany (ratification)[1]	16 October 1998
Iceland (ratification)[1]	21 May 2003
Ireland (accession)[1]	26 April 2001
Italy (accession)[1]	13 October 2006
Luxembourg (accession)[1]	21 November 2005
Mexico (accession)[1]	22 February 2006
New Zealand (ratification)[1,2]	30 July 2001
Norway (ratification)[1,2]	16 December 1999
Saudi Arabia (accession)	2 February 2006
Slovenia (accession)[1]	3 March 2006
South Africa (accession)[1]	23 December 1998
Spain (ratification)[1]	24 March 1999
St Kitts and Nevis (accession)	7 October 2004
Suriname (accession)[1]	11 February 2007
Sweden (ratification)[1,2]	16 October 2000
Switzerland (ratification)[1]	8 September 2000
Tonga (accession)[1]	18 October 2003
Trinidad and Tobago (accession)	6 March 2000
United Kingdom (ratification)[1,2]	15 December 1998
Vanuatu (accession)[1]	18 February 1999

[1] Contracting Party to the London Convention 1972
[2] With a declaration or reservation0

Countries Acceding to the LC 72 or the 1996 Protocol

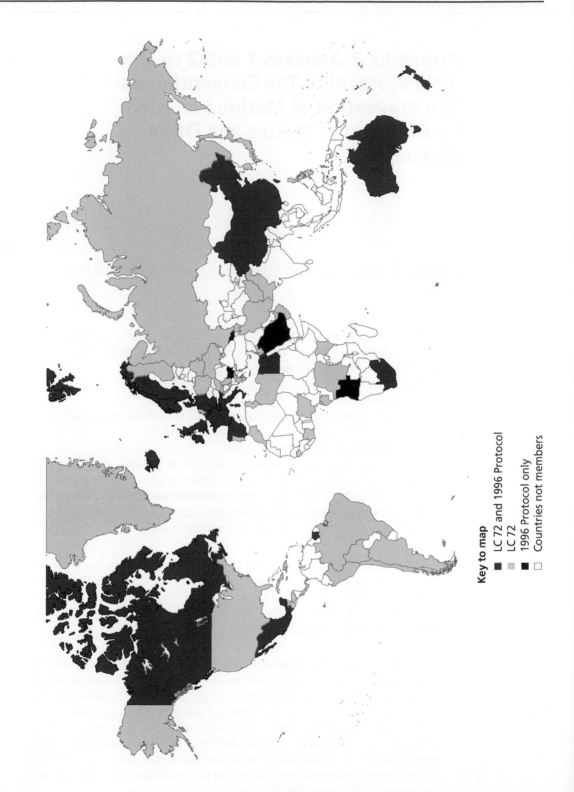

Key to map

- ■ LC 72 and 1996 Protocol
- ▨ LC 72
- ■ 1996 Protocol only
- □ Countries not members

Appendix 2. Annexes 1 and 2 to The 1996 Protocol to The Convention on The Prevention of Marine Pollution by Dumping of Wastes and Other Matter, 1972

ANNEX 1: WASTES OR OTHER MATTER THAT MAY BE CONSIDERED FOR DUMPING

1. The following wastes or other matter are those that may be considered for dumping being mindful of the Objectives and General Obligations of this Protocol set out in articles 2 and 3:
   - dredged material;
   - sewage sludge;
   - fish waste, or material resulting from industrial fish processing operations;
   - vessels and platforms or other artificial structures at sea;
   - inert, inorganic geological material, and
   - organic material of natural origin; and bulky items primarily comprising iron, steel, concrete and similarly unharmful materials for which the concern is physical impact, and limited to those circumstances where such wastes are generated at locations, such as small islands with isolated communities, having no practicable access to disposal options other than dumping.

2. The wastes or other matter listed in paragraphs 1.4 and 1.7 may be considered for dumping, provided that material capable of creating floating debris or otherwise contributing to pollution of the marine environment has been removed to the maximum extent and provided that the material dumped poses no serious obstacle to fishing or navigation.

3. Notwithstanding the above, materials listed in paragraphs 1.1 to 1.7 containing levels of radioactivity greater than de minimis (exempt) concentrations as defined by the IAEA and adopted by Contracting Parties, shall not be considered eligible for dumping; provided further that within 25 years of 20 February 1994, and at each 25 year interval thereafter, Contracting Parties shall complete a scientific study relating to all radioactive wastes and other radioactive matter other than high level wastes or matter, taking into account such other factors as Contracting Parties consider appropriate and shall review the prohibition on dumping of such substances in accordance with the procedures set forth in article 22.

ANNEX 2: ASSESSMENT OF WASTES OR OTHER MATTER THAT MAY BE CONSIDERED FOR DUMPING

GENERAL

1. The acceptance of dumping under certain circumstances shall not remove the obligations under this Annex to make further attempts to reduce the necessity for dumping.

WASTE PREVENTION AUDIT

2. The initial stages in assessing alternatives to dumping should, as appropriate, include an evaluation of:
    – types, amounts and relative hazard of wastes generated;
    – details of the production process and the sources of wastes within that process, and
    – feasibility of the following waste reduction/prevention techniques:
        o product reformulation;
        o clean production technologies;
        o process modification;
        o input substitution, and
        o on-site, closed-loop recycling.

3. In general terms, if the required audit reveals that opportunities exist for waste prevention at source, an applicant is expected to formulate and implement a waste prevention strategy, in collaboration with relevant local and national agencies, which includes specific waste reduction targets and provision for further waste prevention audits to ensure that these targets are being met. Permit issuance or renewal decisions shall assure compliance with any resulting waste reduction and prevention requirements.

4. For dredged material and sewage sludge, the goal of waste management should be to identify and control the sources of contamination. This should be achieved through implementation of waste prevention strategies and requires collaboration between the relevant local and national agencies involved with the control of point and non-point sources of pollution. Until this objective is met, the problems of contaminated dredged material may be addressed by using disposal management techniques at sea or on land.

CONSIDERATION OF WASTE MANAGEMENT OPTIONS

5. Applications to dump wastes or other matter shall demonstrate that appropriate consideration has been given to the following hierarchy of waste management options, which implies an order of increasing environmental impact:
    – re-use;
    – off-site recycling;

  - destruction of hazardous constituents;
  - treatment to reduce or remove the hazardous constituents, and
  - disposal on land, into air and in water.

6. A permit to dump wastes or other matter shall be refused if the permitting authority determines that appropriate opportunities exist to re-use, recycle or treat the waste without undue risks to human health or the environment or disproportionate costs. The practical availability of other means of disposal should be considered in the light of a comparative risk assessment involving both dumping and the alternatives.

CHEMICAL, PHYSICAL AND BIOLOGICAL PROPERTIES

7. A detailed description and characterization of the waste is an essential precondition for the consideration of alternatives and the basis for a decision as to whether a waste may be dumped. If a waste is so poorly characterized that proper assessment cannot be made of its potential impacts on human health and the environment, that waste shall not be dumped.

8. Characterization of the wastes and their constituents shall take into account:
  - origin, total amount, form and average composition;
  - properties: physical, chemical, biochemical and biological;
  - toxicity;
  - persistence: physical, chemical and biological, and
  - accumulation and biotransformation in biological materials or sediments.

ACTION LIST

9. Each Contracting Party shall develop a national Action List to provide a mechanism for screening candidate wastes and their constituents on the basis of their potential effects on human health and the marine environment. In selecting substances for consideration in an Action List, priority shall be given to toxic, persistent and bio-accumulative substances from anthropogenic sources (e.g. cadmium, mercury, organohalogens, petroleum hydrocarbons, and, whenever relevant, arsenic, lead, copper, zinc, beryllium, chromium, nickel and vanadium, organosilicon compounds, cyanides, fluorides and pesticides or their by-products other than organohalogens). An Action List can also be used as a trigger mechanism for further waste prevention considerations.

10. An Action List shall specify an upper level and may also specify a lower level. The upper level should be set so as to avoid acute or chronic effects on human health or on sensitive marine organisms representative of the marine ecosystem. Application of an Action List will result in three possible categories of waste:
  - wastes which contain specified substances, or which cause biological responses, exceeding the relevant upper level shall not be dumped, unless made acceptable for dumping through the use of management techniques or processes;
  - wastes which contain specified substances, or which cause biological responses, below the relevant lower levels should be considered to be of little environmental concern in relation to dumping, and

– wastes which contain specified substances, or which cause biological responses, below the upper level but above the lower level require more detailed assessment before their suitability for dumping can be determined.

DUMP-SITE SELECTION

11. Information required to select a dump-site shall include:
    – physical, chemical and biological characteristics of the water-column and the seabed;
    – location of amenities, values and other uses of the sea in the area under consideration;
    – assessment of the constituent fluxes associated with dumping in relation to existing fluxes of substances in the marine environment, and
    – economic and operational feasibility.

ASSESSMENT OF POTENTIAL EFFECTS

12. Assessment of potential effects should lead to a concise statement of the expected consequences of the sea or land disposal options, i.e. the "Impact Hypothesis". It provides a basis for deciding whether to approve or reject the proposed disposal option and for defining environmental monitoring requirements.

13. The assessment for dumping should integrate information on waste characteristics, conditions at the proposed dump-site(s), fluxes, and proposed disposal techniques and specify the potential effects on human health, living resources, amenities and other legitimate uses of the sea. It should define the nature, temporal and spatial scales and duration of expected impacts based on reasonably conservative assumptions.

14. An analysis of each disposal option should be considered in the light of a comparative assessment of the following concerns: human health risks, environmental costs, hazards, (including accidents), economics and exclusion of future uses. If this assessment reveals that adequate information is not available to determine the likely effects of the proposed disposal option then this option should not be considered further. In addition, if the interpretation of the comparative assessment shows the dumping option to be less preferable, a permit for dumping should not be given.

15. Each assessment should conclude with a statement supporting a decision to issue or refuse a permit for dumping.

MONITORING

16. Monitoring is used to verify that permit conditions are met – compliance monitoring – and that the assumptions made during the permit review and site selection process were correct and sufficient to protect the environment and human health – field monitoring. It is essential that such monitoring programmes have clearly defined objectives.

PERMIT AND PERMIT CONDITIONS

17. A decision to issue a permit should only be made if all impact evaluations are completed and the monitoring requirements are determined. The provisions of the permit shall ensure, as far as practicable, that environmental disturbance and detriment are minimized and the benefits maximized. Any permit issued shall contain data and information specifying:
    - the types and sources of materials to be dumped;
    - the location of the dump-site(s);
    - the method of dumping, and
    - monitoring and reporting requirements.

18. Permits should be reviewed at regular intervals, taking into account the results of monitoring and the objectives of monitoring programmes. Review of monitoring results will indicate whether field programmes need to be continued, revised or terminated and will contribute to informed decisions regarding the continuance, modification or revocation of permits. This provides an important feedback mechanism for the protection of human health and the marine environment.

Appendix 3. Dredged Material Assessment Framework

This summary is based on the "Specific Guidelines for Assessment of Dredged Material" from the International Maritime Organization (IMO). For the full document please see: http://www.imo.org/includes/blastDataOnly.asp/data_id%3D17020/1-DredgedMaterial.pdf.

1. INTRODUCTION

Defines dredging and environmental impacts resulting from dredging activities and the disposal of dredged material under the purview of the London Convention 1972 and the 1996 Protocol and the need for dredging and disposal for capital, maintenance and clean-up dredging. Describes the intentions of the "Generic Guidelines".

A schematic drawing provides a decision tree indicating the stages in apply this guidance. See Figure 5.2 Dredged Material Assessment Framework (DMAF) (adapted from LC 1992) on page 88.

These Guidelines were adopted in 2000 by the Twenty-second Consultative Meeting and are specific to dredged material.

2. WASTE PREVENTION AUDIT

For dredged material, the goal of waste management should be to identify and control the sources of contamination and these are specified.

3. EVALUATION OF DISPOSAL OPTIONS

The results of the physical/chemical/biological characterization will indicate whether the dredged material, in principle, is suitable for disposal at sea. Includes consideration of beneficial uses and management options including treatment, placement on or burial in the sea floor followed by capping and when permits may be issued and when not.

4. DREDGED MATERIAL CHARACTERIZATION

Includes the need for physical characterization of dredged material and when dredged materials are exempted from detailed characterization; the need for chemical characterization and definitions of this; and when to consider biological characterization.

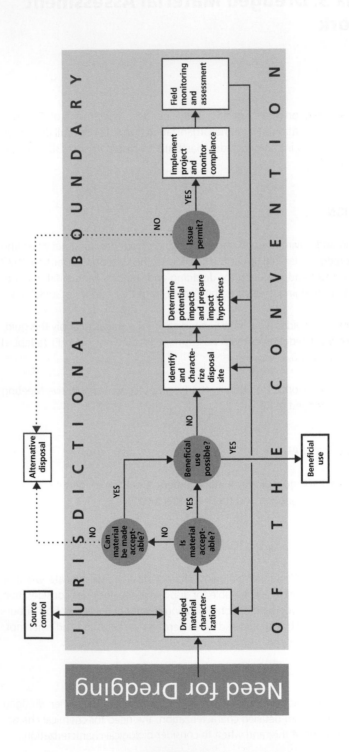

5. ACTION LIST

The Action List provides a screening mechanism for determining whether a material is considered acceptable for dumping. Each Contacting Party is to develop a national Action List. An Action List will specify an upper level so as to avoid acute or chronic effects on human health or on sensitive marine organisms and it may specify a lower level.

6. DUMP-SITE SELECTION

Considers the proper selection of a dump-site at sea, including physical, chemical and biological characteristics; location of amenities, values and other uses of the sea area; assessment of the constituent fluxes; and economic and operational feasibility. Gives guidance on dump-site selection, including nature of the seabed; physical, chemical and biological nature of the water column; and defines the size and the capacity of the dump-site.

Evaluates the potential impacts including increased exposures of organisms to substances that may cause adverse effects; the relative magnitude of the input fluxes; when are critical times of the year for marine life which preclude disposal; and factors effecting contaminant mobility.

7. ASSESSMENT OF POTENTIAL EFFECTS

Should lead to a statement of expected consequences of disposal options, i.e., the "Impact Hypothesis", integrating comprehensive information on waste characteristics and considering amenities and sensitive areas. Link monitoring programmes to the hypotheses, and consider consequences in terms of affected habitats, species; emphasize biological effects and habitat modification. When disposal is repetitive, consider cumulative effects. Each disposal option should be considered; each assessment should conclude with a statement supporting or refusing to support permit issuance. Monitoring should be efficient and cost-effective.

8. MONITORING

Defines compliance monitoring to verify that permit conditions are met and that the assumptions made during the permit review and site selection process were correct. The Impact Hypothesis forms the basis for defining field monitoring and the permitting authority should take account of relevant research information. Predictions within and without the zone of impact should be measured.

9. PERMIT AND PERMIT CONDITIONS

Defines when should a permit be issued; how regulators should consider technological capabilities as well as economic, social and political concerns; and review of permits at regular intervals.

Appendix 4. Other International Agreements and Guidelines

The thirteen regional marine environment protection conventions of the world including their elements relating to sea dumping (listed in alphabetical order by convention short title)

Region	Convention short title	Convention full title and year adopted*	Dumping article
West and Central Africa	Abidjan Convention	Convention for Co-operation in the Protection and Development of the Marine and Coastal Environment of the West and Central African Region 1981	Article 6: Pollution caused by Dumping from Ships and Aircraft
North-East Pacific	Antigua Convention	Convention for Cooperation in the Protection and Sustainable Development of the Marine and Coastal Environment of the North-east Pacific 2002	Article 6: Measures to prevent, reduce, control and remedy pollution and other forms of deterioration of the marine and coastal environment (includes dumping)
Mediterranean Sea	Barcelona Convention	Convention for the Protection of the Marine Environment and Coastal Region of the Mediterranean 1995	Article 5: Pollution caused by Dumping from Ships and Aircraft or Incineration at Sea
Black Sea	Bucharest Convention	Convention on the Protection of the Black Sea against Pollution 1992	Article X: Pollution by Dumping
Wider Caribbean	Cartagena Convention	Convention for the Protection and Development of the Marine Environment of the Wider Caribbean Region 1983	Article 6: Pollution caused by Dumping
Antarctica	CCAMLR	International Convention for the Conservation of Antarctic Marine Living Resources 1980	No specific dumping article but general obligations of environmental protection
Baltic Sea	Helsinki Convention	Convention on the Protection of the Marine Environment of the Baltic Sea Area 1992	Article 11: Prevention of dumping Annex V: Exemptions from the general prohibition of dumping etc.

Dumping protocol or Annex	Secretariat	Further information
No specific dumping protocol	UNEP WACAF/RCU: *West and Central Africa Region/Regional Coordinating Unit* – Abidjan	www.unep.org/regionalseas
No specific dumping protocol	COCATRAM: *Central American Commission for Maritime Transport* – Managua	www.unep.org/regionalseas
Protocol for the Prevention and Elimination of Pollution in the Mediterranean Sea by Dumping from Ships and Aircraft or Incineration at Sea	UNEP MAP: *Mediterranean Action Plan* – Athens	www.unepmap.org
Protocol on the Protection of the Black Sea Marine Environment against Pollution by Dumping	Istanbul Commission: *Commission for the Protection of the Black Sea against Pollution* – Istanbul	www.blackseacommission.org
No specific dumping protocol	UNEP CEP: *Caribbean Environment Programme* – Kingston	www.cep.unep.org
No specific dumping protocol	CCAMLR: *Commission for the Conservation of Antarctic Marine Living Resources* – Hobart	www.ccamlr.org
No specific dumping protocol	Helsinki Commission or HELCOM: *Baltic Marine Environment Protection Commission* – Helsinki	www.helcom.fi

(Continued)

Region	Convention short title	Convention full title and year adopted*	Dumping article
Red Sea and Gulf of Aden	Jeddah Convention	*Regional Convention for the Conservation of the Red Sea and Gulf of Aden Environment* 1982	Article V: *Pollution caused by Dumping from Ships and Aircraft*
Persian Gulf	Kuwait Convention	*Regional Convention for Cooperation on the Protection of the Marine Environment from Pollution* 1978	Article V: *Pollution caused by Dumping from Ships and Aircraft*
East Africa	Nairobi Convention	*Convention for the Protection, Management and Development of the Marine and Coastal Environment of the Eastern African Region* 1985	Article 6: *Pollution caused by Dumping*
Pacific Islands	Noumea Convention	*Convention for the Protection of the Natural Resources and Environment of the South Pacific Region* 1990	Article 10: *Disposal of Wastes*
North-West Atlantic	OSPAR Convention	*Convention for the Protection of the Marine Environment of the North-East Atlantic* 1992	Article 4: *Pollution by Dumping or Incineration*
Caspian Sea	Tehran Convention	*Framework Convention for the Protection of the Marine Environment of the Caspian Sea* 2006	Article 10: *Pollution caused by Dumping*

* Year adopted does not necessarily equate to year of Entry-into-Force

Dumping protocol or Annex	Secretariat	Further information
No specific dumping protocol	PERSGA: *Regional Organization for the Protection of the Red Sea and Gulf of Aden* – Jeddah	www.persga.org
Protocol on the Control of Marine Transboundary Movements and Disposal of Hazardous and Other Wastes	ROPME: *Regional Organization for the Protection of the Marine Environment* – Kuwait	www.ropme.com
No specific dumping protocol	UNEP EAF/RCU: *Eastern African Region/ Regional Coordinating* Unit – Seychelles	www.unep.org/regionalseas
Protocol for the Prevention of Pollution of the South Pacific Region by Dumping (amended in 2004 to align with LP)	SPREP: *Secretariat of the Pacific Regional Environment Programme* – Apia	www.sprep.org
Annex II: *Prevention and elimination of pollution by dumping or incineration*	OSPAR Commission: *Commission for the Protection of the Marine Environment of the North-East Atlantic* – London	www.ospar.org
No specific dumping protocol	Interim - UNEP ROE: *Regional Office for Europe* – Vienna, supported by CEP: *Caspian Environment Programme* – Tehran	www.caspianenvironment.org

Appendix 5. Components of an Environmental Monitoring and Management Plan (EMMP) for Dredging and Reclamation Projects

INTRODUCTION

Environmental habitats in Singapore have recently been afforded very high value status due to international politics (ITLOS 2003), rapid growing public awareness and increasing valuation of coral reefs from a biomedical research viewpoint. Environmental Impact Assessments (EIA) and Environmental Monitoring and Management Plans (EMMP) for marine construction works are carried out in Singapore, with the scale of the EIA being defined on a case-to-case basis during the project planning phase, based on anticipated impacts. The scale of the EMMP is defined by the findings of the EIA.

As a result of the confined nature of Singapore waters and presence of a large number of patch reefs, reclamation and associated dredging activities often take place in very close proximity to coral reefs, seagrass areas and industrial intakes. Recognising the value of these marine habitats and industrial resources, Singapore has established strict Environmental Quality Objectives (EQO) for marine construction activities. In order to document compliance with these EQO's, pro-active Environmental Monitoring and Management Plans (EMMP) based upon feedback monitoring principles are required for marine construction activities to proceed, when these are in close proximity to key environmental receptors.

The components of the EMMP, prior to, during and after completion of dredging and reclamation projects are highlighted in the following section. Thanks go to Mr Chung from JTC, Infrastructure and Planning Department of the Engineering Planning Group, Singapore and to DHI Water & Environment (S) Pte. Ltd., Singapore for the information included here.

OUTLINE OF THE ENVIRONMENTAL MONITORING AND MANAGEMENT PLAN FEEDBACK EMMP

The EMMP is the primary method of control used in Singapore to ensure that reclamation and dredging projects will not cause any adverse impacts to the marine environment. The EMMP is a tool to confirm that impacts have not exceeded the stipulated Environmental Quality Objectives, which are specified for every individual project and dependent on the location of the development area in relation to the location of the sensitive habitats. In the case of unexpected circumstances, the EMMP provides an early warning of impending impacts on a time scale that allows mitigating actions to be taken before any significant impacts to the habitats occur.

For major EMMP projects in Singapore an approach based on feedback methodology is adopted. This is widely regarded as the most effective approach for the control and mitigation of potential environmental impacts arising from dredging and reclamation operations.

The feedback EMMP approach is proactive. It links the results of detailed numerical models of the sediment plumes resulting from the actual dredging works with the results from online sensors and habitat surveys to generate a holistic picture of the limiting sediment spill. The limiting spill, also known as the "spill budget", is the maximum allowable spill that will meet the Environmental Quality Objectives. Hence, provided that the spill, which can be accurately measured, is kept below this limit, there is a strong likelihood that impacts leading to exceedence of the Environmental Quality Objectives will not appear in the field.

The feedback EMMP controls the daily spill against these pre-defined limits. It is a process that can be rigorously enforced with a rapid response time. This ensures that situations that could result in potential impacts are not allowed to develop. The habitat tolerance limits upon which the spill budget is defined are regularly verified via habitat monitoring, and it is possible that the spill budget may be updated (increased or decreased) during the course of the dredging works, based on field monitoring results.

Further, as the feedback EMMP system updates the transport and rate of released spill based on the actual dredging progress, the limiting spill budget can be set for each specific stage of dredging, a level of control that is not possible via traditional static monitoring methods.

The main advantages of the feedback EMMP are, in simple terms:
- Control measurements are targeted at the sediment plume from the dredger. These measurements can be undertaken with a much higher degree of accuracy than is normally associated with fixed suspended sediment (turbidity) sensors at the receptor areas;
- Fixed turbidity sensors at receptor sites cannot differentiate between loading produced from the dredging activities, and natural variability in the background conditions. Control measurements are therefore a very precise tool of measuring sediment spill;
- Sediment plume models are used to keep a running balance of cumulative impact levels based on actual production and measured spill against pre-determined threshold limits. Action can then be taken in advance of any negative impact occurring in the field;
- The use of a sediment plume hindcast model allows complex dredging schedules to be addressed. This results in an accurate assessment of cumulative impacts

and the definitions of spill budgets that are adaptive to changes in construction schedule, equipment numbers and equipment size;

- The models provide a spatial picture at all receptor sites, not just the locations where instrumentation is deployed, as is the case for traditional static monitoring methods. Coverage of the area of interest is thus more rigorous;
- Results of the habitat monitoring can be directly fed back into the model system to update tolerance limits as and when site-specific data are collected. This can allow fine-tuning of production limits, and
- The use of spill control measurements and modelling allows the feedback EMMP to be responsive to changes in conditions (e.g. seasonal effects) and work schedules, which is not possible when utilising traditional impact assessment and static monitoring methods.

Although the feedback EMMP forms a reliable and rigorous method of control, physical mitigation measures such as silt barricades may still be considered as a second level of control to situations where impacts would otherwise develop. However, the feedback EMMP approach ensures that control of spill is monitored and managed at the source and secondary levels of control, like the deployment of silt barricades, are subsequently targeted and their performance is analyzed. Therefore, the success of the environmental management plan is not solely reliant on the performance of physical mitigation measures.

PRE-CONSTRUCTION SCOPE OF WORK

The main EMMP tasks that are carried out prior to a dredging or reclamation project are presented below and summarised in Figure A.1 in a flow chart.

i) Environmental baseline

Appropriate monitoring sites and measurement parameters (e.g. sediment levels) are identified, instrumented and monitored for a statistically significant period of time prior to construction. A baseline period comprised in general a period of 3 months. This task will also include the establishment of the final Environmental Quality Objectives for the project, as set by the relevant authorities, and confirmation of environmental tolerance limits. Online monitoring instrumentation (turbidity sensors, noise meter, current and dissolved oxygen sensors) are generally deployed during the environmental baseline phase of a project;

ii) Elaboration of work plans

Once appointed, the contractor will elaborate its work plan, specifying the distribution of the work in time and space and the procedures that will be used. These plans must include specification of the main dredging equipment (dredger capacity) and production rates;

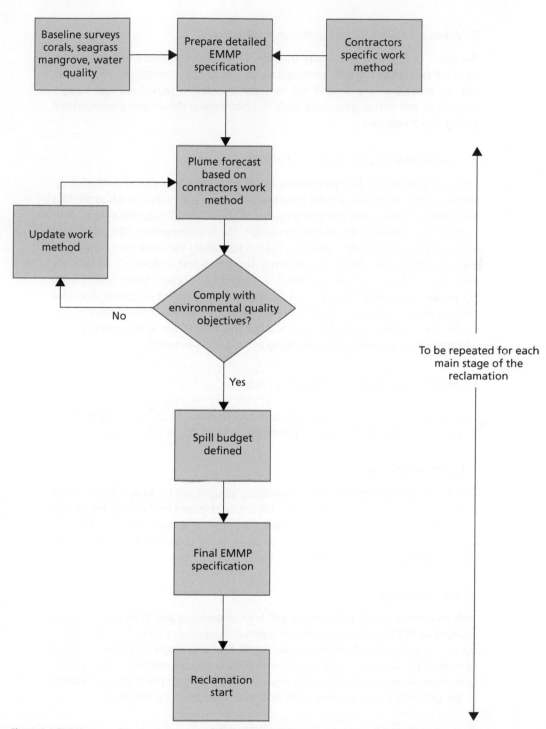

Figure A.1 EMMP scope of work prior to start of dredging and reclamation operations in Singapore.

iii) Assessment of work plans (provisional spill budget)

The effect of performing the work plan specified by the contractor on the environment will be assessed. This is assessed through the use of numerical sediment plume forecast modelling. The models will be used to define a provisional spill budget (amount of spill that is acceptable while still maintaining the agreed environmental quality objectives); and

iv) Revision of work plans (final spill budget)

Depending on the spill budget compared to the contractor's proposed methodology, revision of the work plan may be required to meet the provisional spill budget. This may include, for example, the deployment of Silt Barricades. During this period, close collaboration between the EMMP consultant, the Superintending Officer's Representative (SO Rep) and the contractor is essential. When the work plan is finalised, the sediment plume models will be re-run and a final spill budget will be defined. Once the final spill budget is defined, a Final (Operational) EMMP Specification Document is prepared. This document will specify the detailed execution, response and management process for the EMMP for the entire duration of the project. In particular, the Operational EMMP Specification will include the spill budget against which day-to-day control of the dredging works can be assessed.

SCOPE OF WORK DURING CONSTRUCTION

The main tasks that will be carried out during each phase of a dredging and reclamation project are defined below and presented in Figure A.2 in a flow chart.

i) Control monitoring

The data collected from the online monitoring stations will be analysed on an hour to hour – day to day basis and compared to critical response limits. If the agreed critical limits are exceeded, an environmental management committee will meet to determine if mitigation action is to take place. If no exceedence of critical limits occurs, construction work and regular monitoring continue;

ii) Spill monitoring

Daily monitoring of realised sediment spill from all dredging and reclamation works is assessed via ADCP sediment flux measurements in the dredging areas. The resulting calculation of sediment spill will be compared to the final spill budget for the specific stage of work. If the daily spill budget limits are exceeded, the environmental management committee will meet to determine if mitigation action is to take place. If no exceedence of the spill budget occurs, construction work and spill monitoring continue;

iii) Spill hindcast

The impact of the dredging works on the environment, based on realized production (spill) and tidal conditions are determined on a daily basis through the use of numerical

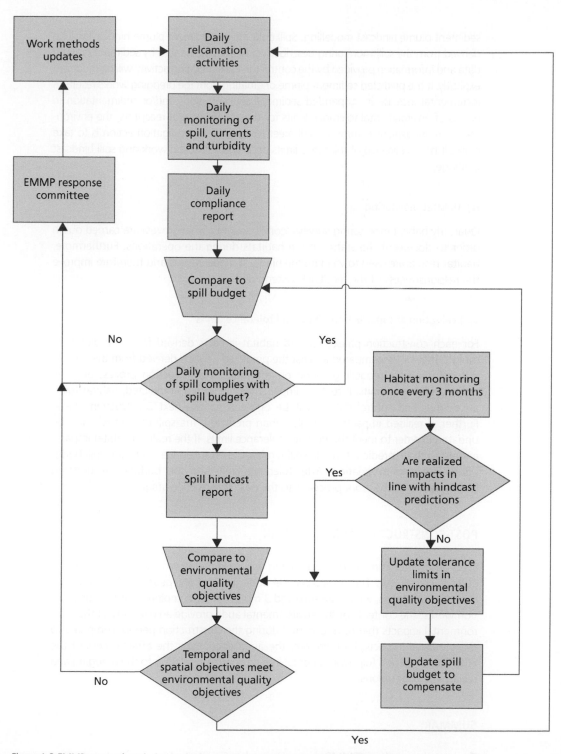

Figure A.2 EMMP scope of work during dredging and reclamation operations in Singapore.

sediment plume hindcast modelling. Spill data for the sediment plume hindcast will be derived from the spill monitoring described above, supplemented by sediment grading data and information provided by the contractor. Each dredging activity will be modelled explicitly. If the predicted sediment plume originating from the dredging works results in incremental increase in suspended sediment concentrations and/or sedimentation in excess of environmental tolerance limits for the various habitat receptors, the environmental management committee will meet to determine if mitigation action is to take place. If no exceedence of tolerance limits occurs, construction work and spill hindcast continue;

iv) Habitat monitoring

Quarterly habitat monitoring surveys (corals, seagrass, mangroves) are carried out in order to document the status of the habitats during the operations. Furthermore, habitat results are used to validate the hindcast model results and therefore improve the responsiveness of the feedback system; and

v) Evaluation of construction phase and tolerance limits

For each construction phase, realised habitat impacts derived from the biological monitoring will be compared against the predicted impacts derived from the numerical spill hindcast modelling based on the realized construction process and the defined tolerance limits. If realised impacts are higher than predicted, habitat tolerance limits and control criteria will be updated for the next construction phase. Further, if realised impacts are higher than predicted, the spill budget will also be updated in order to meet the updated tolerance limits. If the realised habitat impacts equate to those predicted by the spill hindcast, then tolerance limits and spill budgets do not require updating. After tolerance limits and spill budgets are updated (where required) the work proceeds to the next construction stage.

POST-CONSTRUCTION SCOPE OF WORK

Six months after completion of dredging and reclamation works, an environmental audit is carried out. The environmental audit includes a repeat of all the baseline monitoring stations, which are surveyed 3 months after completion of the construction works. The contents of the environmental audit provide an overview of the environmental impacts that have occurred during the construction period, compared to the predicted impacts. Furthermore, the audit documents the effectiveness of the EMMP as a method for environmental impact control in relation to the dredging and reclamation operations.

SUMMARY

To date, the feedback EMMP approach used in Singapore has provided a practical and reliable method for a pro-active management of potential environmental impacts resulting from dredging and reclamation projects. The responsiveness of the system allows unexpected impacts to be mitigated prior to them becoming a serous threat to the environment.

More importantly, the level of documentation provided by the EMMP ensures that developers and contractors are not exposed to unwanted claims concerning environmental degradation as the EMMP allows full segregation of project impacts from other third party disturbances.

The performance of the EMMP methodology has been verified by habitat monitoring surveys. The habitat results are used to validate the hindcast model results and are used to confirm and improve the responsiveness of the feedback system. The EMMP techniques are therefore considered best practice methodology in Singapore and have been successfully adapted to other sites in South East Asia region.

REFERENCE

Doorn-Groen, S.M. and Foster, T.M. (2007). *Environmental Monitoring and Management of Reclamation Works Close to Sensitive Habitats*. Presented at WODCON XVIII World Dredging Conference, Orlando, Florida, USA.

ANNEX B

Case Histories

Project type	Project name and Characteristics	Sources
Capital Dredging for Navigation	**Harwich Haven Approach Channel, United Kingdom** Improved navigation. Dredging affects sediment balance, depleting supply to estuaries. Beneficial use of various dredged materials. Changes to maintenance dredging strategy.	Wallingford HR and Posford Duvivier Environment (1998). 'Harwich Haven Approach Channel Deepening – Environmental Statement', Report EX 3791, January 1998. Allen, R. (1999). 'Reflections made by an Owner – Different Environments, Different Solutions', *Øresund Fixed Link Dredging and Reclamation Conference*, May 1999, Copenhagen, Proceedings pp. 261–266. Challinor, S. (1999). 'Dredging within European Sites of Nature Conservation Importance', *Dredging and Port Construction*, January 1999.
	Channel Deepening Project, Melbourne, Australia Comprehensive environmental assessment for major capital dredging project. Protection of marine mammals, penguins and seagrass	PoMC (2007). Channel Deepening Project. Supplementary Environmental Effects Statement. Port of Melbourne Corporation. www.channelproject.com/publications/ sees_main_report.asp
Capital Dredging for Large Infrastructure	**The Danish Links Projects** Two large road/rail combined bridge/tunnels. Concept of zero solution for change to hydrodynamics. Close monitoring of spill from dredgers in real time.	Møller, J.S. (1989). Denmark's Great Belt Link – Environmental and Hydrodynamic Issues, *American Society of Civil Engineering*, Annual Convention 1989, New Orleans, LA, USA. October 8–11. Fixed Links – The Great Belt, a web site under the Danish Ministry of Transport, September 1999, www.trm.dk/ eng/faste/index.html. (January 2007: http://www.trm.dk/ sw68717.asp)

(Continued)

Project type	Project name and Characteristics	Sources
		Jensen, A. and Lyngby, J.E. (1999). The Øresund Fixed Link – Environmental Management and Monitoring. *Terra et Aqua,* No. 74. March 1999.
		Matthiessen, C.W. (1999). A Cross Border Integration Project on the European Metropolitan Level – Copenhagen and Malmö-Lund united by the Øresund-bridge, *Proceedings of the Øresund Link Dredging & Reclamation Conference in Copenhagen* 26–28 May 1999, Editors Iversen & Mogensen.
		Dynesen, C. (1999) Environmental Management and the Impact on Execution, *Proceedings of the Øresund Link Dredging & Reclamation Conference in Copenhagen* 26–28 May 1999, Editors Iversen & Mogensen.
		Dynesen, C. and Thorkildsen, M. (2001). An Owner's view of hydroinformatics: Its role in realising the bridge and tunnel connection between Denmark and Sweden, *Journal of Hydroinformatics*, Volume 3, April 2001.
Capital Dredging and Reclamation	**Dredging and Reclamation in Hong Kong** The port and airport development strategy (PADS). Exploitation of marine sand sources. Dredging contaminated silt and placement in underwater mud pits. Wave and water flow modelling	Ooms, K. *et al.* (1994). Marine Sand Dredging: Key to the Development of Hong Kong; *Terra et Aqua, No. 54*, March 1994.
		Hodgson, G. (1994). The Environmental Impact of Marine Dredging in Hong Kong *Coastal Management in Tropical Asia,* A Newsletter for Practitioners, Number 2, March 1994.
		Evans, N.C. (1994). Effects of Dredging and Dumping on the Marine Environment of Hong Kong, *Terra et Aqua, No. 57*, December 1994.
		Whiteside, P.G.D. *et al.* (1996). Management of Contaminated Mud in Hong Kong, *Terra et Aqua, No. 65*, December 1996.
		Shaw, J.K. *et al.* (1998). Contaminated Mud in Hong Kong – A case study of Contained Seabed Disposal, *15th World Dredging Congress 1998*, Las Vegas, Nevada, USA.

(Continued)

Project type	Project name and Characteristics	Sources
		Ip, K-L. *et al.* (1994). Wave Modelling for Stage 1 of the Lantau Port development in Hong Kong, *Int. Symposium on Waves – Physical and Numerical Modelling*, Vancouver, Canada, 1994.
		Justesen, P. *et al.* (1996). 2-D & 3-D modelling of Hong Kong Waters, *Proceedings of the Second Intl. Conference on Hydrodynamics*, Hong Kong, 1996, A.A. Balkema/ Rotterdam/ Brookfield.
		Whiteside, P.G.D. *et al.* (1995). Generation and decay of sediment plumes from sand dredging overflow, *Proceedings of the Forurteenth World Dredging Congress*, Amsterdam, The Netherlands, pp. 877–892.
		Government of Hong Kong SAR and its Department homepages – www.info.gov.hk
	Protection of Coral, Seagrass and Mangrove in Singapore Environmental management of dredging and reclamation operations. Prediction and monitoring of effects of dredging and reclamation activities.	Doorn-Groen, S.A. (2007). Environmental Monitoring and Management of Reclamations Works Close to Sensitive Habitats. *Terra et Aqua, No. 108.* (Paper presented at WODCON XVIII, Orlando, FL USA, June 2007).
Capital Dredging and Reclamation for Recreation	**Amager Beach Park Copenhagen, Denmark** Movement of beach line forwards to increase exposure and provide deeper water. Creation of beach lagoon suitable for recreation. Improved flushing and water quality.	Mangor, K. (2004). Shoreline Management Guidelines. DHI Water & Environment. ISBN 87-981950-5-0
		Mangor, K. and Hasløv, D.B. (2007). Coastal Developments in Harmony with Nature and Society, *32nd IAHR Congress on Harmonizing the Demands of Art and Nature in Hydraulics*, July 2007, Venice, Italy.
		Mangor, K. (2006). Presentation of the Amager Beach Park project by Karsten Mangor, Chief Engineer, DHI Water & Environment 20th October 2006 in connection with 36th Meeting of the Environmental Steering Committee, CEDA.
		Owner (2005). Information about the project from the web site of Amager Strandpark I/S at http://www.amager-strand.dk

(Continued)

Project type	Project name and Characteristics	Sources
Capital Dredging for Mining	**Aggregate dredging in United Kingdom and Denmark in 2000**	James, J.W.C. *et al.* (1999). The effective development of offshore aggregates in south-east Asia British Geological Survey Technical report WC/99/9 Overseas Geology Series, prepared for the Department for International Development under the DFID/BGS Knowledge and Research Programme, Project R6840.
		Tender notification from US Dept. of the Interior, Mineral Management Services in Dredging News Online on 7/1-2000 (see www.sandandgravel.com/news/contracts),
		Gaillot, S. and Piégay, H. (1999). Impact of Gravel-Mining on Stream Channel and Coastal Sediment Supply: Example of the Calvi Bay in Corsica (France), *Journal of Coastal Research*, Vol.15, No.3, Published by CERF, Summer 1999.
Capital Dredging for Mining	**Mineral Mining in General**	Davy, A. *et al.* (1998) Environmental assessment of mining projects, Environmental Assessment Sourcebook Update, March 1998, Number 22, Environment Department, The World Bank.
	North Stradbroke Island at Brisbane, Australia	Summary sheets from Consolidated Rutile Limited 'How are CRL products produced?' – The North Stradbroke Island mineral mine, July 1999 – more information on www.consrutile.com.au
	Minerals extracted from sand dunes. Environmental management and rehabilitation.	
	Beenup Mine in Western Australia	Biggs, B. (1998). Acid Drainage in Western Australia, *Groundwork* No. 1, Vol. 2, September 1998.
	Mineral sands – responses to acid drainage.	Price, G. (2000). Environment and External Affairs Superintendent, BHP Titanium Minerals, Australia, supplementary information and updates on Beenup mine, telephone interview, 10 March 2000.
Maintenance Dredging	**New York and New Jersey Harbour, USA**	Asaf Ashar (1997). Impact of Dredging New York Harbor, *Transportation Quarterly, Vol. 51, No. 1*, Winter 1997 (pp. 45–62).
	Stricter placement regulations prevent offshore placement of contaminated materials. Severe environmental constraints. Balance between cost increases and benefit to community.	Karim A. Abood, Susan G. Metzger, Peter Dunlap and William Gay (1998). "A Case Study illustrating the Role of Environmental Remediation in Contaminated Sediment Management in NY/NJ Harbor", *Proceedings of the Fifteenth World Dredging Congress, Volume 2*, Las Vegas,

(Continued)

Project type	Project name and Characteristics	Sources
		Nevada, USA, June 28–July 2, 1998, pp. 677–686. (Several papers covering various aspects of dredging and disposal in NY/NJ were presented at WODCON 1998)
		DHI (1998). "New York 3D Model", numerical model transfer study for the New York/New Jersey Harbor Partnership, a Joint Venture of Lawler, Matusky & Skelly Engineers (USA) and Moffatt & Nichol Engineers (USA), conducted by the Danish Hydraulic Institute, 1998.
	Port of Cochin, India Review of environmental effects of maintenance dredging in the navigation channels.	Balchand, A.N. and Rasheed, K. (2000). Assessment of Short Term Environmental Impacts on Dredging in a Tropical Estuary. *Terra et Aqua, No. 79.*
Maintenance Dredging, Remedial Dredging	**Miami River, Florida, USA** Restoration of navigable channel. Processing of contaminated sediments. Reduction in volume of contaminated and re-use of clean materials.	Taylor, A.S. *et al.* (2006). Deepening, Cleaning and Processing Sediment from the Miami River. *Terra et Aqua, No.103.*
Remedial Dredging, Sanitation Dredging	**Lake Ketelmeer, The Netherlands** Confined Disposal Facility (CDF) for contaminated silt in Rhine delta. Comparison of high accuracy and low turbidity equipment.	Laboyrie, H.P. and Flach, B. (1998). The Handling of Contaminated Dredged Material in The Netherlands; *15th World Dredging Congress* 1998, Las Vegas, Nevada, USA. Heineke, D. and Eversdijk, P.J. (1996). Dutch Design for Storage of Contaminated Spoil, *Land & Water International* 86, 1996. Arts, T. *et al.* (1995). High Accuracy Sanitation Dredging Trials *Terra et Aqua, No. 61,* December 1995. Arts, T. and Kappe, B. (1996). The Sweep Dredge, High Accuracy Dredging Trials Continue *Terra et Aqua, No. 65,* December 1996. Vandycke, S. (1996). New Developments in Environmental Dredging: From Scoop to Sweep Dredge. *Terra et Aqua, No. 65,* December 1996. Roukema, D.C. *et al.* (1998). Realisation of the Ketelmeer Depot, *Terra et Aqua, No. 71,* June 1998.

(Continued)

Project type	Project name and Characteristics	Sources
		IADC/CEDA (1999). Environmental Aspects of Dredging, Guide 5: Reuse, Recycle or Relocate, The Hague, The Netherlands.
Remedial Dredging, Conservation	**Segara Anakan, Indonesia** Deepening of natural lagoon to prevent the spread of mangroves.	Binnie Black and Veatch (2000). Information sheet. Asian Development Bank Completion Report (2006). www.adb.org/Documents/PCRs/INO/22043-INO-PCR.pdf
Remedial Dredging, Restoration	**South Lake, Tunis, Tunisia** Regeneration by dredging of a highly eutrophic lake.	Vandenbroeck, J. and Rafik, B.C. (2001). Restoration and Development Project of South Lake of Tunis and its Shores. *Terra et Aqua Number 85,* December 2001.

Characteristic Features and Functions of Aquatic Systems

C1 GENERAL

The objective of these short introductions to aquatic systems is to describe a few general characteristics which support the discussions of effects presented in the main text. For more detailed explanations and descriptions of the diversity of coastal environments, the reader may find additional information in the References, see for example, Bearman (1989), Pethick (1984) or Viles and Spencer (1995).

Before describing the various aquatic systems some relevant information is given on ecosystems, habitats and zonation.

Ecosystems and habitats

The characteristic space occupied by an individual, a population or a species is called a "habitat". Habitats in the sea are diverse and include such environments as:
- tidal flats
- estuaries
- inter-tidal rocky shores
- water columns
- seabeds.

Within such parts of the sea, most species occupy specific habitats or zones, e.g. pelagic species that live in the open sea or demersal species that inhabit the bottom of the sea. In principle any part of the biosphere, the living part of the world, comprising living and nonliving constituents with a complicated network of mutual interactions and having inputs and outputs with the surrounding biosphere, can be considered as an ecosystem (Baretta et al., 1998). Some parts of the biosphere may have much stronger internal interactions than with the external world, like ponds and lakes for instance.

Marine ecosystems are complicated to define. They are often based on the specific physical nature of the area where the interactions with the outside biosphere may sometimes dominate internal dynamics. Some examples of marine ecosystems are:
- oceanic and coastal rocky and sandy shores;
- estuarine hard substrates; and
- sand and mud flats, including seagrass communities, mangroves and coral reefs.

In the pelagic system, plants and animals are oriented by vertical gradients in light, temperature, and salinity. They are able to exert force against gravity, i.e. prevent themselves from sinking. Pelagic plants are plankton algae and animals are micro-, meso- and macroplankton, and pelagic fish are those living in the water column.

Zonation

On a worldwide scale the geographical distribution of marine biota is fixed by water temperature and light availability, and the tolerance towards changes. At regional or local scale, tidal amplitude, water movement including mixing of salt- and fresh-water, wave action, desiccation and sediment type are important physical factors for zonation and distribution of shore organisms together with biological factors, such as competition and predation.

The availability of light underwater governs the depth distribution of marine plants. In extremely clear water multicellular marine algae have been found at a water depth of 200 m, corresponding to 0.01% of the surface light intensity reaching this depth. In the Mediterranean Sea, the deepest growing kelp occurs at 100 m (Mathieson *et al.*, 1991) and in more turbid waters at much lower depths.

Marine organisms often occur in belts or depth zones because of gradients in environmental factors. This is, for example, easily visible along rocky coasts.

Zonation is determined in different zones by different factors:
- In the littoral zone, particular irradiation, exposure time, desiccation, grazing, and predation are important factors.
- In deep water, temperature is the main factor.
- In estuaries, the salinity gradient is the dominant factor.

Approximately 90% of the yield of the world's fisheries are harvested in continental shelf waters which cover 7% of the oceans surface and less than 0.2% of its volume. The high fishery yield is not explained by the closer proximity of fishing ports to these areas alone (Postma *et al.*, 1988). The high primary productivity, the nutrient supply and recycling of organic matter by the benthic community, favour the abundance of fish compared to ocean waters. Upwelling also plays an important role, for example, for fisheries along the west coast of South America where the continental shelf is narrow. These waters are thus not only important as a food supply but also to local economies.

C2 TIDAL INLETS, ESTUARIES, BAYS AND STRAITS

In coastal areas where the wave action is relatively low and the tidal range is moderate to large, *tidal flats* may develop. These are normally extensive near-horizontal areas and are composed predominantly of silts and clay (mudflats) in the mid flat and high tidal flat part; and coarser material (sand) in the low tidal flat (offshore) part and in the network of tidal channels, which dissects the entire area. The high tidal flats are submerged only at high tide when current speeds fall to zero. During the slack water period, mud settles out of suspension to form the mud flats. In the high tidal flat zone, accumulation of fresh mud sediment normally takes place faster than erosion, which results in sediment accretion. This accretion is enhanced by the cohesive properties of the sediment and by the exposure of this zone for sufficiently long periods to allow for colonisation of land plants. The plant roots help bind the sediment and in turn the plant stems retard the flow, which encourages further deposition of silts and clay and prevents erosion.

Total colonisation of the high tidal flat zone leads to *salt marshes*. In tropical regions mudflats are often colonised by shrubs and trees and are then called *mangrove* or *mangal*. Tidal channels and coastal erosion limit the lateral accretion of salt marshes and mangroves.

Where there is no river mouth, the mangrove forest is usually rather thin. Owing to the very special environmental conditions on tidal flats, with alternating wet and dry periods, strong currents, temperature fluctuations and so on, they are inhabited by a special fauna of inter-tidal animals. Most of these live buried in the sediment.

Although the diversity in tidal flat ecosystems is low, biomass and production can be very high, which attracts other visitors that have adapted to life in tidal areas.

Estuaries

Although the same general principles apply to *estuaries* as to tidal flats, the interaction of tidal currents and river flow modifies the pattern of sediment transport. An estuary is a tidal inlet and may be defined as a semi-enclosed area with open or partially opened connection to the sea, within which seawater is mixed with freshwater from land (an inlet of the sea reaching into a river valley).

Important physical processes are the effects of vertical stratification because of the salinity difference and the longitudinal and vertical mixing by turbulence. This turbulence may be caused by the tides, but also by current shear. Estuaries are classified by:
- their origin (coastal plain estuaries, lagoon estuaries, fjords and so on);
- the type of mixing (saltwedge, partially mixed, and well-mixed estuaries); and
- their tidal range (microtidal – tidal range 0–2 m; mesotidal – tidal range 2–4 m; macrotidal with tidal range more than 4 m).

Estuaries are far from uniform in character and the differences are mainly a result of variations in tidal range and river discharge, which affect the extent to which saline water mixes with freshwater. An estuary is typically divided into a lower, middle and upper estuary. In the lower part, the influence of saltwater is dominant; in the middle part, saltwater and freshwater mix; and in the upper part, freshwater is dominant but subject to tidal influence.

The boundaries are transition zones that shift according to the seasons, the weather and the tides. Tidal flats may form part of the lower estuary zone. Estuarine sediments may show a large variation of sands, silts, clays, shells and other carbonate deposits.

Fine sediments are deposited by aggregation into larger particles with higher settling velocities. Aggregation occurs either by biological processes involving the ingestion and excretion of sediment as faecal pellets, or by flocculation in saline water. The sediment can be supplied by rivers as bed or suspended load, or from the sea by inflow along the bed.

In saltwedge estuaries, most suspended material is river-borne and deposits occur by settling through the halocline, a vertical zone in the saltwater estuary where salinity changes rapidly with depth. At the head of the estuary, where the river meets the salt wedge a coarse sediment bar may build up.

In partially mixed estuaries, river-borne sediments are transported seawards near the surface, and marine-derived suspended sediment is transported landwards near the bed, together with any river-borne sediment that has settled through the water column. Usually a turbidity maximum develops, i.e. a zone of muddy sediments and high concentrations of suspended matter, towards the head of the estuary at the point where transport ceases.

In estuaries with high tidal ranges, the turbidity maximum may reach suspended sediment concentration levels of 10 g/l. In well-mixed estuaries in the Northern Hemisphere marine sediments are deposited on the left-hand bank (looking from the shore seawards), and river-borne sediments on the right-hand bank. The opposite deposition pattern occurs in the Southern Hemisphere. In well-mixed estuaries, salinity hardly varies with depth, although it may vary considerably across the width of the estuary.

In general sedimentation in estuaries may be highly irregular: During exceptional floods or storms more sediments may be deposited in a few days than in many years of normal conditions. However, the delicate balance between sediment supply, deposition and outflow into the sea under normal conditions, sometimes makes it difficult to estimate whether there is export or import of sediment in an estuary. In areas where the sediment discharge from rivers is high and wave and tidal current action is limited, the estuary may rapidly be filled and a delta grows seaward at the expense of the estuary. Deltas may be classified after the relative intensities of river, tidal and wave-dominated processes.

Wetlands

When describing tidally influenced coastline systems and transition zones between terrestrial and aquatic systems another term used, which includes estuaries, is wetlands. This terminology is most frequently used in the USA and is the collective term for marshes, swamps, bogs and similar flat vegetated areas found between water and dry land along the edges of streams, rivers, lakes, and coastlines.

Wetlands are difficult to reduce to a universally recognised definition, because their essential elements are so diverse and variable in character with local differences in climate, soils, topography, hydrology, water chemistry, vegetation and other factors. Depth and duration of inundation can differ greatly between wetland types and may vary from year to year.

Owing to the supply of nutrients in combination with good light conditions, the coastal ecosystems described above are very productive. This implies that estuaries and the like support a variety of habitats interacting within complex food webs. Abundant prey of fish, worms, crabs or clams attract thousands of shorebirds and different types of mammals. Many marine organisms, most commercially valuable fish included, depend on these areas at some point during their life cycles.

The functions of these areas to society and the values obtained from them are many, some of which are mentioned in Table C.1.

Services	Resources
• Transportation – provides natural shelter for sea-going vessels and connects the sea with inland waterways • Coastal defence – a buffer area between land and the sea which reduces flood and storm damages • Water cleaning – the filtration process provided by the salt marsh vegetation improves water quality • Disposal areas – in the past more uncontrolled disposal of waste products and dredged material than today • Recreational – supports a variety of recreational activities including tourism • Educational – the diversity found in estuaries attracts scientists and students within biology, geology, chemistry, physics, history and social issues	• Settlement – land claim for industrial, residential and agricultural development • Fisheries – commercially important fishing grounds and, in tropical zones, breeding habitat for shrimp • Raw materials – aggregate removal; in tropical areas mangrove trees are used for timber and fuel • Energy – tidal power plants

Table C.1 Functions of estuaries to society

The utilisation pattern outlined in Table C.1 involves competing activities. In recognising the values provided by the different uses and functions, the need for careful management and coordination for maintenance and further developments in these areas in response to societal needs becomes clear. This often means seeking a delicate balance between environmental and economic factors, which challenges the designers of new development schemes and the innovators of mitigating measures.

C3 OPEN WAVE EXPOSED COASTS

Characteristic features and functions

The different processes of varying strength and time-scale govern coastal contours or relief are mentioned in Chapter 4. In Figure 4.10 the distribution of coastal types and their frequency of occurrence is shown within various tidal ranges. In this section the focus is on coastal and shallow marine areas, where the patterns of sediment movement and deposition are primarily controlled by wave action.

Although waves and tides are major influences on coastal processes, their action, either combined or separately, does not account for all the fundamentals and dynamics of coastal systems. For example, as highlighted in Annex C4, rivers carrying freshwater, sediments and nutrients to the sea have a significant effect on the sediment transport mechanisms and ecosystems in coastal areas. Sea spray also affects the functioning of coastal bio-geo-morphological systems, as it is an important agent for weathering, as well as a pathway for nutrient transport inland.

An important requirement related to the prediction of the impact of dredging and reclamation activities in coastal environments, and to the identification of suitable solutions, is an understanding of the physics behind coastal problems: Why has the present situation developed? Why are certain solutions applicable and others not? What can be ascribed to natural disturbances such as tectonic and storm events, and what may be related to human impacts? Furthermore, the social and economic dimensions also need to be integrated into the solution.

Erosion

The physical side of many coastal problems is a result of erosion. For example, Mangor (1998) discusses how to establish successful shoreline restoration projects in densely populated areas, which to some extent may already be degraded as a result of applying traditional coastal protection measures. Most coastal erosion problems have their origin in a deficit in the littoral budget for a specific area. Such deficits can have many causes; common problems are, for example, blocking, or reduction of the littoral transport owing to protruding coastal structures, such as port structures, reclaimed areas, inlet regulation works, or coastal protection works; a reduction in the supply of littoral material to a section by river regulation works, sand mining or by protection of neighbouring coastlines; the loss of littoral material caused by placement of maintenance dredged material offshore, and so on.

These causes are well known, and in many cases their damaging effects have been eliminated, at least partly, by enacting regulatory measures. However, in some developed areas, characterised by a highly degraded coastal environment brought on by a historic combination of the above-mentioned causes, sufficient mitigation of these causes is not realistic.

Coastal classification

Mangor (1998) introduces a coastal classification, which may be used to provide some guidelines as regards optimal shoreline restoration measures for different types of coasts (see Figure C.1). The littoral transport for a given coastal environment is mainly dependent on the wave climate in terms of its wave height-direction distribution. The coastal classification is closely related to the variation in transport capacity

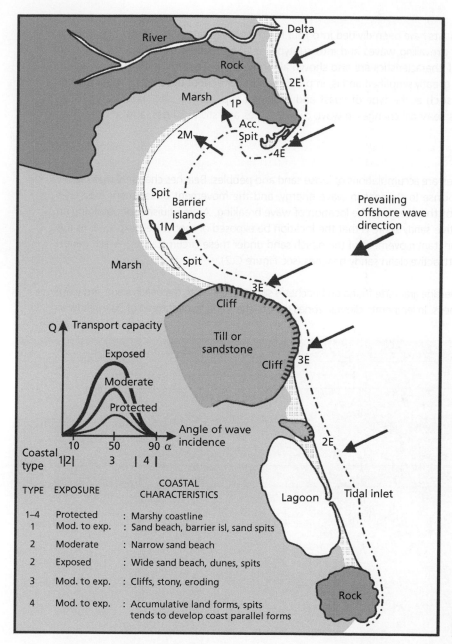

Figure C.1 A coastal classification after Mangor (1998).

as a function of the angle of incidence and the magnitude of the waves, which has been expressed in the form of exposed, moderately exposed and protected coastal areas. The possibility of artificially establishing a practically stable sandy shore is very much dependent on the angle of wave incidence of the prevailing waves and the magnitude of the transport deficit, which is closely related to the transport capacity.

The coasts have been divided into four main types relating to the angle of incidence of the prevailing waves and into subtypes relating to their exposure. The resulting coastal characteristics are also shown along a schematised coastline. This classification is greatly simplified and is, in practice, also dependent upon many other parameters, such as the type of coast and sediment supply from neighbouring areas, as well as seasonal changes in wave climate, tides and storm surges, and so on.

Beaches

Beaches are accumulations of loose sand and pebbles. Beaches change shape rapidly in response to changes in wave energy, and the movement of sediment dissipates some of this energy at the location of wave breaking. A prerequisite for obtaining an attractive sandy shore is that the location be exposed or moderately exposed, as it is the constant movement of the beach sand under these circumstances, which generates attractive clean sandy beaches (see Figure C.2).

The average grain size found on beaches is strongly related to the beach slope and wave steepness. In temperate climatic zones the beaches tend to be formed of pale yellow to

Figure C.2 Wide, clean sandy beaches are often dependent on regular sand replenishment.

brown quartz–rich sands. Around volcanic islands beaches often consist entirely of basaltic or andesitic lava and may be black in appearance. In tropical regions beaches are often comprised of brilliant white sand and gravel-sized fragments of corals and shells.

The following main issues normally expose the coastal zone to competing pressures:
● coast protection and shore protection
● agriculture and fisheries
● habitation, infrastructure, industrial development, public utilities and navigation
● recreation, landscape and environmental preservation, and
● raw material utilisation.

C4 RIVERS

Physical description

Rivers are the arterial system in the global water cycle in which the excess water from precipitation over land is led back to the oceans. The entire river system drains a drainage basin or *catchment area*, which by the topography is bordering neighbouring catchments in watersheds. The amount of water carried by the river, i.e. the *discharge*, often shows a characteristic variation over the year. This is called the *flow regime* and may best be shown in a *hydrograph* (discharge in m^3/s against time).

Discharge

Monsoon-fed rivers have a regular annual hydrograph with a peak during the rainy season and some may almost be empty in the dry season. Rivers in Northern Europe generally show maximum discharge during winter and minimum in the summer. However, as rainfall is governed by weather depressions the hydrographs may often show unpredictable highs and lows. The extent to which individual rainfalls or showers are reflected on the hydrograph depends on the size of the catchment area – a large area will tend to even out their influence.

Erosion

The flowing water exerts a force on the surface that increases with increasing water depth and slope. This initiates erosion of the terrain and the water transports the loosened material further downstream (Figure C.3). Erosion may take place in different ways (Jansen *et al.*, 1979). For example:
● sheet erosion – the wearing away of thin layers of surface soils;
● gully erosion – the removal of soil by concentrated flow;
● stream channel erosion – erosion of the river banks and scouring of the riverbed; and
● landslides and other mass movement of soils are also types of erosion.

Sediments

The material transported by the water thus consists of a combination of sediments and organic matter. Sediments originate from the decomposition of rocks as a result

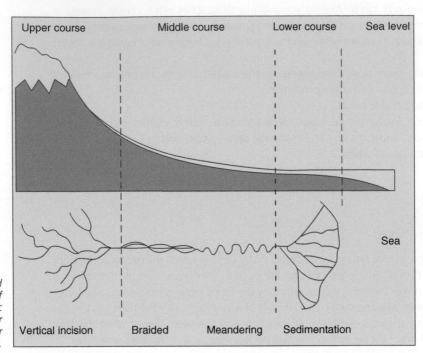

Figure C.3 Plan and longitudinal profile of typical characteristic courses along a river basin (modified after M. Donze, 1990).

of weathering and further crushing in rapid turbulent flow. Large pieces of rock may have the same mineral composition as the mother rock, but sediments such as sand and clays are particles of homogeneous mineral composition.

Quartz is a very durable mineral, which may decrease in size (sand) but still retains its chemical identity. Clays are the weathering products of silicates: feldspars, hornblende, micas and others. Sediments are divided into two groups:
● cohesive (fine sediments such as clay), and
● non-cohesive (sand and coarser sediments).

The sediment is either transported as bed material load or wash load. The bed material load is sediment picked up from the riverbed and moved (rolling or jumping) along the bed (bed load) or in suspension. Wash load refers to sediment, which originates from other parts of the catchment area, is finer than the bed material and is carried through the river in suspension. The quantity of the suspended load is in many cases much higher than the bed load. There is a relation between the water discharge and the bed material load, but there is no relation between the water discharge and the wash load.

Flood plains

Now and then the discharge in a river becomes so large that the water level in the river increases above the level of the banks and overflows the adjacent land. This periodically flooded land, the flood plain, can, in lower reaches of the river, reach quite wide dimensions. In the upper reaches they will usually be narrow if present at all.

With the water goes sediment (and soluble minerals) which deposits and thus builds up these flood plains resulting in very fertile and thus attractive land at some distance from the main course of the river. Sometimes horseshoe-shaped lakes may be found on the flood plain as evidence of abandoned river reaches. This is a typical feature of "meandering" rivers. Rivers channelled by nature's hand are seldom straight; normally, they wind – meander – through the plain.

A river progressing in a single channel is classified as a "meandering" river, and "braided" if it consists of a number of channels. The form it takes depends on the channel slope and the discharge (when running full bank). The same river can thus have straight, braided and meandering reaches.

To understand the often complex behaviour of fluvial systems, and to predict the consequences of modification, both geological aspects related to long-term variations of morphological features, as well as short to mid-term variations in the bed profile caused by a planned use, may need to be considered. For example, a constriction in a river by the building of a bridge may, in the short term, change the shape of the bed and banks several kilometres downstream. In the long term it may not be the most appropriate location for the bridge (Figure C.4).

Figure C.4 The logistics of building a bridge over the dynamic monsoon-fed, braided Jamuna River with its wide flood plains included constructing guide bunds and closure of some river branches.

In combination with simulated scenarios using numerical models, the type of information and parameters characterising the sedimentary processes briefly described above normally form part of investigations of proposed river works.

River ecology

The ecosystems and habitats characterising river systems are not only limited to those constantly "submerged". During summer the river flow may be low, with little or no vegetation and during periods with a high water bed the river may be overgrown. Also insects, which are considered aquatic, have an aerial stage during their life cycle.

Flow velocity
There is a relationship between flow velocity, the nature of the riverbed and distribution of aquatic plants. For example, in very rapid flow, boulders and rocks dominate and the torrential community is often not very varied and appears as cushion-like algae and mosses.

At the other end of the scale where the flow velocity is negligible and the bed is muddy, a littoral community, with vegetation similar to the shore of a lake, is present. Here the ecology is much more varied, but characterised by plants that extend out into the water, like reeds and grasses. Aquatic plants may also provide shelter for the smaller aquatic animals and contribute to the oxygenation of the water. Aquatic fauna may be classified in several groups, often by size and how their skeletons are built, (microscopic unicellular species, groups of worms, species with external skeletons, and those with backbones like fish and so on).

Population densities
However, the description of riverine ecosystems focuses more on "population densities". Like gasses, living organisms tend to fill the space available to them, but they do it by multiplication rather than spreading out thinly, as long as the conditions allow. This is largely determined by the aquatic cycle of organic and inorganic matter.

Animals and plants devoid of chlorophyll, i.e. not capable of photosynthesis, act as consumers, using organic compounds as their primary source of energy. The organic organisms (plants and algae) synthesise their basic nutrients from carbon dioxide and water using energy from sunlight or from inorganic chemical reactions.

Productivity
Detritus can be broken down into mineral salts by saprobic organisms. When the interrelationship of nutrient groups is established, the productivity of the food cycle in an isolated aquatic environment can be determined by the weakest link in the primary cycle. This may be either the rate at which energy is trapped by photosynthesis or the rate at which plant nutrients are made available by saprobic organisms.

A river, however, can hardly be considered an isolated environment because of the loss of nutrients carried away by the flow. Without material brought in from outside, the river will become infertile. Material entering the river may comprise nitrates and sulphates, which can boost the primary cycle. When nutrients are scarce, aquatic life will be limited and there will always be sufficient dissolved oxygen at all depths – "oligotrophic" water, characterised by high transparency (little or no suspended matter) and slow flow.

If nutrients are abundant, aquatic life will develop. In this case, the transparency of the water is limited to an upper shallow layer – "eutrophic" water. In a small body of water this may initiate oxygen depletion caused by detritus falling down to the deeper part, which, in combination with the diurnal variation of the oxygen production in the upper layer, may cause a total de-oxygenation during the night. The consequence may be periods with no aquatic life. Normally, the water in natural streams will fall in between these two extremes.

Socio-economic importance

Rivers provides the "raw material" for drinking water, irrigation, process and cooling water. They are utilised as waterways for navigation and by canalisation they become important infrastructure links. Energy may be furnished to the society through hydropower for instance. Considering all the many different uses and functions of rivers, it is no surprise that some uses create conflicts on a national level and, when rivers cross country borders, on an international level. In these cases river management requires multilateral consultation, because efficient management can only be made for the entire river basin.

Dredging can make an important contribution to obtaining, and maintaining, many of the above-mentioned river uses, for example bed regulation for flood control; to provide sufficient depth in navigation channels and in building new waterways/canals, dredging of backwater zones in reservoirs, sand and gravel mining, clean up operations, i.e. removal of contaminated material, restoring or conservation of river channels and such.

REFERENCES

Baretta-Bekker, Duursma and Kuipers (1998) *Encyclopedia of Marine Sciences, 2nd Edition*, Springer-Verlag Berlin Heidelberg New York.

Bearman, G. (Ed.) (1989) *Waves, Tides and Shallow-water Processes*. The Open University Pergamon Press, Oxford, UK 1987.

Donze, M. (Ed.) (1990) *Shaping the environment: Pollution and dredging in the European Community*, DELWEL Den Haag, The Netherlands.

Jansen, P.Ph *et al.* (1979) *Principles of river engineering – the non-tidal alluvial river*. Rainbow-Bridge Book Co.

Mangor, K. (1998) Coastal restoration considerations. *ICCE '98 26th Coastal Engineering Conference*, Copenhagen, 22–26 June.

Mathieson *et al.* (1991) Intertidal and Littoral Ecosystems, *Ecosystems of the World, Vol. 24*, Elsevier Science Publishers B.V., The Netherlands.

Pethick, J. (1984) *An Introduction to Coastal Geomorphology*, Edward Arnold Publishers, Great Britain.

Postma *et al.* (1988) Continental Shelves, *Ecosystems of the World, Vol. 27*, Elsevier Science Publishers B.V., The Netherlands.

Viles and Spencer (1995) Coastal Problems – *Geomorphology, Ecology and Society at the Coast*, Edward Arnold – Hodder Headline PLC, Great Britain.

Dredged Material Properties

Adequate assessment of sediments is necessary to predict behaviour during and after placement and to evaluate the environmental acceptability of management options. This annex describes the information needed to establish the physical, chemical, biological and engineering characteristics of dredged material. More detailed discussion of the subject and related aspects (e.g. sampling methods and analyses) is given in Chapter 5.

D1 WHAT IS DREDGED MATERIAL?

Aquatic sediments are a mixture of different organic and inorganic matter. The organic matter consists of micro-organisms, the remains of macrophytes and other large-sized organisms, and detritus from decaying material. The inorganic matter consists of mineral grains, small rock particles, clay minerals, precipitates and coatings. Sediments may contain a great variety of contaminants reflecting the industrial, agricultural and municipal discharges of the drainage area. Some sediments have naturally occurring concentrations of various substances, for instance metals from the "parent" rock, which may or may not pose risks to aquatic organisms and human health, depending on bio-availability.

The World Bank (1990) distinguishes five main types of dredged sediments:

1. Material from maintenance dredging of ports, harbours and navigation channels is usually fine-grained (silt and clays) and is often rich in organic matter and contains contaminants.
2. Material from maintenance dredging of sand bars at the entrances to harbours or channels usually consists of fine to coarse-grained sand and is much less contaminated than (1).
3. Material from capital dredging within a port may vary considerably along the vertical profile. The upper layer is usually fine-grained, organic rich and contaminated and the underlying layer is coarse-grained and uncontaminated. In some circumstances, even the deeper layers might contain elevated levels of contaminants from historical discharges.
4. Material from capital dredging of channels or outer harbour areas is likely to be coarse-grained and uncontaminated.
5. Material derived from remediation dredging is usually fine-grained, organic rich and, by definition, highly contaminated. Its relocation requires special control measures.

D2 PHYSICAL PROPERTIES

Information is usually needed on the amount and the physical characteristics of the sediment to be placed. In estimating the total volume to be placed, consideration must be given to possible overdredging and to bulking.

The bulking factor depends on the type of material and the method of dredging and placement. It can vary greatly:
● For mechanically dredged sediments Bray *et al.* (1997) suggest a factor of 10 to 40%.
● For hydraulic dredgers the factor depends on the *in situ* density and the concentrations in the pipeline.
● For hopper dredgers, it depends on the duration of the overflow. Procedure to determine the required initial holding capacity in CDFs and in CAD sites is given in USACE (1987).

As a result of consolidation, changes in volume occur at the placement site as time passes. Recent research shows that in addition to the physical processes, bio-gas production in the dredged material may significantly affect the degree of consolidation (Laboyrie and Flach, 1998). These factors need to be considered when assessing the volume occupied by the dredged material in the long-term.

Several physical classification systems for sediments are in use internationally. Two of the better known ones are the Unified Soil Classification System (USCS) and the PIANC system "Classification of Soils and Rocks to be Dredged" (PIANC, 1984). The following sections are based on a synthesis of these.

The basic physical characteristics are:

Form and composition is a general description based on visual assessment.

Grain size is the basis for most dredged material classifications. A number of samples should be analysed to give a reasonable representation of the material. The particle size distribution is usually described by the percentage by weight that passes each sieve size: Boulders >200 mm; Cobbles <200 mm; Gravel <60 mm; Sand <2.00 mm; Silt <0.063 mm; Clay <0.002 mm. Silts and clays are generally cohesive. They are often described as "mud".

Specific gravity of the solid particles affects the consolidation of placed material and is required in the calculation of void ratio.

Bulk density (or *in situ* density) is important for determining volumes *in situ*, in transport and after placement. Some examples are given in Bray *et al.* (1997).

Plasticity is relevant only to silt and clay. The most commonly used descriptors are the Atterberg liquid limit and plastic limit (LL and PL).

Water content is used to calculate *in situ* void ratio which is consequently used to estimate volume for containment area.

Shear strength is the behaviour of dredged material under load. Shear strength must be determined if the use of dredged material for construction purposes is considered. Rollings and Rollings (1998) stress the significance of this parameter in assessing mound behaviour under the loading of caps.

Water retention characteristics describe the ability of the dredged material to sustain plant life. USACE (1986) gives tables for guidance.

Permeability is a measure of the ease with which water passes through the material. It is determined mainly by the particle size of the material and (for cohesive sediment especially) the degree of consolidation.

Settling behaviour describes the rate at which particles in suspension descend to the bed. It is usually stated as W_{50} which is the median settling velocity of the particles. It is an indicator of the time material will stay in suspension and therefore affects the concentration of solids (and turbidity) in the water. Sedimentation tests provide data to design containment area to meet effluent suspended solids criteria and to provide adequate storage capacity.

Consolidation behaviour describes the rearrangement of sediment particles into a more dense state accompanied by an expulsion of pore water. It is relevant for fine-grained material. Consolidation is an important phenomenon affecting site capacity and, with open-water placement, mound geometry.

Compaction mechanically increases the amount of solids per unit volume of soil. It improves the engineering properties of soils. The degree of compaction required has to be determined before the dredged material is used as a fill for road bases, foundations or embankments.

Organic content contributes to high plasticity, high compressibility, permeability, low strength, enhanced buffering capacity, and immobilisation of contaminants. It affects the material capacity for plant growth.

Note that the dredging process will alter some of the physical properties. For example most dredging processes will reduce bulk density, at least initially.

D3 CHEMICAL PROPERTIES

Commonly tested parameters are the following:

pH is one of the most useful and informative parameters. It is a measure of the concentration and activity of ionised hydrogen. It is used to express the extent to which a medium is acidic or basic. pH is important in evaluating mobility of many metal contaminants. It indicates the kind of analyses or corrective actions needed for determining use or placement.

Calcium Carbonate Equivalent is closely related to pH. It indicates the amount of lime needed to neutralise any acidity present in the dredged material to maintain the required pH. Liming can reduce bioavailability and toxicity in acidic material when aluminium, manganese and other metals are present at elevated concentrations.

Cation Exchange Capacity (CEC) is important because it alters physical properties, causes/corrects acidity and basicity and can purify or alter percolating waters. If CEC is sufficiently high, it can immobilise the heavy metals present thus reducing ground water impact.

Salinity is a measure of the concentration of soluble salts. High salinity can adversely affect the structure of dredged material (decreases the cohesiveness of particles). It can inhibit water and air movement and decrease available nutrient content.

Redox potential (Eh) is analogous to pH. It is a measure of electron activity. It is used to express the degree to which an aquatic medium is oxidising (high Eh) or reducing (low Eh). It is important in determining the stability of various heavy metals and organic species.

Dissolved Oxygen (DO) frequently is the key substance in determining the extent and kinds of life in water ecosystems. Oxygen deficiency is fatal to many aquatic animals.

Biochemical Oxygen Demand (BOD) is the amount of oxygen required during the aerobic decomposition of organic matter in a body of water. Or, a measure of the quantity of oxygen used in the biochemical oxidation of carbonaceous and nitrogenous compounds in a specified time, at a specified temperature and under specified conditions. The standard measurement is made for five days at 20°C and is termed BOD_5. BOD is an indicator of the presence of organic matter in the water.

Total Organic Carbon (TOC) is now recognised as the best means of assessing organic matter content and it is frequently measured. TOC can easily be determined instrumentally as compared to the time-consuming and cumbersome measurement of BOD.

Dissolved Organic Carbon (DOC) is an important parameter for contaminated sediments since many of the contaminants occur in forms bound to DOC (especially PAHs and PCBs). DOC bound forms do not exert toxic effects as opposed to the free forms.

Nutrients (nitrogen and phosphorus compounds) are essential constituents of living organisms. Excess amounts, however, may cause serious water quality problems in sensitive waters (e.g. odour, appearance, fish killings). If vegetation establishment is considered as potential use, adequate levels of nutrients in the dredged material are essential prerequisites.

Carbon:Nitrogen Ratio (C:N) helps to determine if conditions in the dredged material are optimal for the growth of soil microbes and plants.

Potassium availability is important if vegetation establishment is considered as potential use of dredged material.

Contaminants in dredged material raise concern because of their potential to adversely affect aquatic and terrestrial ecosystems and, via the food chain, human health. The list of chemical constituents to be analysed depends on the regulatory requirements and on known or suspected sources of contamination. It is advisable first to conduct a comprehensive analysis on composite samples and to narrow down the set of chemicals that need more detailed investigations. Both total concentrations and pore water concentrations have to be determined. For PAHs and PCBs the pore water concentration can be further broken down to fractions associated to DOC and fractions occurring in free form (only this form is toxic).

Inorganic contaminants (metallic elements) are found in living organisms. Some are essential for life but, in excess, they become toxic. Non-essential elements can give rise to toxic effects even when intakes are slightly higher than normal. Commonly cadmium, chromium, copper, lead, mercury, nickel, and zinc are analysed. Other metals such as arsenic may also be investigated if existing information on contamination makes it necessary.

Organic contaminants (organic micropollutants) include Polycyclic Aromatic Hydrocarbons (PAHs), Polychlorinated Biphenils (PCBs), pesticides (e.g. DDT, dieldrin), and dioxins. Members of this group are definitely toxic. The toxicity varies widely from slightly to extremely toxic depending on the molecule and the exposed species. The need for chemical characterisation of some of the compounds in this group is very case-specific. Decisions should be based on existing sediment chemistry data, information on contamination and applicable priority substance lists.

Organo-metalic compounds are highly toxic and require special attention. They include organo-tin compounds (e.g. TBT) and organic mercury (e.g. methyl mercury). Detailed description of the hazardous compounds in sediments and their environmental significance is given in Donze (1990).

The utility of chemical concentration levels in assessing biological acceptability is limited. Reasons include the uncertainties related to establishing limit concentrations, the lack of considering the interactive effects of multiple contaminants, the possibility of the presence of unknown and untested constituents and the lack of accounting for the physico-chemical conditions dominating at a particular site.

D4 BIOLOGICAL PROPERTIES

Biological characterisation of sediments may involve testing for the presence of microbial constituents and testing for toxicology.

Microbial constituents of concern include pathogens, viruses, yeasts and parasites. Sediments should be tested for these constituents whenever dredging sites are close to sewage discharges, and placement sites are close to sensitive areas such as drinking water intakes, recreation beaches or shellfish beds.

Toxicological characteristics can be assessed by a variety of tests such as acute and chronic toxicity bioassays, bioaccumulation bioassays, biomarkers, microcosm and mesocosm experiments. Although biological tests are not precise predictors of environmental effects they are seen as the best methods available to overcome the limits of chemical characterisation described earlier. Biological tests are conducted with species that are considered sensitive to the contaminants of concern and representative for the specific receiving environment.

Acute toxicity bioassays test the effects of short-term exposure. Toxicity is expressed as median lethal concentration (LC50), i.e. concentration which theoretically would kill 50% of the test organisms within a specified time-span. The duration varies from hours to days.

Chronic toxicity bioassays assess sub-lethal effects which can result from prolonged exposure to relatively low concentrations. Such effects include physiological, pathological, immunological, terratological, mutagenic or carcinogenic effects. Chronic toxicity bioassays may take several weeks.

Toxicity bioassays may be used to evaluate: (a) potential impacts of dissolved and/or suspended sediments on water column organisms (elutriate bioassays) and (b) potential impacts of deposited sediments on benthic organisms (benthic bioassays). In view of possible exposure times elutriate bio-assays should always be acute tests. For benthic bioassays chronic tests would be appropriate but are not practical.

Bioaccumulation bioassays assess bio-availability and the potential for long-term accumulation in aquatic food-webs to levels which might be harmful to top-level consumers, including people, without killing the intermediate organisms. Test species are exposed to sediments for 10 to 28 days, after which concentrations of substances of concern are measured in cell tissue.

Biomarkers may provide early warning of effects at low and sustained concentrations. Most biomarkers are still under development; some however are available for routine use (Murk et al., 1997). Short-term microcosm tests are available to measure community tolerance to toxicants (Gustavson and Wandenberg, 1995). Mecrocosm experiments are time-consuming and expensive. Their use is advised when extrapolation of laboratory results to field condition is difficult.

D5 ENGINEERING PROPERTIES

To evaluate potential uses of dredged material evaluation of engineering properties, in addition to the physical, chemical and biological characterisation described above, will probably be necessary. The importance of evaluating engineering properties relates to the intended use of the material.

Rock always originates from capital dredging which may involve blasting, cutting or ripping. The rock may vary from soft marl to hard granite or basalt with sandstones and coral in between. It may also vary in size depending on how it was dredged and the type of material. Because of size and weight the occurrence of boulders and cobbles tends to improve the stability of foundations. Angularity of particles increases stability. Many engineering uses require rock of a certain size range and it may therefore require sorting or processing (i.e. crushing).

Sand and gravel may be produced in the course of capital or maintenance dredging. Considering that aggregate dredging is an industry in its own right, it is not surprising that this is generally considered to be the most valuable material to arise from a dredging project. The main difference between sand and gravel is that there is less control over the particle size grading of the material.

Gravel and sand have essentially the same engineering properties differing mainly in degree. The defined classification boundary particle size has no engineering significance. They are easy to compact, little affected by moisture and not subject to frost action. Gravels are generally more perviously stable and resistant to erosion and piping than are sands. Well-graded sands and gravels are generally more stable than those which are poorly graded. Irregularity of particles increases the stability slightly. Fine, uniform sand approaches the characteristics of silt, i.e. a decrease in permeability and reduction in stability with increase in moisture.

Silt/mud is inherently unstable, particularly when moisture is increased, and has a tendency to become "quick" when saturated. It is relatively impervious, difficult to compact, highly susceptible to frost heave, easily erodible and subject to piping and boiling. Bulky grains reduce compressibility. Flaky grains (e.g. mica) increase compressibility and produce an elastic silt.

Clay is highly cohesive. The cohesion increases with decrease in moisture content. The permeability of clay is very low. It is difficult to compact when wet and impossible to drain by ordinary means. When consolidated it is resistant to erosion and piping, is not susceptible to frost heave, but is subject to shrinkage and expansion with changes in moisture. The properties are influenced not only by the size and shape (flat, plate-like particles) but also by their mineral composition; i.e. the type of clay mineral and chemical environment (see cationic exchange capacity). In general montmorillonite has the greatest adverse effect on engineering properties and illite and kaolinite the least.

D6 BIO-AVAILABILITY OF CONTAMINANTS *VERSUS* PHYSICO-CHEMICAL CONDITIONS

The bio-availability of heavy metals and organics and consequently their toxicity strongly depends on the form in which they are present in the aquatic environment. These, in turn, are governed by the prevailing physico-chemical conditions (e.g. pH, Eh, salinity, CEC, DO, TOC). The amount and type of clay present and the intensity of biological activity are also important. Changes in conditions induce a complex system of simultaneous, interrelated physico-chemical processes that lead to transformations between the different forms.

Metals may exist in the aquatic environment in four basic, but interactive forms (Figure D.1). Dissolved forms as free metal ions or complex molecules, particulate forms adsorbed to the surface of solid particles or as precipitates (mostly sulphides). In free mode (the most toxic), metals are transported with the water and easily taken up by organisms. The particulate associated forms are relatively inactive. In sediments under natural conditions (anoxic, reduced, near neutral pH) only a small fraction of heavy metals is dissolved; the major portion is bound by sulphides or by structurally complex, large organic compounds. Release of these may be induced by increased salinity, reduced pH and increased redox potential. It should be noted however, that also under aerobic conditions there are mechanisms that remove metals from the dissolved phase: absorption and coprecipitation of metal ions or hydroxides with iron and manganese hydroxides which form coatings on the surface of particles. The release of metal ions by sulphides upon oxidation may be counteracted by these processes.

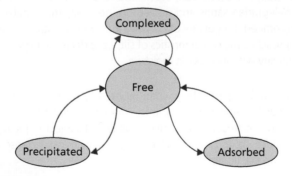

Figure D.1 Basic forms of heavy metals in the aquatic environment and transformation routes (from Goossens and Zwolsman, 1996).

Organic micropollutants may also exist in either dissolved or particulate bound forms. Generally the latter are less bio-available. For most compounds the unionised dissolved forms are the most toxic. The dissociation process depends on the pH: a decrease in pH accelerates it and increases toxicity. However, most chemical changes do not affect the mobilisation of organics.

Contaminated sediments are often in anoxic conditions and the major portion of toxicants is not bio-available. However, the physico-chemical conditions may change during dredging, placement and long-term storage (especially on land, when the deposited material dries out). The environmental assessment of management options should consider the implications of such changes.

D7 SEDIMENT PROPERTIES *VERSUS* DREDGING AND PLACEMENT METHODS

In addition to the original properties at the dredging site, the behaviour of sediments during and after placement will also largely depend on the specific characteristics of the dredging and placement techniques.

Dredging techniques may be subdivided into two categories: mechanical and hydraulic dredging. Mechanical dredgers use mechanical force to dislodge, excavate and lift the sediments (e.g. bucket or clamshell). The material is generally placed and transported to the site of discharge by barges. Barges may have bottom doors or split-hulls and the material can be released within seconds as an instantaneous discharge. Cohesive sediment dredged and placed this way remains intact, close to the *in situ* density through the whole dredging and placement process.

Hydraulic dredgers remove and lift the sediments as a slurry (mixture of sediment solids and large amounts of dredging site water). The slurry is transported to the discharge site via pipeline or in hoppers and released with large amounts of entrained water. Hoppers may release the material through bottom doors or through pump-out operations. Hydraulic dredging and transport break down the original structure of sediments.

Further guidance on the environmental aspects of different types of dredging, transporting and placement plant is given in Chapter 6.

REFERENCES

Bray, R.N. *et al.* (1997). *Dredging; a Handbook for Engineers*. Arnold Publishing, London, UK.

Donze, M. (Ed.) (1990). *Shaping the environment: Pollution and dredging in the European Community*, DELWEL. The Hague, The Netherlands.

Goossens, H. and Zwolsman, J.G. (1996). An evaluation of the behaviour of pollutants during dredging activities. *Terra et Aqua*. Number 62. March 1996. pp 20–28.

Gustavson, K. and Wangberg, S.A. (1995). Tolerance induction and succession in microalgea communities exposed to copper and atrazine. *Aquatic Toxicology.* 32. pp 283–302.

Laboyrie, H. and Flach, B. (1998). The handling of contaminated dredged material in the Netherlands. *Proceedings of World Dredging Congress XV*, Las Vegas, Nevada. WEDA, Vancouver, Washington, USA. pp 513–526.

Murk *et al.* (1996) Chemical-activated luciferase gene expression (CALUX): a novel in vitro bioassay for Ah receptor active compounds in sediments and pore water. *Fund. & Applied Tox.* 33: 149–160.

PIANC (1984) *Classification of Soils to be Dredged*. Report of a Working Group of PTC II, International Navigation Association. Brussels, Belgium.

Rollings, M.E. and Rollings, R. (1998). Consolidation and related geotechnical issues at the 1997 New York Mud Dump Site. *Proceedings of the 15th World Dredging Congress.* Las Vegas, Neveda. WEDA, Vancouver, WA, USA. pp 1–16.

USACE (1986). *Beneficial Uses of Dredged Material.* Engineer Manual No 1110-2-5026. Department of the Army Corps of Engineers. Washington, D.C., USA.

USACE (1987). *Confined disposal of dredged material.* Engineer Manual 1110-2-5027. Office, Chief of Engineers, Washington, D.C., USA.

The World Bank (1990). Environmental Considerations for Port and Harbour Developments. *World Bank Technical paper Number 126.* The World Bank. Washington, D.C., USA.

Index

Printed and bound by CPI Group (UK) Ltd, Croydon, CR0 4YY

24/10/2024

01778288-0002